Game Engine Gems 2

Game Engine Gems 2

Edited by Eric Lengyel

CRC Press
Taylor & Francis Group
Boca Raton London New York

CRC Press is an imprint of the
Taylor & Francis Group, an **informa** business

AN A K PETERS BOOK

CRC Press
Taylor & Francis Group
6000 Broken Sound Parkway NW, Suite 300
Boca Raton, FL 33487-2742

© 2011 by Taylor & Francis Group, LLC
CRC Press is an imprint of Taylor & Francis Group, an Informa business

No claim to original U.S. Government works

ISBN-13: 9781568814377 (hbk)

Visit the Taylor & Francis Web site at
http://www.taylorandfrancis.com

and the CRC Press Web site at
http://www.crcpress.com

Library of Congress Control No. 2010003949

Cover image: Splint/Second screenshot © 2010 Disney.

Contents

Chapter 4   Screen-Space Classification for Efficient Deferred Shading   55

Balor Knight Matthew Ritchie George Parrish

Chapter 5   Delaying OpenGL Calls   75

Patrick Cozzi

Chapter 6   A Framework for GLSL Engine Uniforms   87

Patrick Cozzi

Part II Game Engine Design   249

Part III   Systems Programming   **383**

Preface

The word *gem* has been coined in the fields of computer graphics and game development as a term for describing a short article that focuses on a particular technique, a clever trick, or practical advice that a person working in these fields would find interesting and useful. Several book series containing the word "Gems" in their titles have appeared since the early 1990s, and we continued the tradition by establishing the *Game Engine Gems* series in 2010.

This book is the second volume of the *Game Engine Gems* series, and it comprises a collection of new game engine development techniques. A group of 29 experienced professionals, several of whom also contributed to the first volume, have written down portions of their knowledge and wisdom in the form of the 31 chapters that follow.

The topics covered in these pages vary widely within the subject of game engine development and have been divided into the three broad categories of graphics and rendering, game engine design, and systems programming. The first part of the book presents a variety of rendering techniques and dedicates four entire chapters to the increasingly popular topic of stereoscopic rendering. The second part contains several chapters that discuss topics relating to the design of large components of a game engine. The final part of the book presents several gems concerning topics of a "low-level" nature for those who like to work with the nitty-gritty of engine internals.

Audience

The intended audience for this book includes professional game developers, students of computer science programs, and practically anyone possessing an interest in how the pros tackle specific problems that arise during game engine development. Many of the chapters assume a basic knowledge of computer architecture as well as some knowledge of the high-level design of current-generation game consoles, such as the PlayStation 3 and Xbox 360. The level of mathematics used in the book rarely exceeds that of basic trigonometry and calculus.

The Website

The official website for the *Game Engine Gems* series can be found at the following address:

http://www.gameenginegems.net/

Supplementary materials for many of the gems in this book are posted on this website, and they include demos, source code, examples, specifications, and larger versions of some figures. For chapters that include project files, the source code can be compiled using Microsoft Visual Studio.

Any corrections to the text that may arise will be posted on the website. This is also the location at which proposals will be accepted for the next volume in the *Game Engine Gems* series.

Acknowledgements

Many thanks are due to A K Peters for quickly assuming ownership of the *Game Engine Gems* series after it had lost its home well into the period during which contributing authors had been writing their chapters. Of course, thanks also go to these contributors, who took the transition in stride and produced a great set of gems for this volume. They all worked hard during the editing process so that we would be able to stay on the original schedule.

Part I

Graphics and Rendering

1

Fast Computation of Tight-Fitting Oriented Bounding Boxes

Thomas Larsson
Linus Källberg

Mälardalen University, Sweden

1.1 Introduction

Bounding shapes, or containers, are frequently used to speed up algorithms in games, computer graphics, and visualization [Ericson 2005]. In particular, the oriented bounding box (OBB) is an excellent convex enclosing shape since it provides good approximations of a wide range of geometric objects [Gottschalk 2000]. Furthermore, the OBB has reasonable transformation and storage costs, and several efficient operations have been presented such as OBB-OBB [Gottschalk et al. 1996], sphere-OBB [Larsson et al. 2007], ellipsoid-OBB [Larsson 2008], and ray-OBB [Ericson 2005] intersection tests. Therefore, OBBs can potentially speed up operations such as collision detection, path planning, frustum culling, occlusion culling, ray tracing, radiosity, photon mapping, and other spatial queries.

To leverage the full power of OBBs, however, fast construction methods are needed. Unfortunately, the exact minimum volume OBB computation algorithm given by O'Rourke [1985] has $O(n^3)$ running time. Therefore, more practical methods have been presented, for example techniques for computing a $(1-\varepsilon)$-approximation of the minimum volume box [Barequet and Har-Peled 1999]. Another widely adopted technique is to compute OBBs by using principal component analysis (PCA) [Gottschalk 2000]. The PCA algorithm runs in linear time, but unfortunately may produce quite loose-fitting boxes [Dimitrov et al. 2009]. By initially computing the convex hull, better results are expected since this

3

keeps internal features of the model from affecting the resulting OBB orientation. However, this makes the method superlinear.

The goal of this chapter is to present an alternative algorithm with a simple implementation that runs in linear time and produces OBBs of high quality. It is immediately applicable to point clouds, polygon meshes, or polygon soups, without any need for an initial convex hull generation. This makes the algorithm fast and generally applicable for many types of models used in computer graphics applications.

1.2 Algorithm

The algorithm is based on processing a small constant number of extremal vertices selected from the input models. The selected points are then used to construct a representative simple shape, which we refer to as the *ditetrahedron*, from which a suitable orientation of the box can be derived efficiently. Hence, our heuristic is called the ditetrahedron OBB algorithm, or DiTO for short. Since the chosen number of selected extremal vertices affects the running time of the algorithm as well as the resulting OBB quality, different instances of the algorithm are called DiTO-k, where k is the number of selected vertices.

The ditetrahedron consists of two irregular tetrahedra connected along a shared interior side called the base triangle. Thus, it is a polyhedron having six faces, five vertices, and nine edges. In total, counting also the interior base triangle, there are seven triangles. Note that this shape is not to be confused with the triangular dipyramid (or bipyramid), which can be regarded as two pyramids with equal heights and a shared base.

For most input meshes, it is expected that at least one of the seven triangles of the ditetrahedron will be characteristic of the orientation of a tight-fitting OBB. Let us consider two simple example meshes—a randomly rotated cube with 8 vertices and 12 triangles and a randomly rotated star shape with 10 vertices and 16 triangles. For these two shapes, the DiTO algorithm finds the minimum volume OBBs. Ironically, the PCA algorithm computes an excessively large OBB for the canonical cube example, with a volume approximately two to four times larger than the minimum volume, depending on the orientation of the cube mesh. Similarly, it also computes a loose-fitting OBB for the star shape, with a volume approximately 1.1 to 2.2 times larger than the optimum, depending on the given orientation of the mesh. In Figure 1.1, these two models are shown together with their axis-aligned bounding box (AABB), OBB computed using PCA, and OBB computed using DiTO for a random orientation of the models.

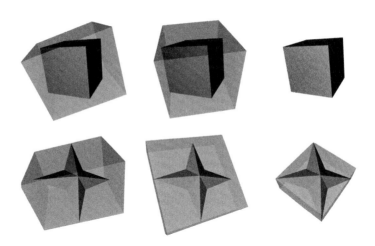

Figure 1.1. Computed boxes for a simple cube mesh (12 triangles) and star mesh (16 triangles). The first column shows the AABB, the second column shows the OBB computed by PCA, and the last column shows the OBB computed by DiTO. The meshes were randomly rotated before the computation.

The OBB is represented here by three orthonormal vectors **u**, **v**, and **w** defining the orientation of the box's axes, the half-extents h_u, h_v, and h_w defining the size of the box, and a midpoint **m** defining the center of the box. In the following, we explain how the algorithm works, and we present our experimental evaluation.

Selecting the Extremal Points

Call the set containing the n vertices of the input model P. The algorithm starts off by selecting a subset S with k representative extremal points from the vertices P, where k is an even number. This is done by finding the vertices with minimum and maximum projection values along $s = k/2$ predefined directions or normal vectors. The points in S are stored systematically so that any extremal point pair $(\mathbf{a}_j, \mathbf{b}_j)$ along a direction $\mathbf{n}_j$ can be retrieved, where $\mathbf{a}_j$ and $\mathbf{b}_j$ are the points with minimum and maximum projection values, respectively. Note that all the points in S are guaranteed to be located on the convex hull of the input mesh. Hopefully, this makes them important for the later determination of the OBB axes. Ideally, the used normal vectors should be uniformly distributed in direction space. However, to be able to optimize the projection calculations that are calculated by simple dot products, normals with many 0s and 1s may be preferable, given that they sample the direction space in a reasonable manner.

Clearly, the DiTO-k algorithm relies on the choice of an appropriate normal set N_s, and simply by choosing a different normal set a new instance of DiTO-k is created. In the experiments described later, five normal sets are used, yielding five algorithm instances. The normal sets are listed in Table 1.1. The normals in N_6, used in DiTO-12, are obtained from the vertices of a regular icosahedron with the mirror vertices removed. Similarly, the normals in N_{10}, used in DiTO-20, are taken from the vertices of a regular dodecahedron. The normal set $N_{16} = N_6 \cup N_{10}$ is used in DiTO-32. The normals in N_7 and N_{13}, used in DiTO-14 and DiTO-26, are not uniformly distributed, but they are still usually regarded as good choices for computing k-DOPs [Ericson 2005]. Therefore, they are also expected to work well in this case.

N_6	N_{10}	N_7	N_{13}
$(0,1,a)$	$(0,a,1+a)$	$(1,0,0)$	$(1,0,0)$
$(0,1,-a)$	$(0,a,-1-a)$	$(0,1,0)$	$(0,1,0)$
$(1,a,0)$	$(a,1+a,0)$	$(0,0,1)$	$(0,0,1)$
$(1,-a,0)$	$(a,-1-a,0)$	$(1,1,1)$	$(1,1,1)$
$(a,0,1)$	$(1+a,0,a)$	$(1,1,-1)$	$(1,1,-1)$
$(a,0,-1)$	$(1+a,0,-a)$	$(1,-1,1)$	$(1,-1,1)$
	$(1,1,1)$	$(1,-1,-1)$	$(1,-1,-1)$
	$(1,1,-1)$		$(1,1,0)$
	$(1,-1,1)$		$(1,-1,0)$
	$(1,-1,-1)$		$(1,0,1)$
			$(1,0,-1)$
			$(0,1,1)$
			$(0,1,-1)$

Table 1.1. Efficient normal sets N_6, N_{10}, N_7, and N_{13} used for DiTO-12, DiTO-20, DiTO-14, and DiTO-26, respectively, with the value $a = (\sqrt{5}-1)/2 \approx 0.61803399$. The normals in N_6 and N_{10} are uniformly distributed.

Finding the Axes of the OBB

To determine the orientation of the OBB, candidate axes are generated, and the best axes found are kept. To measure the quality of an orientation, an OBB covering the subset S is computed. In this way, the quality of an orientation is determined in $O(k)$ time, where $k \ll n$ for complex models. For very simple models, however, when $n \leq k$, the candidate OBBs are of course computed using the entire point set P. To measure the quality, the surface areas of the boxes are used, rather than their volumes. This effectively avoids the case of comparing boxes with zero volumes covering completely flat geometry. Furthermore, it makes sense to reduce the surface area as much as possible for applications such as ray casting. Of course, if preferred, a quality criterion based on the volume can be used in the algorithm instead.

The best axes found are initialized to the standard axis-aligned base, and the best surface area is initialized to the surface area of the AABB covering P. Then the algorithm proceeds by calculating what we call the *large base triangle*. The first edge of this triangle is given by the point pair with the furthest distance among the s point pairs in S, that is, the point pair that satisfies

$$\max \|\mathbf{a}_j - \mathbf{b}_j\|.$$

Call these points $\mathbf{p}_0$ and $\mathbf{p}_1$. Then a third point $\mathbf{p}_2$ is selected from S that lies furthest away from the infinite line through $\mathbf{p}_0$ and $\mathbf{p}_1$. An example of a constructed large base triangle is shown on the left in Figure 1.2.

The base triangle is then used to generate three different candidate orientations, one for each edge of the triangle. Let $\mathbf{n}$ be the normal of the triangle, and $\mathbf{e}_0 = \mathbf{p}_1 - \mathbf{p}_0$ be the first edge. The axes are then chosen as

$$
\begin{aligned}
\mathbf{u}_0 &= \mathbf{e}_0 / \|\mathbf{e}_0\|, \\
\mathbf{u}_1 &= \mathbf{n} / \|\mathbf{n}\|, \\
\mathbf{u}_2 &= \mathbf{u}_0 \times \mathbf{u}_1.
\end{aligned}
$$

The axes are chosen similarly for the other two edges of the triangle. For each computed set of axes, an approximation of the size of the resulting OBB is computed by projecting the points in S on the axes, and the best axes found are kept. In Figure 1.3, an example of the three considered OBBs for the base triangle is shown.

Next, the algorithm proceeds by constructing the ditetrahedron, which consists of two connected tetrahedra sharing the large base triangle. For this, two additional points $\mathbf{q}_0$ and $\mathbf{q}_1$ are computed by searching S for the points furthest

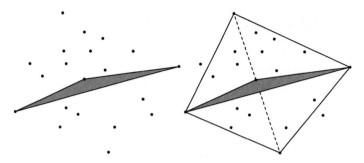

Figure 1.2. Illustration of how the large base triangle spanning extremal points (left) is extended to tetrahedra in two directions by finding the most distant extremal points below and above the triangle surface (right).

above and below the plane of the base triangle. An example of a ditetrahedron constructed in this way is shown on the right in Figure 1.2. This effectively generates six new triangles, three top triangles located above the plane of the base triangle and three bottom triangles located below the base triangle. For each one of these triangles, candidate OBBs are generated in the same way as already described above for the large base triangle, and the best axes found are kept.

After this, all that remains is to define the final OBB appropriately. A final pass through all n vertices in P determines the true size of the OBB, that is, the smallest projection values s_u, s_v, and s_w, as well as the largest projection values l_u,

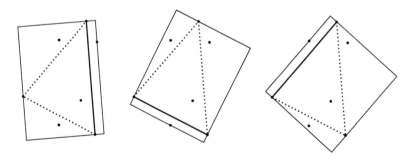

Figure 1.3. The three different candidate orientations generated from the normal and edges of the large base triangle. In each case, the box is generated from the edge drawn with a solid line.

l_v, and l_w of P along the determined axes $\mathbf{u}$, $\mathbf{v}$, and $\mathbf{w}$. The final OBB parameters besides the best axes found are then given by

$$h_u = \frac{l_u - s_u}{2},$$

$$h_v = \frac{l_v - s_v}{2},$$

$$h_w = \frac{l_w - s_w}{2},$$

$$\mathbf{m} = \frac{l_u + s_u}{2}\mathbf{u} + \frac{l_v + s_v}{2}\mathbf{v} + \frac{l_w + s_w}{2}\mathbf{w}.$$

The parameters h_u, h_v, and h_w are the half-extents, and $\mathbf{m}$ is the midpoint of the box. Note that $\mathbf{m}$ needs to be computed in the standard base, rather than as the midpoint in its own base.

A final check is also made to make sure that the OBB is still smaller than the initially computed AABB; otherwise, the OBB is aligned with the AABB instead. This may happen in some cases, since the final iteration over all n points in P usually grows the OBB slightly compared to the best-found candidate OBB, whose size only depends on the subset S.

This completes our basic presentation of the DiTO algorithm. Example source code in C/C++ for the DiTO-14 algorithm is available, which shows how to efficiently implement the algorithm and how the low-level functions work.

Handling Detrimental Cases

There are at least three cases of detrimental input that the algorithm needs to detect and handle appropriately. The first case is when the computation of the first long edge in the base triangle results in a degenerate edge; that is, the two points are located at the same spot. In this case, the OBB is aligned with the already computed AABB, and the algorithm is aborted.

The second case arises when the computation of the third point in the base triangle results in a degenerate triangle; that is, the point is collinear with the endpoints of the first long edge. In this case, one of the OBB axes is aligned with the already-found first long edge in the base triangle, and the other two axes are chosen arbitrarily to form an orthogonal base. The dimensions of the OBB are then computed, and the algorithm is terminated.

The third detrimental case is when the construction of a tetrahedron (either the upper or the lower) fails because the computed fourth point lies in the plane of the already-found base triangle. When this happens, the arising triangles of the degenerate tetrahedron are simply ignored by the algorithm; that is, they are not used in the search for better OBB axes.

1.3 Evaluation

To evaluate the DiTO algorithm, we compared it to three other methods referred to here as AABB, PCA, and brute force (BF). The AABB method simply computes an axis-aligned bounding box, which is then used as an OBB. While this method is expected to be extremely fast, it also produces OBBs of poor quality in general.

The PCA method was first used to compute OBBs by Gottschalk et al. [1996]. It works by first creating a representation of the input model's shape in the form of a covariance matrix. High-quality OBB axes are then assumed to be given by the eigenvectors of this matrix. As an implementation of the PCA method, we used code from Gottschalk et al.'s RAPID source package. This code works on the triangles of the model and so has linear complexity in the input size.

The naive BF method systematically tests $90 \times 90 \times 90$ different orientations by incrementing Euler angles one degree at a time using a triple-nested loop. This method is of course extremely slow, but in general it is expected to create OBBs of high quality which is useful for comparison to the other algorithms. To avoid having the performance of BF break down completely, only 26 extremal points are used in the iterations. This subset is selected initially in the same way as in the DiTO algorithm.

All the algorithms were implemented in C/C++. The source code was compiled using Microsoft Visual Studio 2008 Professional Edition and run single-threaded using a laptop with an Intel Core2 Duo T9600 2.80 GHz processor and 4 GB RAM. The input data sets were triangle meshes with varying shapes and varying geometric complexity. The vertex and triangle counts of these meshes are summarized in Table 1.2. Screenshots of the triangle meshes are shown in Figure 1.4.

To gather statistics about the quality of the computed boxes, each algorithm computes an OBB for 100 randomly generated rotations of the input meshes. We then report the average surface area A_{avg}, the minimum and maximum surface areas A_{min} and A_{max}, as well as the average execution time t_{avg} in milliseconds. The results are given in Table 1.3.

The BF algorithm is very slow, but it computes high-quality OBBs on average. The running times lie around one second for each mesh since the triple-nested loop acts like a huge hidden constant factor. The quality of the boxes var-

Model	Vertices	Triangles
Pencil	1234	2448
Teddy	1598	3192
Frog	4010	7964
Chair	7260	14,372
Bunny	32,875	65,536
Hand	327,323	654,666

Table 1.2. The number of vertices and triangles for the polygon meshes used for evaluation of the algorithms.

ies slightly due to the testing of somewhat unevenly distributed orientations arising from the incremental stepping of the Euler angles. The quality of boxes computed by the other algorithms, however, can be measured quite well by comparing them to the sizes of the boxes computed by the BF method.

The DiTO algorithm is very competitive. For example, it runs significantly faster than the PCA algorithm, although both methods are fast linear algorithms. The big performance difference is mainly due to the fact that the PCA method needs to iterate over the polygon data instead of iterating over the list of unique vertices. For connected triangle meshes, the number of triangles is roughly twice the number of vertices, and each triangle has three vertices. Therefore, the total size of the vertex data that the PCA method processes is roughly six times as large as the corresponding data for the DiTO algorithm.

Figure 1.4. Visualizations of the triangle meshes used for evaluation of the algorithms.

Pencil

Method	A_{avg}	A_{min}	A_{max}	t_{avg}
AABB	1.4155	0.2725	2.1331	0.02
PCA	0.2359	0.2286	0.2414	0.51
BF	0.2302	0.2031	0.2696	1009
DiTO-12	0.2316	0.1995	0.2692	0.11
DiTO-14	0.2344	0.1995	0.2707	0.09
DiTO-20	0.2306	0.1995	0.2708	0.14
DiTO-26	0.2331	0.1995	0.2707	0.15
DiTO-32	0.2229	0.1995	0.2744	0.20

Chair

Method	A_{avg}	A_{min}	A_{max}	t_{avg}
AABB	7.1318	4.6044	8.6380	0.07
PCA	4.7149	4.7139	4.7179	1.87
BF	3.6931	3.6106	4.6579	1047
DiTO-12	3.8261	3.6119	4.1786	0.35
DiTO-14	3.8094	3.6129	4.2141	0.25
DiTO-20	3.8213	3.6164	3.9648	0.37
DiTO-26	3.8782	3.6232	4.0355	0.35
DiTO-32	3.8741	3.6227	3.9294	0.49

Teddy

Method	A_{avg}	A_{min}	A_{max}	t_{avg}
AABB	3.9655	3.5438	4.3102	0.02
PCA	4.0546	4.0546	4.0546	0.60
BF	3.3893	3.3250	3.5945	1043
DiTO-12	3.7711	3.5438	4.0198	0.14
DiTO-14	3.7203	3.5438	3.9577	0.12
DiTO-20	3.7040	3.5438	3.8554	0.19
DiTO-26	3.7193	3.5438	3.8807	0.16
DiTO-32	3.7099	3.5438	3.8330	0.22

Bunny

Method	A_{avg}	A_{min}	A_{max}	t_{avg}
AABB	5.7259	4.7230	6.4833	0.19
PCA	5.2541	5.2540	5.2541	8.76
BF	4.6934	4.5324	4.9091	1041
DiTO-12	4.9403	4.5635	5.7922	1.13
DiTO-14	4.9172	4.5810	5.6695	0.98
DiTO-20	4.8510	4.5837	5.5334	1.55
DiTO-26	4.7590	4.5810	5.3967	1.42
DiTO-32	4.7277	4.6552	5.1037	2.04

Frog

Method	A_{avg}	A_{min}	A_{max}	t_{avg}
AABB	4.6888	3.0713	5.7148	0.07
PCA	2.6782	2.6782	2.6782	1.14
BF	2.7642	2.6582	3.5491	1037
DiTO-12	2.7882	2.6652	3.0052	0.28
DiTO-14	2.7754	2.6563	2.9933	0.24
DiTO-20	2.7542	2.6602	2.9635	0.40
DiTO-26	2.7929	2.6579	3.0009	0.36
DiTO-32	2.7685	2.6538	2.9823	0.44

Hand

Method	A_{avg}	A_{min}	A_{max}	t_{avg}
AABB	2.8848	2.4002	3.2693	1.98
PCA	2.5066	2.5062	2.5069	86.6
BF	2.3071	2.2684	2.4531	1067
DiTO-12	2.3722	2.2946	2.5499	11.8
DiTO-14	2.3741	2.2914	2.5476	10.0
DiTO-20	2.3494	2.2805	2.4978	15.5
DiTO-26	2.3499	2.2825	2.5483	14.5
DiTO-32	2.3372	2.2963	2.4281	20.6

Table 1.3. The average, minimum, and maximum area as well as the average execution time in ms over 100 random orientations of the input meshes.

The DiTO algorithm also produces oriented boxes of relatively high quality. For all meshes except Frog, the DiTO algorithm computes OBBs with smaller surface areas than the PCA method does. For some of the models, the difference is significant, and for the Teddy model, the PCA method computes boxes that are actually looser fitting than the naive AABB method does. The DiTO algorithm, however, is in general more sensitive than the PCA method to the orientation of the input meshes as can be seen in the minimum and maximum area columns.

Among the included DiTO instances, there seems to be a small quality improvement for increasing k values for some of the models. DiTO-32 seems to compute the best boxes in general. The quality difference of the computed boxes, however, is quite small in most cases. Therefore, since DiTO-14 is approximately twice as fast as DiTO-32, it is probably the preferable choice when speed is a prioritized factor.

Bounding Volume Hierarchy Quality

We also tried the different OBB fitting algorithms for OBB hierarchy construction to evaluate them on different levels of geometric subdivision. On models with smooth curvature, the geometry enclosed on the lower levels of the hierarchy tend to get an increasing flatness with each level, and we wanted to compare the algorithms on this type of input. It turns out that our algorithm handles this better than the other algorithms we tested. This is visualized using the Bunny model in Figure 1.5.

For each one of the six test models, we applied the OBB fitting algorithms during the hierarchy construction. These test runs have been done only once for each model and OBB algorithm, with the model kept in its original local coordinate system, due to the longer construction times for whole hierarchies. Table 1.4 shows the total surface areas of the resulting hierarchies.

Although the RAPID source package also includes a hierarchy builder, we extracted the functions involved in fitting the OBBs and plugged them into our own hierarchy builder since it uses a more elaborate strategy for partitioning primitives that generally creates better hierarchy structures, albeit at a higher computational cost. The table does not show execution times because the time for hierarchy construction is influenced too much by factors other than the time to fit the bounding volumes, such as the strategy used to find clusters in the geometry to build a good tree structure. However, the construction times were about the same with all the algorithms. Note also that the BF algorithm is not included since it gives unreasonably long construction times.

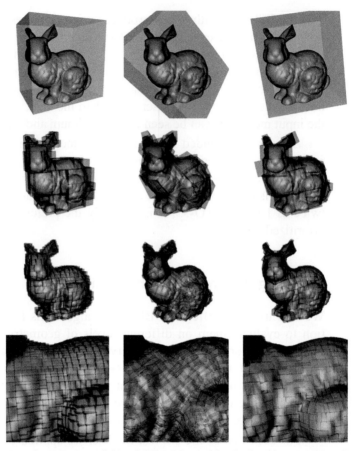

Figure 1.5. Levels 0, 6, 9, and 12 of OBB hierarchies built using AABBs (leftmost column), PCA (middle column), and DiTO-20 (rightmost column). As can be seen in the magnified pictures in the bottom row, PCA and DiTO both produce OBBs properly aligned with the curvature of the model, but the boxes produced by PCA have poor mutual orientations with much overlap between neighboring boxes.

The algorithm used for building the tree structure is a top-down algorithm, where the set of primitives is recursively partitioned into two subsets until there is only one primitive left, which is then stored in a leaf node. Before partitioning the primitives in each step, the selected OBB fitting procedure is called to create an OBB to store in the node. This means that the procedure is called once for every node of the tree. To partition the primitives under a node, we use a strategy that tries to minimize the tree's total surface area, similar to that used by Wald et al. [2007] for building AABB hierarchies.

Model	AABB	PCA	DiTO-12	DiTO-14	DiTO-20	DiTO-26	DiTO-32
Pencil	9.64676	4.05974	3.03241	3.03952	3.03098	3.0111	3.00679
Teddy	90.6025	86.2482	71.9596	74.071	74.3019	73.1202	74.7392
Frog	56.1077	52.2178	42.7492	41.9447	41.6065	41.2272	42.0487
Chair	81.3232	92.9097	64.5399	66.9878	64.2992	65.5366	64.8454
Bunny	170.71	171.901	125.176	122.108	119.035	119.791	119.172
Hand	112.625	117.241	50.8327	49.8038	50.0446	48.8985	49.7918

Table 1.4. Total surface areas of OBB hierarchies built using the different OBB fitting algorithms.

As Table 1.4 shows, the DiTO algorithms create better trees than the other two algorithms do. An implication of the increasing flatness on the lower hierarchy levels is that the first base triangle more accurately captures the spatial extents of the geometry, and that the two additional tetrahedra get small heights. It is therefore likely that the chosen OBB most often is found from the base triangle and not from the triangles in the tetrahedra. The PCA fitter frequently produces poor OBBs, and in three cases (Chair, Bunny, and Hand), produces even worse OBBs than the AABB method does. It is also never better than any of the DiTO versions.

Interesting to note is that there is a weak correspondence between the number of extremal directions used in DiTO and the tree quality. This can be partly explained by the fact that the directions included in a smaller set are not always included in a larger set, which, for example, is the case in DiTO-14 versus DiTO-12. This means that for some models, fewer directions happen to give better results than a larger number of directions. Another part of the explanation is that the tested OBBs are only fitted to the set of extracted extremal points S, which means that good-quality OBBs might be missed because worse OBBs get better surface areas on the selected extremal points. All this suggests that execution time can be improved by using fewer extremal directions (see Table 1.3), while not much OBB quality can be gained by using more.

Note that the AABB hierarchies sometimes have somewhat undeservedly good figures because the models were kept in their local coordinate systems during the construction. This gives the AABB an advantage in, for example, Pencil and Chair, where much of the geometry is axis-aligned.

1.4 Optimization Using SIMD Instructions

By using data-parallelism at the instruction level, the execution speed of the DiTO algorithm can be improved substantially. As an initial case study, a new version of DiTO-14 has been written that utilizes Intel's Streaming SIMD Extensions (SSE). For this version of the algorithm, the vertices are pre-stored as groups of four vertices, which can be packed componentwise into three full SSE registers, one for each coordinate component. For example, the first four vertices

$$\mathbf{p}_0 = (x_0, y_0, z_0),$$
$$\mathbf{p}_1 = (x_1, y_1, z_1),$$
$$\mathbf{p}_2 = (x_2, y_2, z_2),$$
$$\mathbf{p}_3 = (x_3, y_3, z_3),$$

are stored in the first vertex group as three arrays

$$\mathbf{X}_0 = (x_0, x_1, x_2, x_3),$$
$$\mathbf{Y}_0 = (y_0, y_1, y_2, y_3),$$
$$\mathbf{Z}_0 = (z_0, z_1, z_2, z_3).$$

Usually, this kind of data structure is referred to as structure of arrays (SoA).

There are two main passes in the algorithm, and they are the two loops over all input vertices. These two passes can be implemented easily using SSE. First, consider the last loop over all input vertices that determines the final dimensions of the OBB by computing minimum and maximum projection values on the given axes. This operation is ideal for an SSE implementation. By using the mentioned SoA representation, four dot products are computed simultaneously. Furthermore, the branches used in the scalar version to keep track of the most extremal projection values are eliminated by using the far more efficient `minps` and `maxps` SSE instructions.

Similarly, the first loop over all points of the model can also benefit a lot from using SSE since extremal projections along different axes are also computed. However, in this case the actual extremal points along the given directions are wanted as outputs in addition to the maximum and minimum projection values. Therefore, the solution is slightly more involved, but the loop can still be converted to quite efficient SSE code using standard techniques for branch elimination.

As shown in Table 1.5, our initial SIMD implementation of DiTO-14 is significantly faster than the corresponding scalar version. Speed-ups greater than

Model	t	t_v	s
Pencil	0.09	0.035	2.6
Teddy	0.12	0.04	3.0
Frog	0.24	0.08	3.0
Chair	0.25	0.08	3.1
Bunny	0.98	0.21	4.7
Hand	10.0	1.99	5.0

Table 1.5. The execution time t for the scalar version of DiTO-14 versus t_v for the vectorized SSE version. All timings are in ms, and the speed-up factors s are listed in the last column.

four times are achieved for the most complex models, Bunny and Hand. To be as efficient as possible for models with fewer input points, the remaining parts of the algorithm have to be converted to SSE as well. This is particularly true when building a bounding volume hierarchy, since most of the boxes in the hierarchy only enclose small subsets of vertices.

1.5 Discussion and Future Work

The assumption that the constructed ditetrahedron is a characterizing shape for a tight-fitting OBB seems valid. According to our experiments, the proposed algorithm is faster and gives better OBBs than do algorithms based on the PCA method. Also, when building hierarchies of OBBs, our method gives more accurate approximations of the geometry on the lower hierarchy levels, where the geometry tends to become flatter with each level. In addition, our method is more general since it requires no knowledge of the polygon data.

We have not found any reliable data indicating which instance of the DiTO algorithm is best. The tested variations produce quite similar results in terms of surface area, although in some cases, there seems to be a small quality advantage with increasing sampling directions. For construction of OBB trees, however, it may be unnecessarily slow to use many sampling directions since n is less than k for a large number of nodes at the lower hierarchy levels.

Although the presented heuristics work fine for fast OBB computation, it would still be interesting to try to improve the algorithm further. Perhaps the constructed ditetrahedron can be utilized more intelligently when searching for good

OBB axes. As it is now, we have only considered the triangles of this shape one at a time. Furthermore, the construction of some simple shape other than the ditetrahedron may be found to be more advantageous for determining the OBB axes.

Finally, note that DiTO can be adapted to compute oriented bounding rectangles in two dimensions. The conversion is straightforward. In this case, the large base triangle simplifies to a base line segment, and the ditetrahedron simplifies to a *ditriangle* (i.e., two triangles connected by the base line segment). There are better algorithms available in two dimensions such as the rotating calipers method, which runs in $O(n)$ time, but these methods require the convex hull of the vertices to be present [Toussaint 1983].

References

[Barequet and Har-Peled 1999] Gill Barequet and Sariel Har-Peled. "Efficiently Approximating the Minimum-Volume Bounding Box of a Point Set in Three Dimensions." *SODA '99: Proceedings of the Tenth Annual ACM-SIAM Symposium on Discrete Algorithms*, 1999, pp. 98–91.

[Dimitrov et al. 2009] Darko Dimitrov, Christian Knauer, Klaus Kriegel, and Günter Rote. "Bounds on the Quality of the PCA Bounding Boxes." *Computational Geometry: Theory and Applications* 42 (2009), pp. 772–789.

[Ericson 2005] Christer Ericson. *Real-Time Collision Detection*. San Francisco: Morgan Kaufmann, 2005.

[Gottschalk et al. 1996] Stefan Gottschalk, Ming Lin, and Dinesh Manocha. "OBBTree: A Hierarchical Structure for Rapid Interference Detection." *Proceedings of SIGGRAPH 1996*, ACM Press / ACM SIGGRAPH, Computer Graphics Proceedings, Annual Conference Series, ACM, pp. 171–180.

[Gottschalk 2000] Stefan Gottschalk. "Collision Queries using Oriented Bounding Boxes." PhD dissertation, University of North Carolina at Chapel Hill, 2000.

[Larsson et al. 2007] Thomas Larsson, Tomas Akenine-Möller, and Eric Lengyel. "On Faster Sphere-Box Overlap Testing." *Journal of Graphics Tools* 12:1 (2007), pp. 3–8.

[Larsson 2008] Thomas Larsson. "An Efficient Ellipsoid-OBB Intersection Test." *Journal of Graphics Tools* 13:1 (2008), pp. 31–43.

[O'Rourke 1985] Joseph O'Rourke. "Finding Minimal Enclosing Boxes." *International Journal of Computer and Information Sciences* 14:3 (June 1985), pp. 183–199.

[Toussaint 1983] Godfried Toussaint. "Solving Geometric Problems with the Rotating Calipers." *Proceedings of IEEE Mediterranean Electrotechnical Conference 1983*, pp. 1–4.

[Wald et al. 2007] Ingo Wald, Solomon Boulos, and Peter Shirley. "Ray Tracing Deformable Scenes Using Dynamic Bounding Volume Hierarchies." *ACM Transactions on Graphics* 26:1 (2007).

2

Modeling, Lighting, and Rendering Techniques for Volumetric Clouds

Frank Kane
Sundog Software, LLC

Pregenerated sky box textures aren't sufficient for games with varying times of day, or games where the camera may approach the clouds. Rendering clouds as real 3D objects, as shown in Figure 2.1, is a challenging task that many engines shy away from, but techniques exist to produce realistic results with good performance. This gem presents an overview of the procedural generation of cloud layers, the simulation of light transport within a cloud, and several volumetric

Figure 2.1. Volumetric clouds at dusk rendered using splatting. (*Image from the Silver-Lining SDK, courtesy of Sundog Software, LLC.*)

rendering techniques that may be used to produce realistically lit, volumetric clouds that respond to changes in lighting conditions and grow over time.

2.1 Modeling Cloud Formation

Before we render a cloud, we need to know its size, shape, and position. A scene may contain hundreds of clouds. While requiring a level designer to manually place and shape each one maximizes artistic control, achieving realistic results in this manner is time-consuming and challenging. Fortunately, procedural techniques exist to model the size, shape, and position of clouds within a cloud layer. A designer may simply define a bounding region for clouds of a specific type and let the simulation handle the rest.

Growing Clouds with Cellular Automata

While you could attempt to simulate the growth of clouds using fluid dynamics, doing this at the scale of an entire cloud layer would be slow and overly complex (although it has been done [Kajiya and Herzen 1984].) Clouds are complex natural phenomena, and attempting a rigorous physical simulation of their growth and the transport of light within them is computationally prohibitive. Rather, we seek techniques that produce results consistent with a viewer's expectations of what a cloud should look like, without overthinking the underlying physical properties.

One such shortcut is the use of cellular automata to grow clouds. You might remember cellular automata from the game *Life*, where very simple rules about a virtual cell's neighbors can produce complex colonies of cells that grow and shrink over time. Work from Nagel and Raschke [1992] and Dobashi et al. [2000] applying this same idea to the formation and growth of clouds is summarized here.

There is some physical basis to this technique; in general, we know that clouds form when a humid pocket of air rises and cools, causing a phase transition that turns its water vapor into water droplets. We also know that clouds tend to form vertically and horizontally, but generally don't grow downward.

We start by defining our cloud layer as an axis-aligned bounding box divided into cubic regions that represent cells of air. Later, this three-dimensional array of cells will become the *voxels* that are volumetrically rendered. Each cell consists of three states, each represented by a single bit:

- ▪ VAPOR_BIT, indicates whether the cell contains enough water vapor to form a cloud.

- PHASE_TRANSITION_BIT, indicates that the phase transition from vapor to droplets is ready to occur in this cell.
- HAS_CLOUD_BIT, indicates that the cell contains water droplets, and should be rendered as a part of a cloud.

The rules may be summarized simply:

1. A cell acquires the VAPOR_BIT if it currently has the VAPOR_BIT (which is initially set randomly throughout the cloud volume) and doesn't have the PHASE_TRANSITION_BIT.
2. A cell acquires the PHASE_TRANSITION_BIT if it does not have the PHASE_TRANSITION_BIT and it has the VAPOR_BIT, and one of its neighbors beside or below it also has the PHASE_TRANSITION_BIT.
3. A cell acquires the HAS_CLOUD_BIT if it has the HAS_CLOUD_BIT or the PHASE_TRANSITION_BIT.

Additionally, there is some random probability that a cell may spontaneously acquire humidity or phase transition or that it may lose the HAS_CLOUD_BIT as the cloud evaporates. As the cellular automaton continues to be iterated, this results in clouds randomly changing shape over time rather than reaching a steady state. We find that a ten percent probability of acquiring vapor or losing cloud status, and a 0.1 percent probability of spontaneously hitting a phase transition, produces satisfying results. This logic for a single iteration of the automaton is implemented in Listing 2.1, which assumes the existence of a 3D cells array for which each cell includes a character for its states and floating-point values for its phase transition, vapor, and extinction probabilities.

```
#define HAS_CLOUD_BIT             0x01
#define PHASE_TRANSITION_BIT     0x02
#define VAPOR_BIT                0x04

for (int i = 0; i < cellsAcross; i++)
{
    for (int j = 0; j < cellsDeep; j++)
    {
        for (int k = 0; k < cellsHigh; k++)
        {
            char phaseStates = 0;
```

```
if (i + 1 < cellsAcross)
    phaseStates |= cells[i + 1][j][k]->states;
if (j + 1 < cellsDeep)
    phaseStates |= cells[i][j + 1][k]->states;
if (k + 1 < cellsHigh)
    phaseStates |= cells[i][j][k + 1]->states;
if (i - 1 >= 0)
    phaseStates |= cells[i - 1][j][k]->states;
if (j - 1 >= 0)
    phaseStates |= cells[i][j - 1][k]->states;
if (k - 1 >= 0)
    phaseStates |= cells[i][j][k - 1]->states;
if (i - 2 >= 0)
    phaseStates |= cells[i - 2][j][k]->states;
if (i + 2 < cellsAcross)
    phaseStates |= cells[i + 2][j][k]->states;
if (j - 2 >= 0)
    phaseStates |= cells[i][j - 2][k]->states;
if (j + 2 < cellsDeep)
    phaseStates |= cells[i][j + 2][k]->states;
if (k - 2 >= 0)
    phaseStates |= cells[i][j][k - 2]->states;

bool phaseActivation =
    (phaseStates & PHASE_TRANSITION_BIT) != 0;

bool thisPhaseActivation = (cells[i][j][k]->states &
    PHASE_TRANSITION_BIT) != 0;

// Set whether this cell is in a phase transition state.
double rnd = random();        // Uniform within 0.0 - 1.0.

bool phaseTransition = ((!thisPhaseActivation) &&
    (cells[i][j][k]->states & VAPOR_BIT) && phaseActivation)
|| (rnd < cells[i][j][k]->phaseTransitionProbability);

if (phaseTransition)
    cells[i][j][k]->states |= PHASE_TRANSITION_BIT;
else
    cells[i][j][k]->states &= ~PHASE_TRANSITION_BIT;
```

```
        // Set whether this cell has acquired humidity.
        rnd = random();

        bool vapor = ((cells[i][j][k]->states & VAPOR_BIT) &&
            !thisAct) || (rnd < cells[i][j][k]->vaporProbability);

        if (vapor)
            cells[i][j][k]->states |= VAPOR_BIT;
        else
            cells[i][j][k]->states &= ~VAPOR_BIT;

        // Set whether this cell contains a cloud.
        rnd = random();

        bool hasCloud = ((cells[i][j][k]->states & HAS_CLOUD_BIT)
                || thisAct)
                && (rnd > cells[i][j][k]->extinctionProbability);

        if (hasCloud)
            cells[i][j][k]->states |= HAS_CLOUD_BIT;
        else
            cells[i][j][k]->states &= ~HAS_CLOUD_BIT;
    }
  }
}
```

Listing 2.1. Cellular automata for cloud formation.

You may have noticed that the probabilities for spontaneous acquisition of the vapor or phase transition states, as well as the probability for cloud extinction, are actually stored per cell rather than being applied globally. This is how we enforce the formation of distinct clouds within the automaton; each cloud within the layer is defined by an ellipsoid within the layer's bounding volume. Within each cloud's bounding ellipsoid, the phase and vapor probabilities approach zero toward the edges and the extinction probability approaches zero toward the center. Simply multiply the extinction probability by

$$\frac{x^2}{a^2} + \frac{y^2}{b^2} + \frac{z^2}{c^2},$$

where (x, y, z) is the position of the cell relative to the ellipsoid center, and

(a, b, c) are the semiaxis lengths of the ellipsoid. Subtract this expression from one to modulate the phase and vapor probabilities with distance from the ellipsoid center. As an optimization, the cellular automaton may be limited to cells contained by these ellipsoids, or each ellipsoid may be treated as independent cellular automata to eliminate the storage and rendering overhead of cells that are always empty. If you're after cumulus clouds with flattened bottoms, using hemiellipsoids as bounding volumes for the clouds instead of ellipsoids is also more efficient.

You may grow your simulated clouds by placing a few random phase transition seeds at the center of each ellipsoid and iterating over the cellular automaton a few times. The resulting 3D array of cloud states may then be stored for rendering, or you may continue to iterate at runtime, smoothing the cloud states in the time domain to produce real-time animations of cloud growth. In reality, however, clouds change their shape very slowly—their growth and extinction is generally only noticeable in time-lapse photography.

Simulating the Distribution of Clouds

We discussed using bounding ellipsoids to contain individual clouds within our simulation, but how do we position and size these ellipsoids? Some approaches leave the modeling of clouds entirely to artists or level designers [Wang 2004], but procedural approaches exist to make the generation of realistic cloud volumes easier. The Plank exponential model [Plank 1969] is one such technique, based on the analysis of experimental data of cloud size distributions over Florida.

Plank found an exponential relationship between cloud sizes and their density in a region of cumulus clouds; further, he found there is an upper bound of cumulus cloud size at any given time of day, and there are fewer of these large clouds than smaller clouds.

His algorithm may be implemented by iteratively calling the `GetNextCloud()` method shown in Listing 2.2 until it returns false. `GetNextCloud()` is assumed to be a method of a class that is initialized with the desired cloud coverage, area, and minimum and maximum cloud sizes. For the constants referenced in the code, we use an `alpha` value of 0.001, `chi` of 0.984, `nu` of 0.5, and `beta` of −0.10. We use a minimum cloud size of 500 meters, a maximum of 5000 meters, and an `epsilon` of 100 meters.

```
bool GetNextCloud(double& width, double& depth, double& height)
{
    while (currentN >= targetN)
```

```
{
    currentD -= epsilon;
    if (currentD <= GetMinimumSize()) return (false);

    currentN = 0;
    targetN = (int) (((2.0 * GetDesiredArea() * epsilon * alpha *
        alpha * alpha * GetDesiredCoverage()) / (PI * chi)) *
        exp(-alpha * (currentD)));
}

if (currentD <= GetMinimumSize()) return (false);

// select random diameter within currentD += bandwidth
double variationW = random() * (epsilon);
double variationH = random() * (epsilon);

width = currentD - epsilon * 0.5 + variationW;
depth = currentD - epsilon * 0.5 + variationH;

double D = (width + depth) * 0.5;
double hOverD = nu * pow(D / GetMaximumSize(), beta);
height = D * hOverD;

currentN++;
return (true);
}
```

Listing 2.2. The Plank exponential cloud size distribution model.

2.2 Cloud Lighting Techniques

Clouds are not the simple translucent objects you might think they are—
somehow, these collections of water droplets and air turn opaque and white and
end up brighter than the sky behind them. As light enters a cloud, it is deflected
many times by water vapor particles (which are highly reflective) in directions
that are not entirely random. This is known as *multiple scattering*, and our task is
to find a way to approximate it without tracing every ray of light as it bounces
around inside the cloud.

As it turns out, for thin slabs of clouds (<100 m), *single scattering* accounts
for most of the radiative transfer [Bouthors 2008]. That is, most of the light only

bounces off a single cloud droplet at this distance. This means we may approximate light transport through a cloud by dividing it up into chunks of 100 meters or less, computing how much light is scattered and absorbed by single scattering within each chunk, and passing the resulting light as the incident light into the next chunk. This technique is known as *multiple forward scattering* [Harris 2002]. It benefits from the fact that computing the effects of single scattering is a well-understood and simple problem, while the higher orders of scattering may only be solved using computationally expensive Monte Carlo ray tracing techniques. These higher orders of scattering are increasingly diffuse, meaning we can reasonably approximate the missing scattered light with a simple ambient term.

As you might guess, the chunks of cloud we just described map well to the voxels we generated from the cellular automaton above. Multiple forward scattering computes the color and transparency of a given voxel by shooting a ray from the light source toward the voxel and iteratively compositing the scattering and extinction from each voxel we pass through. Essentially, we accumulate the scattering and absorption of light on a voxel-by-voxel basis, producing darker and more opaque voxels the deeper we go into the cloud.

To compute the transparency of a given voxel in isolation, we need to compute its *optical depth* τ [Blinn 1982], given by

$$\tau = n\pi p^2 T.$$

Here, n is the number of water droplets per unit volume, p is the effective radius of each droplet, and T is the thickness of the voxel. Physically realistic values of p in cumulus clouds are around 0.75 μm, and n is around 400 droplets per cubic centimeter [Bouthors 2008]. The extinction of light α within this voxel is then

$$\alpha = 1 - e^{-\tau}.$$

This informs us as to the transparency of the voxel, but we still need to compute its color due to forward scattering. The voxel's color $\mathbf{C}$ is given by

$$\mathbf{C} = \frac{a\tau \mathbf{L} P(\cos\Theta)}{4\pi}.$$

Here, a is the albedo of the water droplets, which is very high—between 0.9 and 1.0. $\mathbf{L}$ is the light color incident on the voxel (which itself may be physically simulated [Kane 2010]), and $P(\cos\Theta)$ is the *phase function* of the cloud, which is a function of the dot product between the view direction and the light direction.

The phase function is where things get interesting. Light has a tendency to scatter in the forward direction within a cloud; this is what leads to the bright "silver lining" you see on clouds that are lit from behind. The more accurate a phase function you use, the more realistic your lighting effects are.

A simple phase function is the *Rayleigh function* [Rayleigh 1883]. Although it is generally used to describe the scattering of atmospheric molecules and is best known as the reason the sky is blue, it turns out to be a reasonable approximation of scattering from cloud droplets under certain cloud densities and wavelengths [Petty 2006] and has been used successfully in both prior research [Harris 2002] and commercial products.[1] The Rayleigh function is given by

$$P(\cos\Theta) = \frac{3}{4}(1 - \cos^2\Theta).$$

The Rayleigh function is simple enough to execute in a fragment program, but one problem is that it scatters light equally in the backward direction and the forward direction. For the larger particles that make up a typical cloud, the *Henyey-Greenstein function* [Henyey and Greenstein 1941] provides a better approximation:

$$P(\cos\Theta) = \frac{1 - g^2}{\left(1 + g^2 - 2g\cos\Theta\right)^{3/2}}.$$

The parameter g describes the asymmetry of the function, and is typically high (~ 0.99). Positive values of g produce forward scattering, and negative values produce backward scattering. Since a bit of both actually occurs, more sophisticated implementations actually use a *double-lobed* Henyey-Greenstein function. In this case, two functions are evaluated—one with a positive value of g and one with a negative value, and they are blended together, heavily favoring the positive (forward) scattering component.

The ultimate phase function is given by *Mie theory*, which simulates actual light waves using Maxwell's equations in three-dimensional space [Bohren and Huffman 1983]. It is dependent on the droplet size distribution within the cloud, and as you can imagine, it is very expensive to calculate. However, a free tool called MiePlot[2] is available to perform offline solutions to Mie scattering, which may be stored in a texture to be looked up for a specific set of conditions. Mie scattering not only gives you silver linings but also wavelength-dependent

---

[1] For example, SilverLining. See http://www.sundog-soft.com/.
[2] See http://www.philiplaven.com/mieplot.htm.

effects such as fogbows and glories. It has been applied in real-time successfully by either chopping the function's massive forward peak [Bouthers et al. 2006] or restricting it to phase angles where wavelength-dependent effects occur, and using simpler phase functions for other angles [Petty 2006].

If you rely exclusively on multiple forward scattering of the incident sunlight on a cloud, your cloud will appear unnaturally dark. There are other light sources to consider—skylight, reflected light from the ground, and the light from higher-order scattering should not be neglected. We approximate these contributions with an ambient term; more sophisticated implementations may use hemisphere lighting techniques [Rost and Licea-Kane 2009] to treat skylight from above and light reflected from the ground below independently.

Tone mapping and gamma correcting the final result are also vitally important for good image quality. We use a gamma value of 2.2 together with the simplest form of the Reinhard tone-mapping operator [Reinhard et al. 2002] with good results:

$$\mathbf{L}_d = \frac{\mathbf{L}}{1 + \mathbf{L}}.$$

For added realism, you'll also want to simulate atmospheric perspective effects on distant clouds. Exponentially blending the clouds into the sky with distance is a simple approach that's generally "good enough," although more rigorous approaches are available [Preetham et al. 1999].

2.3 Cloud Rendering Techniques

At this point, we have a means of generating a 3D volume of voxels representing a collection of clouds and an algorithm for lighting these voxels. Visualizing the volumetric data described by these voxels may be achieved through a variety of techniques.

Volumetric Splatting

The simplest technique is called *splatting* and is illustrated in Figure 2.2. Each voxel is represented by a billboard that represents an individual cloud puff. Mathematically, this texture should represent a Gaussian distribution, but adding some wispy detail and randomly rotating it produces visually appealing results. Figure 2.3 illustrates how a single texture representing a cloud puff is used to generate a realistic scene of cumulus clouds.

Figure 2.2. Volumetric cloud data rendered using splatting. (*Image courtesy of Sundog Software, LLC.*)

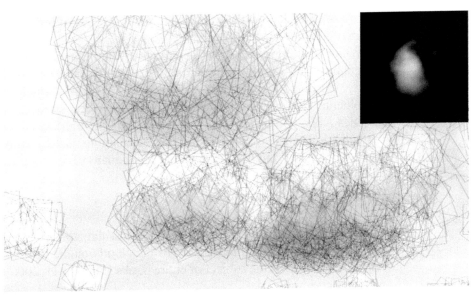

Figure 2.3. Wireframe overlay of splatted clouds with the single cloud puff texture used (inset). (*Image courtesy Sundog Software, LLC.*)

Lighting and rendering are achieved in separate passes. In each pass, we set the blending function to (ONE, ONE_MINUS_SRC_ALPHA) to composite the voxels together. In the lighting pass, we set the background to white and render the voxels front to back from the viewpoint of the light source. As each voxel is rendered, the incident light is calculated by multiplying the light color and the frame buffer color at the voxel's location prior to rendering it; the color and transparency of the voxel are then calculated as above, stored, and applied to the billboard. This technique is described in more detail by Harris [2002].

When rendering, we use the color and transparency values computed in the lighting pass and render the voxels in back-to-front order from the camera.

To prevent breaking the illusion of a single, cohesive cloud, we need to ensure the individual billboards that compose it aren't perceptible. Adding some random jitter to the billboard locations and orientations helps, but the biggest issue is making sure all the billboards rotate together in unison as the view angle changes. The usual trick of axis-aligned billboards falls apart once the view angle approaches the axis chosen for alignment. Our approach is to use two orthogonal axes against which our billboards are aligned. As the view angle approaches the primary axis (pointing up and down), we blend toward using our alternate (orthogonal) axis instead.

To ensure good performance, the billboards composing a cloud must be rendered as a vertex array and not as individual objects. Instancing techniques and/or geometry shaders may be used to render clouds of billboards from a single stream of vertices.

While splatting is fast for sparser cloud volumes and works on pretty much any graphics hardware, it suffers from fill rate limitations due to high depth complexity. Our lighting pass also relies on pixel read-back, which generally blocks the pipeline and requires rendering to an offscreen surface in most modern graphics APIs. Fortunately, we only need to run the lighting pass when the lighting conditions change. Simpler lighting calculations just based on each voxel's depth within the cloud from the light direction may suffice for many applications, and they don't require pixel read-back at all.

Volumetric Slicing

Instead of representing our volumetric clouds with a collection of 2D billboards, we can instead use a real 3D texture of the cloud volume itself. Volume rendering of 3D textures is the subject of entire books [Engel et al. 2006], but we'll give a brief overview here.

The general idea is that some form of simple proxy geometry for our volume is rendered using 3D texture coordinates relative to the volume data. We then get the benefit of hardware bilinear filtering in three dimensions to produce a smooth image of the 3D texture that the proxy geometry slices through.

In volumetric slicing, we render a series of camera-facing polygons that divide up the volume at regular sampling intervals from the volume's farthest point from the camera to its nearest point, as shown in Figure 2.4. For each polygon, we compute the 3D texture coordinates of each vertex and let the GPU do the rest.

It's generally easiest to do this in view coordinates. The only hard part is computing the geometry of these slicing polygons; if your volume is represented by a bounding box, the polygons that slice through it may have anywhere from three to six sides. For each slice, we must compute the plane-box intersection and turn it into a triangle strip that we can actually render. We start by transforming the bounding box into view coordinates, finding the minimum and maximum z coordinates of the transformed box, and dividing this up into equidistant slices along the view-space z-axis. The Nyquist-Shannon sampling algorithm [Nyquist 1928] dictates that we must sample the volume at least twice per voxel diameter to avoid sampling artifacts.

For each sample, we test for intersections between the x-y slicing plane in view space and each edge of the bounding box, collecting up to six intersection points. These points are then sorted in counterclockwise order around their center and tessellated into a triangle strip.

The slices are then rendered back to front and composited using the blending function (ONE, ONE_MINUS_SRC_ALPHA). As with splatting, the lighting of each voxel needs to be done in a separate pass. One technique is to orient the slicing planes halfway between the camera and the light source, and accumulate the voxel colors to a pixel buffer during the lighting pass. We then read from this pixel buffer to obtain the light values during the rendering pass [Ikits et al. 2007].

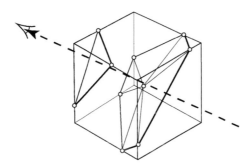

Figure 2.4. Slicing proxy polygons through a volume's bounding box.

Unfortunately, half-angle slicing requires sampling the volume even more frequently, and we're still left with massive demands on fill rate as a result. In the end, we really haven't gained much over splatting.

GPU Ray Casting

Real-time ray casting may sound intimidating, but in many ways it's the simplest and most elegant technique for rendering clouds. It does involve placing most of the computing in a fragment program, but if optimized carefully, high frame rates are achievable with precise per-fragment lighting. Figure 2.5 shows a dense, 60 square kilometer stratocumulus cloud layer rendering at over 70 frames per second on consumer-grade hardware using GPU ray casting.

The general idea is to just render the bounding box geometry of your clouds (with back-face culling enabled) and let the fragment processor do the rest. For each fragment of the bounding box, our fragment program shoots a ray through it from the camera and computes the ray's intersection with the bounding volume. We then sample our 3D cloud texture along the ray within the volume from front to back, compositing the results as we go.

Figure 2.5. Volumetric cloud data rendered from a single bounding box using GPU ray casting. (*Image courtesy Sundog Software, LLC.*)

The color of each sample is determined by shooting another ray to it from the light source and compositing the lighting result using multiple forward scattering—see Figure 2.6 for an illustration of this technique.

It's easy to discard this approach, thinking that for every fragment, you need to sample your 3D texture hundreds of times within the cloud, and then sample each sample hundreds of more times to compute its lighting. Surely that can't scale! But, there's a dirty little secret about cloud rendering that we can exploit to keep the actual load on the fragment processor down: clouds aren't really all that transparent at all, and by rendering from front to back, we can terminate the processing of any given ray shortly after it intersects a cloud.

Recall that we chose a voxel size on the order of 100 meters for multiple forward scattering-based lighting because in larger voxels, higher orders of scattering dominate. Light starts bouncing around shortly after it enters a cloud, making the cloud opaque as soon as light travels a short distance into it. The mean free path of a cumulus cloud is typically only 10 to 30 meters [Bouthors et al. 2008]—beyond that distance into the cloud, we can safely stop marching our ray into it. Typically, only one or two samples per fragment really need to be lit, which is a wonderful thing in terms of depth complexity and fill rate.

The first thing we need to do is compute the intersection of the viewing ray for the fragment with the cloud's axis-aligned bounding box. To simplify our calculations in our fragment program, we work exclusively in 3D texture coordinates relative to the bounding box, where the texture coordinates of the box range from 0.0 to 1.0. An optimized function for computing the intersection of a ray with this unit cube is shown in Listing 2.3.

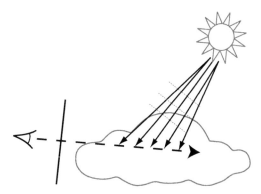

Figure 2.6. GPU ray casting. We shoot a ray into the volume from the eye point, terminating once an opacity threshold is reached. For each sample along the ray, we shoot another ray from the light source to determine the sample's color and opacity.

```
void getVolumeIntersection(in vec3 pos, in vec3 dir, out float tNear,
        out float tFar)
{
    // Intersect the ray with each plane of the box.
    vec3 invDir = 1.0 / dir;
    vec3 tBottom = -pos * invDir;
    vec3 tTop = (1.0 - pos) * invDir;

    // Find min and max intersections along each axis.
    vec3 tMin = min(tTop, tBottom);
    vec3 tMax = max(tTop, tBottom);

    // Find largest min and smallest max.
    vec2 t0 = max(tMin.xx, tMin.yz);
    tNear = max(t0.x, t0.y);
    t0 = min(tMax.xx, tMax.yz);
    tFar = min(t0.x, t0.y);

    // Clamp negative intersections to 0.
    tNear = max(0.0, tNear);
    tFar = max(0.0, tFar);
}
```

Listing 2.3. Intersection with a ray and unit cube.

With the parameters of our viewing ray in hand, the rest of our fragment program becomes straightforward. We sample this ray in front-to-back order from the camera; if a sample contains a cloud voxel, we then shoot another ray from the light source to determine the sample's color. We composite these samples together until an opacity threshold is reached, at which point we terminate the ray early. Listing 2.4 illustrates this technique.

```
uniform sampler3D    volumeData;
uniform sampler3D    noiseData;
uniform vec3         cameraTexCoords;        // The camera in texture coords
uniform vec3         lightTexCoords;         // The light dir in tex coords
uniform float        extinction;             // 1 - exp(optical depth)
uniform float        constTerm;              // (albedo * optical depth)/4*pi
uniform vec3         viewSampleDimensions;   // Size of view sample
```

```
                                             // in texcoords
uniform vec3      lightSampleDimensions;  // Size of light sample,
                                          // texcoords
uniform vec3      skyLightColor;          // RGB sky light component
uniform vec3      multipleScatteringTerm; // RGB higher-order
                                          // scattering term
uniform vec4      lightColor;             // RGBA direct sun light color

void main()
{
    vec3 texCoord = gl_TexCoord[0].xyz;

    vec3 view = texCoord - cameraTexCoords;
    vec3 viewDir = normalize(view);

    // Find the intersections of the volume with the viewing ray.
    float tminView, tmaxView;
    getVolumeIntersection(cameraTexCoords, viewDir,
        tminView, tmaxView);

    vec4 fragColor = vec4(0, 0, 0, 0);

    // Compute the sample increments for the main ray and
    // the light ray.
    vec3 viewSampleInc = viewSampleDimensions * viewDir;
    float viewInc = length(viewSampleInc);
    vec3 lightSampleInc = lightSampleDimensions * lightTexCoords;

    // Ambient term to account for skylight and higher orders
    // of scattering.
    vec3 ambientTerm = skyLightColor + multipleScatteringTerm;

    // Sample the ray front to back from the camera.
    for (float t = tminView; t <= tmaxView; t += viewInc)
    {
        vec3 sampleTexCoords = cameraTexCoords + viewDir * t;

        // Look up the texture at this sample, with fractal noise
        // applied for detail.
        float texel = getCloudDensity(sampleTexCoords);
        if (texel == 0) continue;
```

```
// If we encountered a cloud voxel, compute its lighting.
// It's faster to just do 5 samples, even if it overshoots,
// than to do dynamic branching and intersection testing.

vec4 accumulatedColor = lightColor;
vec3 samplePos = sampleTexCoords + lightSampleInc * 5;

vec3 scattering = lightColor * constTerm;

for (int i = 0; i < 5; i++)
{
    float lightSample = texture3D(volumeData, samplePos).x;

    if (lightSample != 0)
    {
        // Multiple forward scattering:
        vec4 srcColor;
        srcColor.xyz = accumulatedColor.xyz * scattering
            * phaseFunction(1.0);
        srcColor.w = extinction;
        srcColor *= lightSample;

        // Composite the result:
        accumulatedColor = srcColor + (1.0 - srcColor.w)
            * accumulatedColor;
    }

    samplePos -= lightSampleInc;
}

vec4    fragSample;

// Apply our phase function and ambient term.
float cosa = dot(viewDir, lightTexCoords);
fragSample.xyz = accumulatedColor.xyz * phaseFunction(cosa)
    + ambientTerm;
fragSample.w = extinction;

// apply texture and noise
fragSample *= texel;
```

```
    // "Under operator" for compositing:
    fragColor = fragColor + (1.0 - fragColor.w) * fragSample;

    // Early ray termination!
    if (fragColor.w > 0.95) break;
  }

  // Apply tone-mapping and gamma correction to final result.
  toneMap(fragColor);

  gl_FragColor = fragColor * gl_Color;
}
```

Listing 2.4. GPU ray casting of a cloud.

Note that the texture lookup for each sample isn't just a simple `texture3D` call—it calls out to a `getCloudDensity()` function instead. If you rely on the 3D volume data alone, your clouds will look like nicely shaded blobs. The `get-CloudDensity()` function needs to add in procedural noise for realistic results—we upload a 32^3-texel RGB texture of smoothed random noise, and apply it as a displacement to the texture coordinates at a couple of octaves to produce fractal effects. Perlin noise [Perlin 1985] would also work well for this purpose. An example implementation is shown in Listing 2.5; the `noiseOffset` uniform vector is used to animate the noise over time, creating turbulent cloud animation effects.

```
float getCloudDensity(in vec3 texCoord)
{
   vec3 r = viewSampleDimensions.xyz * 32.0;

   vec3 perturb = vec3(0, 0, 0);
   vec3 uvw = ((texCoord + noiseOffset) / viewSampleDimensions) / 256.0;
   perturb += 1.0 * texture3D(noiseData, 2.0 * uvw).xyz - 0.5;
   perturb += 0.5 * texture3D(noiseData, 4.0 * uvw).xyz - 0.25;

   return (texture3D(volumeData, texCoord + perturb * r).x);
}
```

Listing 2.5. Adding fractal noise to your clouds.

Unfortunately, adding high-frequency noise to the cloud impacts performance. We restrict ourselves to two octaves of noise because filtered 3D texture lookups are expensive, and high frequency noise requires us to increase our sampling rate to at least double the noise frequency to avoid artifacts. This may be mitigated by fading out the higher octaves of noise with distance from the camera and adaptively sampling the volume such that only the closer samples are performed more frequently. The expense of adding procedural details to ray-casted clouds is the primary drawback of this technique compared to splatting, where the noise is just part of the 2D billboards used to represent each voxel.

Intersections between the scene's geometry and the cloud volume need to be handled explicitly with GPU ray-casting; you cannot rely on the depth buffer, because the only geometry you are rendering for the clouds is its bounding box. While this problem is best just avoided whenever possible, it may be handled by rendering the depth buffer to a texture, and reading from this texture as each sample along the viewing ray is computed. If the projected sample is behind the depth buffer's value, it is discarded.

Viewpoints inside the volume also need special attention. The intersection code in Listing 2.3 handles this case properly, but you will need to detect when the camera is inside the volume, and flip the faces of the bounding box being rendered to represent it.

Three-dimensional textures consume a lot of memory on the graphics card. A $256 \times 256 \times 32$ array of voxels each represented by a single byte consumes two megabytes of memory. While VRAM consumption was a show-stopper on early 3D cards, it's easily handled on modern hardware. However, addressing that much memory at once can still be slow. Swizzling the volume texture by breaking it up into adjacent, smaller bricks can help with cache locality as the volume is sampled, making texture lookups faster.

Fog and blending of the cloud volume is omitted in Listing 2.4 but should be handled for more realistic results.

Hybrid Approaches

Although GPU ray casting of clouds can be performant on modern graphics hardware, the per-fragment lighting calculations are still expensive. A large number of computations and texture lookups may be avoided by actually performing the lighting on the CPU and storing the results in the colors of the voxels themselves to be used at rendering time. Recomputing the lighting in this manner results in a pause in framerate whenever lighting conditions change but makes rendering the cloud extremely fast under static lighting conditions. By eliminating the lighting calculations from the fragment processor, we're just left with finding the first intersection of the view ray with a cloud and terminating the ray early—we've now rendered our cloud volume with an effective depth complexity close to one!

Volumetric slicing may also benefit from a hybrid approach. For example, a fragment program may be employed to perform lighting of each fragment using GPU ray casting, while still relying on the slice geometry to handle compositing along the view direction.

Other approaches render a cloud as a mesh [Bouthors et al. 2008], again taking advantage of the low mean free path of cumulus clouds. This allows more precise lighting and avoids the intersection problems introduced by proxy geometry in volume rendering.

Ultimately, choosing a cloud rendering technique depends on the trade-offs you're willing to make between hardware compatibility, physical realism, and performance—fortunately, there are a variety of techniques to choose from.

References

[Blinn 1982] James F. Blinn. "Light Reflection Functions for Simulation of Clouds and Dusty Surfaces." *Computer Graphics (Proceedings of SIGGRAPH 82)* 16:3, ACM, pp. 21–29.

[Bohren and Huffman 1983] Craig F. Bohren and Donald R. Huffman. *Absorption and Scattering of Light by Small Particles*. New York: Wiley-Interscience, 1983.

[Bouthers et al. 2006] Antoine Bouthors, Fabrice Neyret, and Sylvain Lefebvre. "Real-Time Realistic Illumination and Shading of Stratiform Clouds." *Eurographics Workshop on Natural Phenomena*, 2006, pp. 41–50.

[Bouthors 2008] Antoine Bouthors. "Real-Time Realistic Rendering of Clouds." PhD dissertation, Grenoble Universitès, 2008.

[Bouthors et al. 2008] Antoine Bouthors, Fabrice Neyret, Nelson Max, Eric Bruneton, and Cyril Crassin. "Interactive Multiple Anisotropic Scattering in Clouds." *ACM Symposium on Interactive 3D Graphics and Games (I3D)*, 2008.

[Dobashi et al. 2000] Yoshinori Dobashi, Kazufumi Kaneda, Hideo Yamashita, Tsuyoshi Okita, and Tomoyuki Nishita. "A Simple, Efficient Method for Realistic Animation of Clouds." *Proceedings of SIGGRAPH 2000*, ACM Press / ACM SIGGRAPH, Computer Graphics Proceedings, Annual Conference Series, ACM, pp. 19–28.

[Engel et al. 2006] Klaus Engel, Markus Hadwiger, Joe M. Kniss, Christof Rezk-Salama, and Daniel Weiskopf. *Real-Time Volume Graphics*. Wellesley, MA: A K Peters, 2006.

[Harris 2002] Mark Harris. "Real-Time Cloud Rendering for Games." *Game Developers Conference*, 2002.

[Henyey and Greenstein 1941] L. Henyey and J. Greenstein. "Diffuse Reflection in the Galaxy." *Astrophysics Journal* 93 (1941), p. 70.

[Ikits et al. 2007] Milan Ikits, Joe Kniss, Aaron Lefohn, and Charles Hansen. "Volume Rendering Techniques." *GPU Gems*, edited by Randima Fernando. Reading, MA: Addison-Wesley, 2007.

[Kajiya and Herzen 1984] James T. Kajiya and Brian P. Von Herzen. "Ray Tracing Volume Densities." *Computer Graphics* 18:3 (July 1984), ACM, pp. 165–174.

[Kane 2010] Frank Kane. "Physically-Based Outdoor Scene Lighting." *Game Engine Gems 1*, edited by Eric Lengyel. Sudbury, MA: Jones and Bartlett, 2010.

[Nagel and Raschke 1992] K. Nagel and E. Raschke. "Self-Organizing Criticality in Cloud Formation?" *Physica A* 182:4 (April 1992), pp. 519–531.

[Nyquist 1928] Harry Nyquist. "Certain Topics in Telegraph Transmission Theory." *AIEE Transactions* 47 (April 1928), pp. 617–644.

[Perlin 1985] Ken Perlin. "An Image Synthesizer." *Computer Graphics (Proceedings of Siggraph 85)* 19:3, ACM, pp. 287–296.

[Petty 2006] Grant W. Petty. *A First Course in Atmospheric Radiation*. Madison, WI: Sundog Publishing, 2006.

[Plank 1969] Vernon G. Plank. "The Size Distribution of Cumulus Clouds in Representative Florida Populations." *Journal of Applied Meteorology* 8 (1969), pp. 46–67.

[Preetham et al. 1999] Arcot J. Preetham, Peter Shirley, and Brian Smits. "A Practical Analytic Model for Daylight." *Proceedings of SIGGRAPH 1999*, ACM Press / ACM SIGGRAPH, Computer Graphics Proceedings, Annual Conference Series, ACM, pp. 91–100.

[Rayleigh 1883] Lord Rayleigh. "Investigation of the Character and Equilibrium of an Incompressible Heavy Fluid of Variable Density." *Proceedings of the London Mathematical Society* 14 (1883), pp. 170–177.

[Reinhard et al. 2002] Erik Reinhard, Michael Stark, Peter Shirley, and James Ferwerda. "Photographic Tone Reproduction for Digital Images." *Proceedings of SIGGRAPH 2002*, ACM Press / ACM SIGGRAPH, Computer Graphics Proceedings, Annual Conference Series, ACM, pp. 267–276.

[Rost and Licea-Kane 2009] Radni J. Rost and Bill Licea-Kane. *OpenGL Shading Language*, Third Edition. Reading, MA: Addison-Wesley Professional, 2009.

[Wang 2004] Niniane Wang. "Realistic and Fast Cloud Rendering." *Journal of Graphics Tools* 9:3 (2004), pp. 21–40.

3

Simulation of Night-Vision and Infrared Sensors

Frank Kane

Sundog Software, LLC

Many action games simulate infrared (IR) and night-vision goggles (NVG) by simply making the scene monochromatic, swapping out a few textures, and turning up the light sources. We can do better. Rigorous simulations of IR and NVG sensors have been developed for military training and simulation applications, and we can apply their lessons to game engines. The main differences between visible, IR, and near-IR wavelengths are easily modeled. Sensors may also include effects such as light blooms, reduced contrast, blurring, atmospheric transmittance, and reduced resolution that we can also simulate, adding to the realism.

3.1 The Physics of the Infrared

The world of the infrared is a very different place from the world of visible light—you're not just seeing reflected sunlight, you're seeing how objects radiate heat. Accurately representing an IR scene requires understanding some basic thermodynamics.

Fortunately, the bit we need isn't very complicated—we can get by with just an understanding of the *Stefan-Boltzmann law*. It tells us that the black body radiation j^* of an object is given by

$$j^* = \varepsilon \sigma T^4.$$

Here, T is the absolute temperature of the object (in Kelvins), ε is the *thermal emissivity* of the material, and σ is the Stefan-Boltzmann constant, 5.6704×10^{-8} $\text{J} \cdot \text{s}^{-1} \cdot \text{m}^{-2} \cdot \text{K}^{-4}$. If the ambient temperature and temperature of the objects in your scene remain constant, the radiation emitted may be precomputed and baked into special IR versions of your texture maps. Figure 3.1 illustrates texture-based IR simulation of a tank (note the treads and engine area are hot). If you want to simulate objects cooling off over time, just store the emissivity in your materials and/or textures and compute the equation above in a vertex program as the temperature varies.

Table 3.1 lists emissivity values for some common materials at 8 μm. Note that most organic materials have high emissivity and behave almost like ideal black bodies, while metals are more reflective and have lower IR emissions.

It is the subtle differences in emissivity that distinguish different materials in your scene that are at the same ambient temperature; although the differences may be small, they are important for adding detail. For objects that emit heat of their own (i.e., the interior of a heated house or living organisms), the changes in temperature are the main source of contrast in your IR scene. Stop thinking about modeling your materials and textures in terms of RGB colors, but rather in terms of emissivity and absolute temperature. An alternate set of materials and/or textures is required.

To add additional detail to the IR scene, there is a physical basis to blending in the visible light texture as well—in a pinch, this lets you repurpose the visible-

Figure 3.1. Simulated image of physically based IR fused with visible lights. (*Image courtesy of SDS International.*)

Material	Emissivity	Material	Emissivity
Aluminum	0.20	Paint	0.84 (white)–0.99 (black)
Asphalt	0.95	Paper	0.95
Cast iron	0.70–0.90	Plastic	0.95
Clay	0.95	Rusted iron	0.60–0.90
Cloth	0.95	Sand	0.90
Glass	0.85	Snow	0.90
Granite	0.90	Soil	0.90–0.98
Grass	0.98	Steel	0.80–0.90 (cold-rolled), 0.25 (polished)
Gravel	0.95	Tin	0.10–0.30
Human skin	0.95	Water	0.93
Ice	0.98	Wood	0.90–0.95

Table 3.1. Emissivity of common materials.

light textures of your objects as detail for the thermal information. There is a range of about 0.15 in the emissivity between white and black objects; light colors reflect more heat, and dark colors absorb it. Your fragment shader may convert the RGB values of the visible-light textures to monochromatic luminance values and perturb the final emissivity of the fragment accordingly. Listing 3.1 illustrates a snippet of a fragment shader that might approximate the thermal radiation of a fragment given its visible color and knowledge of its underlying material's emissivity and temperature. This is a valid approach only for objects that do not emit their own heat; for these objects, emissivities and temperatures should be encoded directly in specialized IR textures, rather than blending the visible-light texture with vertex-based thermal properties.

```
uniform sampler2D    visibleTexture;
uniform sampler2D    thermalTexture;
uniform float        level;
uniform float        gain;
uniform float        stefanBoltzmannConstant;
uniform float        emissivityBlendFactor;
```

```
varying float          materialEmissivity;
varying float          materialTemperature;

void main()
{
    vec3 texCoords = gl_TexCoord[0].xy;
    vec3 visibleColor = texture2D(visibleTexture, texCoords).xyz;

    // Convert color to luminance.
    float visibleLuminance = dot(vec3(0.2126, 0.7152, 0.0722),
                                 visibleColor);

    // Convert luminance to thermal emissivity.
    float emissivityFromColor = (1.0 - visibleLuminance) * 0.15 + 0.84;

    // Blend the material-based emissivity with the color-based.
    float finalEmissivity = mix(materialEmissivity,
        emissivityFromColor, emissivityBlendFactor);

    // Stefan-Boltzmann equation:
    float radiation = finalEmissivity * stefanBoltzmannConstant *
        materialTemperature * materialTemperature *
        materialTemperature * materialTemperature;

    // Apply auto-gain control.
    float mappedRadiation = (radiation * gain) + level;

    // In a "white-hot system," we're done:
    gl_FragColor = vec4(mappedRadiation, mappedRadiation,
                        mappedRadiation, 1.0);
}
```

Listing 3.1. A simplified fragment shader for IR sensor simulation using a hybrid of visible-light textures with material-based thermal properties.

Atmospheric transmittance should also be applied to IR scenes, just as we apply fog to simulate visibility effects in visible light. In the IR band, however, visibility is largely ruled by the humidity in the air and not by the particles in it. Water in the air absorbs heat quite effectively, although IR still provides superior visibility in foggy conditions. Military-grade simulations apply rigorous atmospheric transmittance models for this purpose such as LOWTRAN[1] and MOD-TRAN,[2] but we can get by with a simpler approximation. Atmospheric transmittance from water vapor for a wavelength λ in micrometers is approximated [Bonjean et al. 2006] by

$$\tau_\lambda = \left(\tau_{\omega\lambda}\right)^{m\omega/20}.$$

Here, ω is the depth of precipitable water in millimeters; this is a measure of the air's humidity. Typical values range from 10 to 60 mm. The value $\tau_{\omega\lambda}$ is given by:

$$\tau_{\omega\lambda} = 10^{-0.0075/\lambda^2}.$$

The value of m represents the *air mass*, which is a measure of how much atmosphere the heat source has passed through before reaching the sensor. This is a unitless value, normalized to one for a vertical path through the atmosphere at sea level. More sophisticated implementations for "serious games" may also take atmospheric scattering and scattering from dust into account.

3.2 Simulating Infrared Sensor Effects

The Stefan-Boltzmann law gives you the simulated radiation being emitted from an object in your scene, but that's not directly useful for rendering. You need to map these radiation values into monochromatic luminance values that are displayable—real sensors have a feature called *auto gain control* that does this. Typically, they compute the range of radiation values in the visible scene and map these values linearly from 0.0 to 1.0. The range being mapped can be quite high—the night sky's temperature is near absolute zero, while a missile's exhaust could be several thousand Kelvins [Thomas 2003].

The actual colors used to represent this radiation vary by the type of sensor. Most forward-looking infrared (FLIR) sensors used by the military represent IR in black and white, and are configurable between "black-hot" and "white-hot" polarities (shown in Figure 3.2). Thermal imaging sensors may blend between red for hot spots in the image and blue for cold spots. These are all easy effects to achieve in a fragment program.

In addition to modeling gain and polarity, there are many imperfections in IR sensors that we can also emulate. Older CRT-based devices were prone to effects

---

[1] See http://www1.ncdc.noaa.gov/pub/software/lowtran/.
[2] See http://www.modtran.org/.

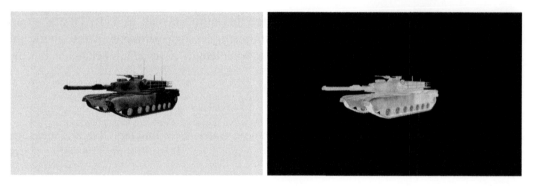

Figure 3.2. Black-hot (left) and white-hot (right) simulated IR images of a tank. (*Images courtesy of SDS International.*)

from *capacitive coupling*, which produced dark streaks surrounding hot spots in the final image. This can be emulated by using a convolution filter [Bhatia and Lacy 1999] or, more simply, by applying a textured quad representing the streaks over the hotter objects.

More important is simulating the resolution of the device you are emulating; your monitor resolution is likely much higher than the resolution of the IR sensor. Furthermore, IR images tend to lose high-frequency details—your simulated image should be blurred to simulate this effect as well. Both effects may be achieved by rendering the simulated image to an offscreen surface. To simulate blurring, render the image to a texture smaller than the sensor resolution, and let hardware bilinear filtering blur it when you scale it up on a quad rendered at the sensor resolution. That quad could in turn be rendered to another texture, which is scaled up to the monitor resolution using GL_NEAREST filtering to pixelate it and simulate the lower resolution of the sensor.

Persistence is another effect we may simulate easily by using an accumulation buffer [Shreiner 2006] or velocity-depth-gradient buffers [Lengyel 2010]. This is similar to motion blur, where hot spots tend to smear a bit across the display as they move.

Images from sensors also include noise, often significantly; in older IR sensors, this would manifest itself as visible streaks in the scan lines of the final image. A Gaussian distribution of random noise used to stretch out pixels by a given distance over the image accomplishes this effect. For more modern sensors, random noise could be added to the image on a per-fragment basis, at the point where the simulated image is being rendered at the sensor's resolution. Random noise increases with the gain of the image.

Fixed-pattern noise reflects variances in the individual elements of the sensor that remain static over time; this may be implemented through an overlay added to the image at the sensor's resolution. You may also wish to simulate dead elements of the sensor, where a few random pixels remain statically cold.

3.3 Night-Vision Goggle Simulation

NVGs are often described as *image intensification* devices, but modern NVGs do not simply amplify visible light—they actually operate in the near-IR part of the spectrum with very little overlap with the visible spectrum at all. This is because most of the natural radiation at night occurs within the near-IR band. This doesn't mean that all the thermal effects discussed above also apply to NVG; near-IR behaves more like visible light and has little to do with heat emission. As such, simulating night vision is simpler from a physical standpoint. Figure 3.3 shows one example of a simulated NVG scene.

We could get by with converting the visible-light textures to luminance values as we did in Listing 3.1 and mapping this to the monochromatic green colors typical in NVGs. There is one material-based effect worth simulating, however, and that is the *chlorophyll effect* [Hogervorst 2009]. The leaves of plants and

Figure 3.3. Simulated NVG viewpoint with depth-of-field effects. (*Image courtesy of SDS International.*)

trees containing chlorophyll are highly reflective in the near-IR band. As a result, they appear much brighter than you would expect in modern NVGs. Simply apply a healthy boost to ambient light when rendering the plants and trees in your night-vision scene to accomplish this.

The same techniques for simulating auto gain, sensor resolution, and random "shot noise" discussed for IR sensors apply to NVGs as well. It's especially important to increase the noise as a function of gain; a moonless night has significant noise compared to a scene illuminated by a full moon. Many NVGs also have noticeable fixed-pattern noise in the form of honeycomb-shaped sensor elements that may be added to the final output. Figure 3.4 compares the same scene under low-light and high-light conditions, with noise varying accordingly. Note also the loss of contrast under low-light conditions.

There are anomalies specific to NVGs worth simulating as well. One is depth of field; NVGs are focused just like a pair of binoculars—as a result, much of the resulting scene is blurry. Figure 3.3 illustrates depth of field in a simulated NVG scene; note that the distant buildings are blurred. Depth of field may be implemented by jittering the view frustum about the focal point a few times and accumulating the results [Shreiner 2006].

The prominence of stars on a clear night should also be simulated, and this can also be seen in Figure 3.4. While there are only about 8000 stars visible on a clear night to the naked eye, many more are visible with NVGs. Individual stars may saturate entire pixel elements in the sensor.

Light blooms are an important effect specific to NVGs. Bright light sources in a scene saturate the sensor; for point light sources, this manifests itself as bright halos surrounding the light. These halos may be simulated by rendering

Figure 3.4. Identical scene rendered under high ambient light (left) and low ambient light (right). (*Images courtesy of SDS International.*)

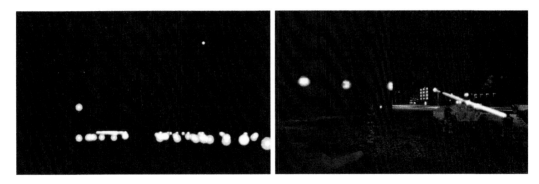

Figure 3.5. Light blooms and glow from light sources in simulated NVG scenes. (*Images courtesy of SDS International.*)

billboards over the position of point lights in the scene. These halos typically cover a viewing angle of 1.8 degrees, irrespective of the distance of the light from the viewer [Craig et al. 2005]. For non-point-light sources, such as windows into illuminated buildings, the halo effect manifests itself as a glow radiating 1.8 degrees from the shape of the light. These effects are illustrated in Figure 3.5.

Finally, you may want to use a stencil or overlay to restrict the field of view to a circular area, to match the circular collectors at the end of the NVG tubes.

References

[Bhatia and Lacy 1999] Sanjiv Bhatia and George Lacy. "Infra-Red Sensor Simulation." I/ITSEC, 1999.

[Bonjean et al. 2006] Maxime E. Bonjean, Fabian D. Lapierre, Jens Schiefele, and Jacques G. Verly. "Flight Simulator with IR and MMW Radar Image Generation Capabilities." *Enhanced and Synthetic Vision 2006 (Proceedings of SPIE)* 6226.

[Craig et al. 2005] Greg Craig, Todd Macuda, Paul Thomas, Rob Allison, and Sion Jennings. "Light Source Halos in Night Vision Goggles: Psychophysical Assessments." *Proceedings of SPIE* 5800 (2005), pp. 40–44.

[Hogervorst 2009] Maarten Hogervorst. "Toward Realistic Night Vision Simulation." SPIE Newsroom, March 27, 2009. Available at http://spie.org/x34250.xml? ArticleID=x34250.

[Lengyel 2010] Eric Lengyel. "Motion Blur and the Velocity-Depth-Gradient Buffer." *Game Engine Gems 1*, edited by Eric Lengyel. Sudbury, MA: Jones and Bartlett, 2010.

[Shreiner 2006] Dave Shreiner. *OpenGL Programming Guide*, Seventh Edition. Reading, MA: Addison-Wesley, 2006.

[Thomas 2003] Lynn Thomas. "Improved PC FLIR Simulation Through Pixel Shaders." *IMAGE 2003* Conference.

4

Screen-Space Classification for Efficient Deferred Shading

Balor Knight
Matthew Ritchie
George Parrish
Black Rock Studio

4.1 Introduction

Deferred shading is an increasingly popular technique used in video game rendering. Geometry components such as depth, normal, material color, etc., are rendered into a geometry buffer (commonly referred to as a G-buffer), and then deferred passes are applied in screen space using the G-buffer components as inputs.

A particularly common and beneficial use of deferred shading is for faster lighting. By detaching lighting from scene rendering, lights no longer affect scene complexity, shader complexity, batch count, etc. Another significant benefit of deferred lighting is that only relevant and visible pixels are lit by each light, leading to less pixel overdraw and better performance.

The traditional deferred lighting model usually includes a fullscreen lighting pass where global light properties, such as sun light and sun shadows, are applied. However, this lighting pass can be very expensive due to the number of onscreen pixels and the complexity of the lighting shader required.

A more efficient approach would be to take different shader paths for different parts of the scene according to which lighting calculations are actually required. A good example is the expensive filtering techniques needed for soft shadow edges. It would improve performance significantly if we only performed this filter on the areas of the screen that we know are at the edges of shadows.

This can be done using dynamic shader branches, but that can lead to poor performance on current game console hardware.

Swoboda [2009] describes a technique that uses the PlayStation 3 SPUs to analyze the depth buffer and classify screen areas for improved performance in post-processing effects, such as depth of field. Moore and Jefferies [2009] describe a technique that uses low-resolution screen-space shadow masks to classify screen areas as in shadow, not in shadow, or on the shadow edge for improved soft shadow rendering performance. They also describe a fast multisample antialiasing (MSAA) edge detection technique that improves deferred lighting performance.

These works provided the background and inspiration for this chapter, which extends things further by classifying screen areas according to the global light properties they require, thus minimizing shader complexity for each area. This work has been successfully implemented with good results in *Split/Second*, a racing game developed by Disney's Black Rock Studio. It is this implementation that we cover in this chapter because it gives a practical real-world example of how this technique can be applied.

4.2 Overview of Method

The screen is divided into 4×4 pixel tiles. For every frame, each tile is classified according to the minimum global light properties it requires. The seven global light properties used on *Split/Second* are the following:

1. *Sky*. These are the fastest pixels because they don't require any lighting calculations at all. The sky color is simply copied directly from the G-buffer.
2. *Sun light*. Pixels facing the sun require sun and specular lighting calculations (unless they're fully in shadow).
3. *Solid shadow*. Pixels fully in shadow don't require any shadow or sun light calculations.
4. *Soft shadow*. Pixels at the edge of shadows require expensive eight-tap percentage closer filtering (PCF) unless they face away from the sun.
5. *Shadow fade*. Pixels near the end of the dynamic shadow draw distance fade from full shadow to no shadow to avoid pops as geometry moves out of the shadow range.
6. *Light scattering*. All but the nearest pixels have a light scattering calculation applied.
7. *Antialiasing*. Pixels at the edges of polygons require lighting calculations for both 2X MSAA fragments.

We calculate which light properties are required for each 4×4 pixel tile and store the result in a 7-bit classification ID. Some of these properties are mutually exclusive for a single pixel, such as sky and sunlight, but they can exist together when properties are combined into 4×4 pixel tiles.

Once we've generated a classification ID for every tile, we then create an index buffer for each ID that points to the tiles with that ID and render it using a shader with the minimum lighting code required for those light properties.

We found that a 4×4 tile size gave the best balance between classification computation time and shader complexity, leading to best overall performance. Smaller tiles meant spending too much time classifying the tiles, and larger tiles meant more lighting properties affecting each tile, leading to more complex shaders. A size of 4×4 pixels also conveniently matches the resolution of our existing screen-space shadow mask [Moore and Jefferies 2009], which simplifies the classification code, as explained later. For *Split/Second*, the use of 4×4 tiles adds up to 57,600 tiles at a resolution of 1280×720. Figures 4.1 and 4.2 show screenshots from the *Split/Second* tutorial mode with different global light properties highlighted.

Figure 4.1. A screenshot from *Split/Second* with soft shadow edge pixels highlighted in green.

Figure 4.2. A screenshot from *Split/Second* with MSAA edge pixels highlighted in green.

4.3 Depth-Related Classification

Tile classification in *Split/Second* is broken into two parts. We classify four of the seven light properties during our screen-space shadow mask generation, and we classify the other three in a per-pixel pass. The reason for this is that the screen-space shadow code is already generating a one-quarter resolution (320×180) texture, which perfectly matches our tile resolution of 4×4 pixels, and it is also reading depths, meaning that we can minimize texture reads and shader complexity in the per-pixel pass by extending the screen-space shadow mask code to perform all depth-related classification.

Moore and Jefferies [2009] explain how we generate a one-quarter resolution screen-space shadow mask texture that contains three shadow types per pixel: pixels in shadow, pixels not in shadow, and pixels near the shadow edge. This work results in a texture containing zeros for pixels in shadow, ones for pixels not in shadow, and all other values for pixels near a shadow edge. By looking at this texture for each screen-space position, we can avoid expensive PCF for all areas except those near the edges of shadows that we want to be soft.

For tile classification, we extend this code to also classify light scattering and shadow fade since they're both calculated from depth alone, and we're already reading depth in these shaders to reconstruct world position for the shadow projections.

```
float shadowType = CalcShadowType(worldPos, depth);

float lightScattering = (depth > scatteringStartDist) ? 1.0 : 0.0;
float shadowFade = (depth > shadowFadeStartDist) ? 1.0 : 0.0;

output.color = float4(shadowType, lightScattering, shadowFade, 0.0);
```

Listing 4.1. Classifying light scattering and shadow fade in the first-pass shadow mask shader.

Recall that the shadow mask is generated in two passes. The first pass calculates the shadow type per pixel at one-half resolution (640×360) and the second pass conservatively expands the pixels marked as near shadow edge by downsampling to one-quarter resolution. Listing 4.1 shows how we add a simple light scattering and shadow fade classification test to the first-pass shader.

Listing 4.2 shows how we extend the second expand pass to pack the classification results together into four bits so they can easily be combined with the per-pixel classification results later on.

```
// Read 4 texels from 1st pass with sample offsets of 1 texel.
#define OFFSET_X (1.0 / 640.0)
#define OFFSET_Y (1.0 / 360.0)

float3 rgb = tex2D(tex, uv + float2(-OFFSET_X, -OFFSET_Y)).rgb;
rgb += tex2D(tex, uv + float2(OFFSET_X, -OFFSET_Y)).rgb;
rgb += tex2D(tex, uv + float2(-OFFSET_X, OFFSET_Y)).rgb;
rgb += tex2D(tex, uv + float2(OFFSET_X, OFFSET_Y)).rgb;

// Pack classification bits together.
#define RAW_SHADOW_SOLID        (1 << 0)
#define RAW_SHADOW_SOFT         (1 << 1)
#define RAW_SHADOW_FADE         (1 << 2)
#define RAW_LIGHT_SCATTERING    (1 << 3)

float bits = 0.0;

if (rgb.r == 0.0)
    bits += RAW_SHADOW_SOLID / 255.0;
else if (rgb.r < 4.0)
    bits += RAW_SHADOW_SOFT / 255.0;
```

```
if (rgb.b != 0.0)
    bits += RAW_SHADOW_FADE / 255.0;

if (rgb.g != 0.0)
    bits += RAW_LIGHT_SCATTERING / 255.0;

// Write results to red channel.
output.color = float4(bits, 0.0, 0.0, 0.0);
```

Listing 4.2. Packing classification results together in the second-pass shadow mask shader. Note that this code could be simplified under shader model 4 or higher because they natively support integers and bitwise operators.

4.4 Pixel Classification

It helps to explain how this pass works by describing the *Split/Second* G-buffer format (see Table 4.1). Moore and Jefferies [2009] explain how we calculate a per-pixel MSAA edge bit by comparing the results of centroid sampling against linear sampling. We pack this into the high bit of our motion ID byte in the G-buffer. For classifying MSAA edges, we extract this MSAA edge bit from both MSAA fragments and also compare the normals of each of the fragments to catch situations in which there are no polygon edges (e.g., polygon intersections).

The motion ID is used for per-pixel motion blur in a later pass, and each object type has its own ID. For the sky, this ID is always zero, and we use this value to classify sky pixels.

For sun light classification, we test normals against the sun direction (unless it's a sky pixel). Listing 4.3 shows how we classify MSAA edge, sky, and sunlight from both G-buffer 1 fragments.

Buffer	Red	Green	Blue	Alpha
Buffer 0	Albedo red	Albedo green	Albedo blue	Specular amount
Buffer 1	Normal x	Normal y	Normal z	Motion ID + MSAA edge
Buffer 2	Prelit red	Prelit green	Prelit blue	Specular power

Table 4.1. The *Split/Second* G-buffer format. Note that each component has an entry for both 2X MSAA fragments.

```
// Separate motion IDs and MSAA edge fragments from normals.
float2 edgeAndID_frags = float2(gbuffer1_frag0.w, gbuffer1_frag1.w);

// Classify MSAA edge (marked in high bit).
float2 msaaEdge_frags = (edgeAndID_frags > (128.0 / 255.0));
float mssaEdge = any(msaaEdge_frags);

float3 normalDiff = gbuffer1_frag0.xyz - gbuffer1_frag1.xyz;
mssaEdge += any(normalDiff);

// Classify sky (marked with motion ID of 0 - MSAA edge bit
// will also be 0).
float2 sky_frags = (edgeAndID_frags == 0.0);
float sky = any(sky_frags);

// Classify sunlight (except in sky).
float2 sunlight_frags;
sunlight_frags.x = sky_frags.x ? 0.0 : -dot(normal_frag0, sunDir);
sunlight_frags.y = sky_frags.y ? 0.0 : -dot(normal_frag1, sunDir);
float sunlight = any(sunlight_frags);

// Pack classification bits together.
#define RAW_MSAA_EDGE    (1 << 4)
#define RAW_SKY          (1 << 5)
#define RAW_SUN_LIGHT    (1 << 6)

float bits = msaaEdge ? (RAW_MSAA_EDGE / 255.0) : 0.0;
bits += sky ? (RAW_SKY / 255.0) : 0.0;
bits += sunlight ? (RAW_SUN_LIGHT / 255.0) : 0.0;
```

Listing 4.3. Classifying MSAA edge, sky, and sun light. This code could also be simplified in shader model 4 or higher.

4.5 Combining Classification Results

We now have per-pixel classification results for MSAA edge, sky, and sunlight, but we need to downsample each 4×4 pixel area to get a per-tile classification ID. This is as simple as ORing each 4×4 pixel area of the pixel classification results together. We also need to combine these results with the depth-related classification results to get a final classification ID per tile. Both these jobs are done in a very different way on each platform in order to make the most of their particular strengths and weaknesses, as explained in Section 4.9.

4.6 Index Buffer Generation

Once both sets of classification results are ready, a GPU callback triggers index buffer generation for each classification ID. There is one preallocated index buffer containing exactly enough indices for all tiles. On the Xbox 360, we use the RECT primitive type, which requires three indices per tile, and on the PlayStation 3, we use the the QUAD primitive type, which requires four indices per tile. The index buffer references a prebuilt vertex buffer containing a vertex for each tile corner. At a tile resolution of 4×4 pixels, this equates to 321×181 vertices at a screen resolution of 1280×720.

Index buffer generation is performed in three passes. The first pass iterates over every tile and builds a table containing the number of tiles using each classification ID, as shown in Listing 4.4. The second pass iterates over this table and builds a table of offsets into the index buffer for each classification ID, as shown in Listing 4.5. The third pass fills in the index buffer by iterating over every tile, getting the current index buffer offset for the tile's classification ID, writing new indices for that tile to the index buffer, and advancing the index buffer pointer. An example using the QUAD primitive is shown in Listing 4.6. We now have a final index buffer containing indices for all tiles and a table of starting indices for each classification ID. We're ready to render!

```
#define SHADER_COUNT        128
#define TILE_COUNT          (320 * 180)

unsigned int shaderTileCounts[SHADER_COUNT];
for (int shader = 0; shader < SHADER_COUNT; shader++)
{
    shaderTileCount[shader] = 0;
}

for (int tile = 0; tile < TILE_COUNT; tile++)
{
    unsigned char id = classificationData[tile];
    shaderTileCount[id]++;
}
```

Listing 4.4. This code counts the number of tiles using each classification ID.

```
unsigned int      *indexBufferPtrs[SHADER_COUNT];
int                indexBufferOffsets[SHADER_COUNT];
int                currentIndexBufferOffset = 0;


for (int shader = 0; shader < SHADER_COUNT; shader++)
{
    // Store shader index buffer ptr.
    indexBufferPtrs[shader] = indexBufferStart +
            currentIndexBufferOffset;

    // Store shader index buffer offset.
    indexBufferOffsets[shader] = currentIndexBufferOffset;

    // Update current offset.
    currentIndexBufferOffset += shaderTileCounts[shader] * INDICES_PER_PRIM;
}
```

Listing 4.5. This code builds the index buffer offsets. We store a pointer per shader for index buffer generation and an index per shader for tile rendering.

```
#define TILE_WIDTH      320
#define TILE_HEIGHT     180

for (int y = 0; y < TILE_HEIGHT; y++)
{
    for (int x = 0; x < TILE_WIDTH; x++)
    {
        int tileIndex = y * TILE_WIDTH + x;
        unsigned char id = classificationData[tileIndex];
        unsigned int index0 = y * (TILE_WIDTH + 1) + x;

        *indexBufferPtrs[id]++ = index0;
        *indexBufferPtrs[id]++ = index0 + 1;
        *indexBufferPtrs[id]++ = index0 + TILE_WIDTH + 2;
        *indexBufferPtrs[id]++ = index0 + TILE_WIDTH + 1;
    }
}
```

Listing 4.6. This code builds the index buffer using the QUAD primitive.

4.7 Tile Rendering

To render the tiles, we'd like to simply loop over each classification ID, activate the shaders, then issue a draw call to render the part of the index buffer we calculated earlier. However, it's not that simple because we want to submit the index buffer draw calls before we've received the classification results and built the index buffers. This is because draw calls are submitted on the CPU during the render submit phase, but the classification is done later on the GPU. We solve this by submitting each shader activate, then inserting the draw calls between each shader activate later on when we've built the index buffers and know their starting indices and counts. This is done in a very platform-specific way and is explained in Section 4.9.

4.8 Shader Management

Rather than trying to manage 128 separate shaders, we opted for a single uber-shader with all lighting properties included, and we used conditional compilation to remove the code we didn't need in each case. This is achieved by prefixing the uber-shader with a fragment defining just the properties needed for each shader. Listing 4.7 shows an example for a shader only requiring sunlight and soft shadow. The code itself is not important and just illustrates how we conditionally compile out the code we don't need.

```
// Fragment defining light properties.
#define SUN_LIGHT
#define SOFT_SHADOW

// Uber-shader starts here.
...

// Output color starts with prelit.
float3 oColor = preLit;

// Get sun shadow contribution.
#if defined(SOFT_SHADOW) && defined(SUN_LIGHT)

    float sunShadow = CascadeShadowMap_8Taps(worldPos, depth);

#elif defined(SUN_LIGHT) && !defined(SOLID_SHADOW)
```

```
    float sunShadow = 1.0;

#else

    float sunShadow = 0.0;

#endif

// Fade sun shadow.
#if (defined(SOLID_SHADOW) || defined(SOFT_SHADOW)) &&
        defined(SHADOW_FADE) && defined(SUN_LIGHT)

    sunShadow = lerp(sunShadow, 1.0,
        saturate(depth * shadowFadeScale + shadowFadeOffset));

#endif

// Apply sunlight.
#if defined(SUN_LIGHT) && !defined(SOLID_SHADOW)

    float3     sunDiff, sunSpec;

    Global_CalcDirectLighting(normal, view, sunShadow, specIntAndPow,
        sunDiff, sunSpec);
    oColor += (albedo * sunDiff) + sunSpec;

#endif

// Apply light scattering.
#ifdef LIGHT_SCATTERING

    float3 colExtinction, colInscattering;
    LightScattering(view, depth, lightDir, colExtinction,
        colInscattering);
    oColor = oColor * colExtinction + colInscattering;

#endif
```

Listing 4.7. This example shader code illustrates how we generate a shader for sunlight and soft shadow only.

4.9 Platform Specifics

Xbox 360

On the Xbox 360, downsampling pixel classification results and combining with the depth-related classification results is performed inside the pixel classification shader, and the final 7-bit classification IDs are written to a one-quarter resolution buffer in main memory using the `memexport` API. We use `memexport` rather than rendering to texture so we can output the IDs as nonlinear blocks, as shown in Figure 4.3. This block layout allows us to speed up index buffer generation by coalescing neighboring tiles with the same ID, as explained in Section 4.10. Another benefit of using `memexport` is that it avoids a resolve to texture. Once we've written out all final IDs to CPU memory, a GPU callback wakes up a CPU thread to perform the index buffer generation.

Before we can allow tile rendering to begin, we must make sure that the CPU index buffer generation has finished. This is done by inserting a GPU block that waits for a signal from the CPU thread (using asynchronous resource locks). We insert other GPU jobs before the block to avoid any stalls.

We use Xbox Procedural Synthesis (XPS) callbacks for tile rendering as they allow us to dynamically generate draw calls inside the callback. We insert an XPS callback after each shader activate during the CPU render submit, then submit each draw call in the XPS callback using the index buffer offsets and counts we calculated during index buffer generation.

Figure 4.4 shows how it all fits together, particularly the classification flow between GPU and CPU. The dotted arrow represents other work that we do to keep the GPU busy while the CPU generates the index buffer.

0	1	2	3	16	17	18	19
4	5	6	7	20	21	22	23
8	9	10	11	24	25	26	27
12	13	14	15	28	29	30	31

Figure 4.3. Xbox 360 tile classification IDs are arranged in blocks of 4×4 tiles, giving us 80×45 blocks in total. The numbers show the memory offsets, not the classification IDs.

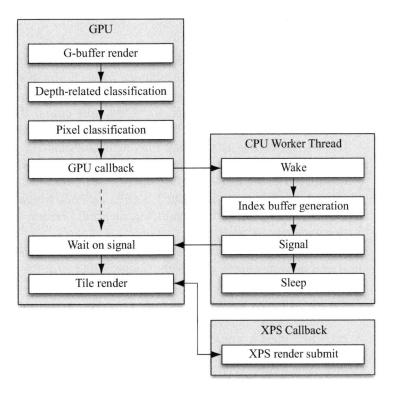

Figure 4.4. Xbox 360 classification flow.

PlayStation 3

On the PlayStation 3, the pixel classification pass is piggybacked on top of an existing depth and normal restore pass as an optimization to avoid needing a specific pass. This pass creates non-antialiased, full-width depth and normal buffers for later non-antialiased passes, such as local lights, particles, post-processing, etc., and we write the classification results to the unused w component of the normal buffer.

Once we've rendered the normals and pixel classification to a full-resolution texture, we then trigger a series of SPU downsample jobs to convert this texture into a one-quarter resolution buffer containing only the pixel classification results. Combination with the depth-related classification results is performed later on during the index buffer generation because those results aren't ready yet. This is due to the fact that we start the depth-related classification work on the GPU at the same time as these SPU downsample jobs to maximize parallelization between the two.

We spread the work across four SPUs. Each SPU job takes 64×64 pixels of classification data (one main memory frame buffer tile), ORs each 4×4 pixel area together to create a 16×16 block of classification IDs, and streams them back to main memory. Figure 4.5 shows how output IDs are arranged in main memory. We take advantage of this block layout to speed up index buffer generation by coalescing neighboring tiles with the same ID, as explained in Section 4.10. Using 16×16 tile blocks also allows us to send the results back to main memory in a single DMA call. Once this SPU work and the depth related classification work have both finished, a GPU callback triggers SPU jobs to combine both sets of classification results together and perform the index buffer generation and draw call patching.

The first part of tile rendering is to fill the command buffer with a series of shader activates interleaved with enough padding for the draw calls to be inserted later on, once we know their starting indices and counts. This is done on the CPU during the render submit phase.

Index buffer generation and tile rendering is spread across four SPUs, where each SPU runs a single job on a quarter of the screen. The first thing we do is combine the depth-related classification with the pixel classification. Remember that we couldn't do it earlier because the depth-related classification is rendered on the GPU at the same time as the pixel classification downsample jobs are running on the SPUs. Once we have final 7-bit IDs, we can create the final draw calls. Listings 4.5 and 4.6 show how we calculate starting indices and counts for each shader, and we use these results to patch the command buffer with each draw call.

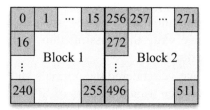

Figure 4.5. PlayStation 3 tile classification IDs are arranged in blocks of 16×16 tiles, giving us 20×12 blocks in total. The numbers show the memory offsets, not the classification IDs.

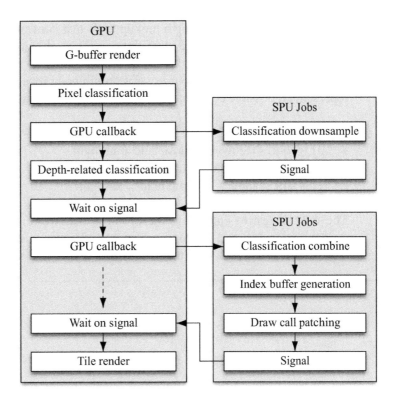

Figure 4.6. PlayStation 3 classification flow.

Figure 4.6 shows how it all fits together, particularly the classification flow between GPU and SPU jobs. The dotted arrow represents other work that we do to keep the GPU busy while the SPU generates the index buffer.

4.10 Optimizations

Reducing Shader Count

We realized that some of the 7-bit classification combinations are impossible, such as sun light and solid shadow together, no sunlight and soft shadow together, etc., and we were able to optimize these seven bits down to five by collapsing the four sun and shadow bits into two. This reduced the number of shaders from 128 to 32 and turned out to be a very worthwhile optimization.

We do this by prebuilding a lookup table that converts a 7-bit raw classification ID into a 5-bit optimized classification ID, as shown in Listing 4.8. Each ID is passed through this lookup table before being used for index buffer generation.

These two bits give us four possible combined properties, which is enough to represent all possible combinations in the scene as follows:

- 00 = solid shadow.
- 01 = solid shadow + shadow fade + sunlight.
- 10 = sun light (with no shadows).
- 11 = soft shadow + shadow fade + sunlight.

The only caveat to collapsing these bits is that we're now always calculating shadow fade when soft shadows are enabled. However, this extra cost is negligible and is far outweighed by the benefits of reducing the shader count to 32.

```
#define LIGHT_SCATTERING      (1 << 0)
#define MSAA_EDGE             (1 << 1)
#define SKY                   (1 << 2)
#define SUN_0                 (1 << 3)
#define SUN_1                 (1 << 4)

unsigned char      output[32];

for (int iCombo = 0; iCombo < 128; iCombo++)
{
    // Clear output bits.
    unsigned char bits = 0;

    // Most combos are directly copied.
    if (iCombo & RAW_LIGHT_SCATTERING) bits |= LIGHT_SCATTERING;
    if (iCombo & RAW_MSAA_EDGE) bits |= MSAA_EDGE;
    if (iCombo & RAW_SKY) bits |= SKY;

    // If in solid shadow.
    if (iCombo & RAW_SHADOW_SOLID)
    {
        // Set bit 0 if fading to sun.
        if ((iCombo & RAW_SHADOW_FADE) &&
            (iCombo & RAW_SUN_LIGHT))
                bits |= SUN_0;
```

```
    }
    else if (iCombo & RAW_SUN_LIGHT)     // else if in sun
    {
        // Set bit 1.
        bits |=  SUN_1;

        // Set bit 0 if in soft shadow.
        if (iCombo & RAW_SHADOW_SOFT)
            bits |= SUN_0;
    }

    // Write output.
    output[iCombo] = bits;
}
```

Listing 4.8. This builds a lookup table to convert from raw to optimized material IDs.

Tile Coalescing

We mentioned earlier that we take advantage of the tile block layout to speed up index buffer generation by coalescing adjacent tiles. This is done by comparing each entry in a block row, which is very fast because they're contiguous in memory. If all entries in a row are the same, we join the tiles together to make a single quad. Figure 4.7 shows an example for the Xbox 360 platform. By coalescing a row of tiles, we only have to generate one primitive instead of four. On the PlayStation 3, the savings are even greater because a block size is 16×16 tiles. We extend this optimization even further on the Xbox 360 and coalesce an entire block into one primitive if all 16 IDs in the block are the same.

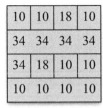

Figure 4.7. Coalescing two rows of tiles within a single block on the Xbox 360.

4.11 Performance Comparison

Using the scene shown in Figure 4.8, we compare performance between a naive implementation that calculates all light properties for all pixels and our tile classification method. The results are shown in Table 4.2. The pixel classification cost on the PlayStation 3 is very small because we're piggybacking onto an existing pass, and the extra shader cost is minimal.

Figure 4.8. The *Split/Second* scene used for our performance comparisons.

Task	PlayStation 3 Naive	PlayStation 3 Classification	Xbox 360 Naive	Xbox 360 Classification
Depth-related classification	0.00	1.29	0.00	0.70
Pixel classification	0.00	~0.00	0.00	0.69
Global light pass	13.93	4.82	8.30	2.69
Total	13.93	6.11	8.30	4.08

Table 4.2. Performance comparisons for a naive implementation versus our tile classification method. All numbers are times measured in milliseconds.

References

[Swoboda 2009] Matt Swoboda. "Deferred Lighting and Post Processing on Play-Station 3." *Game Developers Conference*, 2009.

[Moore and Jefferies 2009] Jeremy Moore and David Jefferies. "Rendering Techniques in Split/Second." *Advanced Real-Time Rendering in 3D Graphics and Games*, ACM SIGGRAPH 2009 course notes.

5

Delaying OpenGL Calls

Patrick Cozzi

Analytical Graphics, Inc.

5.1 Introduction

It is a well known best practice to write an abstraction layer over a rendering API such as OpenGL. Doing so has numerous benefits, including improved portability, flexibility, performance, and above all, ease of development. Given OpenGL's use of global state and selectors, it can be difficult to implement clean abstractions for things like shader uniforms and frame buffer objects. This chapter presents a flexible and efficient technique for implementing OpenGL abstractions using a mechanism that delays OpenGL calls until they are finally needed at draw time.

5.2 Motivation

Since its inception, OpenGL has relied on context-level global state. For example, in the fixed-function days, users would call glLoadMatrixf() to set the entries of one of the transformation matrices in the OpenGL state. The particular matrix modified would have been selected by a preceding call to glMatrix-Mode(). This pattern is still in use today. For example, setting a uniform's value with glUniform1f() depends on the currently bound shader program defined at some earlier point using glUseProgram().

The obvious downside to these selectors (e.g., the current matrix mode or the currently bound shader program) is that global state is hard to manage. For example, if a virtual call is made during rendering, can you be certain the currently bound shader program did not change? Developing an abstraction layer over OpenGL can help cope with this.

```
std::string vertexSource = // ...
std::string fragmentSource = // ...
ShaderProgram program(vertexSource, fragmentSource);

Uniform *diffuse = program.GetUniformByName("diffuse");
Uniform *specular = program.GetUniformByName("specular");

diffuse->SetValue(0.5F);
specular->SetValue(0.2F);
context.Draw(program, /* ... */);
```

Listing 5.1. Using an OpenGL abstraction layer.

Our goal is to implement an efficient abstraction layer that does not expose selectors. We limit our discussion to setting shader uniforms, but this technique is useful in many other areas, including texture units, frame buffer objects, and vertex arrays. To further simplify things, we only consider scalar floating-point uniforms. See Listing 5.1 for an example of what code using such an abstraction layer would look like.

In Listing 5.1, the user creates a shader program, sets two floating-point uniforms, and eventually uses the program for drawing. The user is never concerned with any globals like the currently bound shader program.

In addition to being easy to use, the abstraction layer should be efficient. In particular, it should avoid needless driver CPU validation overhead by eliminating redundant OpenGL calls. When using OpenGL with a language like Java or C#, eliminating redundant calls also avoids managed to native code round-trip overhead. With redundant calls eliminated, the code in Listing 5.2 only results in two calls to `glUniform1f()` regardless of the number of times the user sets uniform values or issues draw calls.

```
diffuse->SetValue(0.5F);
specular->SetValue(0.2F);
context.Draw(program, /* ... */);

diffuse->SetValue(0.5F);
specular->SetValue(0.2F);
context.Draw(program, /* ... */);
context.Draw(program, /* ... */);
```

Listing 5.2. An abstraction layer should filter out these redundant calls.

5.3 Possible Implementations

Now that we know what we are trying to achieve, let's briefly consider a few naive implementations. The simplest implementation for assigning a value to a uniform is to call glUniform1f() every time a user provides a value. Since the user isn't required to explicitly bind a program, the implementation also needs to call glUseProgram() to ensure the correct uniform is set. This is shown in Listing 5.3.

```
void SetValue(float value)
{
    glUseProgram(m_handle);
    glUniform1f(m_location, value);
}
```

Listing 5.3. Naive implementation for setting a uniform.

This implementation results in a lot of unnecessary calls to glUseProgram() and glUniform1f(). This overhead can be minimized by keeping track of the uniform's current value and only calling into OpenGL if it needs to be changed, as in Listing 5.4.

```
void SetValue(float value)
{
    if (m_currentValue != value)
    {
        m_currentValue = value;
        glUseProgram(_handle);
        glUniform1f(_location, value);
    }
}
```

Listing 5.4. A first attempt at avoiding redundant OpenGL calls.

Although epsilons are usually used to compare floating-point values, an exact comparison is used here. In some cases, the implementation in Listing 5.4 is sufficient, but it can still produce redundant OpenGL calls. For example, "thrashing" occurs if a user sets a uniform to 1.0F, then to 2.0F, and then back to 1.0F before finally issuing a draw call. In this case, `glUseProgram()` and `glUniform1f()` would be called three times each. The other downside is that `glUseProgram()` is called each time a uniform changes. Ideally, it would be called only once before all of a program's uniforms change.

Of course, it is possible to keep track of the currently bound program in addition to the current value of each uniform. The problem is that this becomes difficult with multiple contexts and multiple threads. The currently bound program is context-level global state, so each uniform instance in our abstraction needs to be aware of the current thread's current context. Also, tracking the currently bound program in this fashion is error prone and susceptible to thrashing when different uniforms are set for different programs in an interleaved manner.

5.4 Delayed Calls Implementation

In order to come up with a clean and efficient implementation for our uniform abstraction, observe that it doesn't matter what value OpenGL thinks a uniform has until a draw call is issued. Therefore, it is not necessary to call into OpenGL when a user provides a value for a uniform. Instead, we keep a list of uniforms that were changed on a per-program basis and make the necessary `glUniform1f()` calls as part of a draw command. We call the list of changed uniforms the program's *dirty list*. We *clean* the list by calling `glUniform1f()` for each uniform and then clear the list itself. This is similar to a cache where dirty cache lines are flushed to main memory and marked as clean.

An elegant way to implement this delayed technique is similar to the *observer pattern* [Gamma et al. 1995]. A shader program "observes" its uniforms. When a uniform's value changes, it notifies its observer (the program), which adds the uniform to the dirty list. The dirty list is then cleaned as part of a draw command.

The observer pattern defines a one-to-many dependency between objects. When an object changes, its dependents are notified. The object that changes is called the *subject* and its dependents are called *observers*. For us, the situation is simplified: each uniform is a subject with only one observer—the program. Since we are interested in using this technique for abstracting other areas of OpenGL, we introduce the two generic interfaces shown in Listing 5.5.

The shader program class will implement `ICleanableObserver`, so it can add a uniform to the dirty list when it is notified that the uniform changed. The uniform class will implement `ICleanable` so it can call `glUniform1f()` when the dirty list is cleaned. These relationships are shown in Figure 5.1.

```
class ICleanable
{
    public:

        virtual ~ICleanable() {}
        virtual void Clean() = 0;
};

class ICleanableObserver
{
    public:

        virtual ~ICleanableObserver() {}
        virtual void NotifyDirty(ICleanable *value) = 0;
};
```

Listing 5.5. Interfaces used for implementing the observer pattern.

Let's first consider how to implement the class ShaderProgram. As we saw in Listing 5.1, this class represents a shader program and provides access to its uniforms. In addition, it cleans the dirty list before draw commands are issued. The relevant parts of its implementation are shown in Listing 5.6. The shader program keeps two sets of uniforms: one set for all the active uniforms, which is accessed by uniform name (m_uniforms), and another set for just the dirty uniforms (m_dirtyUniforms). The dirty uniforms set is a std::vector of ICleanable pointers, since the only operation that will be applied to them is calling

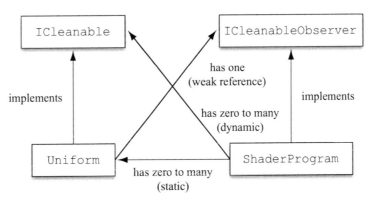

Figure 5.1. Relationship between shader programs and uniforms.

their `Clean()` method. The constructor is responsible for creating, compiling, and linking shader objects, as well as iterating over the program's active uniforms, to populate `m_uniforms`.

A user accesses a particular uniform by calling `GetUniformByName()`, which has a straightforward implementation that uses the `find()` method of `std::map` to look up the uniform. This method should not be called every time the uniform is updated because of the string-based map search. Instead, the method should be called once, and the returned `Uniform` object should be reused every frame to modify the uniform, similar to what is done in Listings 5.1 and 5.2.

The most important methods in the `ShaderProgram` class are `NotifyDirty()` and `Clean()`. As we will see when we look at the implementation for the `Uniform` class, `NotifyDirty()` is called when a uniform wants to notify the program that it is dirty. In response, the program adds the uniform to the dirty list. It is the uniform's responsibility to make sure it doesn't redundantly notify the program and be put on the dirty list multiple times. Finally, before a draw call, the shader's `Clean()` method needs to be called. The method iterates over each dirty uniform, which in turn makes the actual OpenGL call to modify the uniform's value. The dirty list is then cleared since no uniforms are dirty.

```cpp
class ShaderProgram : public ICleanableObserver
{
public:

    ShaderProgram(const std::string& vertexSource,
        const std::string& fragmentSource)
    {
        m_handle = glCreateProgram();
        // ... Create, compile, and link shader objects.

        // Populate m_uniforms with program's active uniforms by
        // calling glGetActiveUniform to get the name and location
        // for each uniform.
    }

    virtual ~ShaderProgram()
    {
        // Delete shader objects, program, and m_uniforms.
    }
```

```
    Uniform *GetUniformByName(const std::string name)
    {
        std::map<std::string, Uniform *>::iterator i =
            m_uniforms.find(name);
        return ((i != m_uniforms.end()) ? i->second : 0);
    }

    void Clean()
    {
        std::for_each(m_dirtyUniforms.begin(),
            m_dirtyUniforms.end(), std::mem_fun(&ICleanable::Clean));
        m_dirtyUniforms.clear();
    }

    // ICleanableObserver implementation.
    virtual void NotifyDirty(ICleanable *value)
    {
        m_dirtyUniforms.push_back(value);
    }

private:

    GLuint                              m_handle;
    std::vector<ICleanable *>           m_dirtyUniforms;
    std::map<std::string, Uniform *>    m_uniforms;
};
```

Listing 5.6. Partial implementation for a shader program abstraction.

The other half of the implementation is the code for Uniform, which is shown in Listing 5.7. A uniform needs to know its OpenGL location (m_location), its current value (m_value), if it is dirty (m_dirty), and the program that is observing it (m_observer). When a program creates a uniform, the program passes the uniform's location and a pointer to itself to the uniform's constructor. The constructor initializes the uniform's value to zero and then notifies the shader program that it is dirty. This has the effect of initializing all uniforms to zero. Alternatively, the uniform's value can be queried with glGetUniform(), but this has been found to be problematic on various drivers.

The bulk of the work for this class is done in SetValue() and Clean(). When the user provides a clean uniform with a new value, the uniform marks itself as dirty and notifies the program that it is now dirty. If the uniform is already dirty or the user-provided value is no different than the current value, the program is not notified, avoiding adding duplicate uniforms to the dirty list. The `Clean()` function synchronizes the uniform's value with OpenGL by calling `glUniform1f()` and then marking itself clean.

```cpp
class Uniform : public ICleanable
{
    public:

        Uniform(GLint location, ICleanableObserver *observer) :
            m_location(location), m_currentValue(0.0F), m_dirty(true),
            m_observer(observer)
        {
            m_observer->NotifyDirty(this);
        }

        float GetValue() const
        {
            return (m_currentValue);
        }

        void SetValue(float value)
        {
            if ((!m_dirty) && (m_currentValue != value))
            {
                m_dirty = true;
                m_observer->NotifyDirty(this);
            }

            m_currentValue = value;
        }

        // ICleanable implementation.
        virtual void Clean()
        {
            glUniform1f(m_location, m_currentVvalue);
            m_dirty = false;
        }
```

```
    private:

        GLint                   m_location;
        GLfloat                 m_currentValue;
        bool                    m_dirty;
        ICleanableObserver      *m_observer;
};
```

Listing 5.7. Implementation for a scalar floating-point uniform abstraction.

The final piece of the puzzle is implementing a draw call that cleans a shader program. This is as simple as requiring the user to pass a `ShaderProgram` instance to every draw call in your OpenGL abstraction (you're not exposing a separate method to bind a program, right?), then calling `glUseProgram()`, followed by the program's `Clean()` method, and finally calling the OpenGL draw function. If the draw calls are part of a class that represents an OpenGL context, it is also straightforward to factor out redundant `glUseProgram()` calls.

5.5 Implementation Notes

Our implementation is efficient in that it avoids redundant OpenGL calls and uses very little CPU. Once the `std::vector` has been "primed," adding a uniform to the dirty list is a constant time operation. Likewise, iterating over it is efficient because only dirty uniforms are touched. If no uniforms changed between one draw call and the next, then no uniforms are touched. If the common case in your engine is that most or all uniforms change from draw call to draw call, consider removing the dirty list and just iterating over all uniforms before each draw.

If you are using reference counting when implementing this technique, keep in mind that a uniform should keep a weak reference to its program. This is not a problem in garbage-collected languages.

Also, some methods, including `ShaderProgram::Clean()`, `ShaderProgram::NotifyDirty()`, and `Uniform::Clean()`, should not be publicly accessible. In C++, this can be done by making them private or protected and using the somewhat obscure `friend` keyword. A more low-tech option is to use a naming convention so clients know not to call them directly.

5.6 Improved Flexibility

By delaying OpenGL calls until draw time, we gain a great deal of flexibility. For starters, calling `Uniform::GetValue()` or `Uniform::SetValue()` does not require a current OpenGL context. For games with multiple contexts, this can minimize bugs caused by incorrect management of the current context. Likewise, if you are developing an engine that needs to play nice with other libraries using their own OpenGL context, `Uniform::SetValue()` has no context side effects and can be called anytime, not just when your context is current.

Our technique can also be extended to minimize managed to native code round-trip overhead when using OpenGL with languages like Java or C#. Instead of making fine-grained `glUniform1f()` calls for each dirty uniform, the list of dirty uniforms can be passed to native C++ code in a single coarse-grained call. On the C++ side, `glUniform1f()` is called for each uniform, thus eliminating the per-uniform round trip. This can be taken a step further by making all the required OpenGL calls for a draw in a single round trip.

5.7 Concluding Remarks

An alternative to our technique is to use direct state access (DSA) [Kilgard 2009], an OpenGL extension that allows updating OpenGL state without previously setting global state. For example, the following two lines,

```
glUseProgram(m_handle);
glUniform1f(m_location, value);
```

can be combined into one:

```
glProgramUniform1fEXT(m_handle, m_location, m_currentValue);
```

As of this writing, DSA is not a core feature of OpenGL 3.3, and as such, is not available on all platforms, although `glProgramUniform*()` calls are mirrored in the separate shader objects extension [Kilgard et al. 2010] which has become core functionality in OpenGL 4.1.

Delaying selector-based OpenGL calls until draw time has a lot of benefits, although there are some OpenGL calls that you do not want to delay. It is important to allow the CPU and GPU to work together in parallel. As such, you would not want to delay updating a large vertex buffer or texture until draw time because this could cause the GPU to wait, assuming it is not rendering one or more frames behind the CPU.

Finally, I've had great success using this technique in both commercial and open source software. I've found it quick to implement and easy to debug. An excellent next step for you is to generalize the code in this chapter to support all uniform types (vec2, vec3, etc.), uniform buffers, and other areas of OpenGL with selectors. Also consider applying this technique to higher-level engine components, such as when the bounding volume of a model in a spatial data structure changes.

Acknowledgements

Thanks to Kevin Ring and Sylvain Dupont from Analytical Graphics, Inc., and Christophe Riccio from Imagination Technologies for reviewing this chapter.

References

[Gamma et al. 1995] Erich Gamma, Richard Helm, Ralph Johnson, and John M. Vlissides. *Design Patterns*. Reading, MA: Addison-Wesley, 1995.

[Kilgard 2009] Mark Kilgard. EXT_direct_state_access OpenGL extension, 2009. Available at http://www.opengl.org/registry/specs/EXT/direct_state_access.txt.

[Kilgard et al. 2010] Mark Kilgard, Greg Roth, and Pat Brown. ARB_separate_shader_objects OpenGL extension, 2010. Available at http://www.opengl.org/registry/specs/ARB/separate_shader_objects.txt.

6

A Framework for GLSL Engine Uniforms

Patrick Cozzi
Analytical Graphics, Inc.

6.1 Introduction

The OpenGL 3.x and 4.x core profiles present a clean, shader-centric API. Many veteran developers are pleased to say goodbye to the fixed-function pipeline and the related API entry points. The core profile also says goodbye to the vast majority of GLSL built-in uniforms, such as `gl_ModelViewMatrix` and `gl_ProjectionMatrix`. This chapter addresses the obvious question: what do we use in place of GLSL built-in uniforms?

6.2 Motivation

Our goal is to design a framework for commonly used uniforms that is as easy to use as GLSL built-in uniforms but does not have their drawback: global state. GLSL built-in uniforms were easy to use because a shader could just include a built-in uniform, such as `gl_ModelViewMatrix`, and it would automatically pick up the current model-view matrix, which may have been previously set with calls like the following:

```
glMatrixMode(GL_MODELVIEW);
glLoadMatrixf(modelViewMatrix);
```

A shader could even use built-in uniforms derived from multiple GL states, such as `gl_ModelViewProjectionMatrix`, which is computed from the current model-view and projection matrices (see Listing 6.1(a)). Using built-in uniforms makes it easy to use both fixed-function and shader-based rendering code in the same application. The drawback is that the global state is error prone and hard to manage.

```
#version 120

void main()
{
    gl_Position = gl_ModelViewProjectionMatrix * gl_Vertex;
}
```

Listing 6.1(a). Pass-through vertex shader using GLSL built-in uniforms.

```
#version 330

in vec4      position;
uniform mat4  u_ModelViewProjectionMatrix;

void main()
{
    gl_Position = u_ModelViewProjectionMatrix * position;
}
```

Listing 6.1(b). Pass-through vertex shader using our engine uniforms framework.

Our replacement framework should be just as easy to use as built-in uniforms, but without relying on global state. Shader authors should be able to define and use engine uniforms as shown in Listing 6.1(b), and C++ code should automatically identify and set them.

In Listing 6.1(b), u_ModelViewProjectionMatrix is an engine uniform that serves as a replacement for gl_ModelViewProjectionMatrix. You can use any naming convention you'd like; in this case, the prefix u_ stands for "uniform." Our framework should support any number of these engine uniforms, and it should be easy to add new ones.

In addition to ease of use, the goal of our framework is to avoid global state. The solution to this is to introduce a State object that is passed to draw methods. This object should contain all the necessary state to set our engine uniforms before issuing an OpenGL draw call. Listing 6.2 shows a very minimal interface for this class. The user can set the model, view, and projection matrices; the model-view and model-view-projection are derived state (similar to gl_ModelViewProjectionMatrix). This class uses a matrix type called Matrix44, which is assumed to have the standard matrix goodies: static methods for affine trans-

```
class State
{
    public:

        const Matrix44& GetModel() const;
        void SetModel(const Matrix44& value);

        const Matrix44& GetView() const;
        void SetView(const Matrix44& value);

        const Matrix44& GetProjection() const;
        void SetProjection(const Matrix44& value);

        const Matrix44& GetModelView() const;
        const Matrix44& GetModelViewProjection() const;
};
```

Listing 6.2. Minimal interface for state used to automatically set engine uniforms.

formations, operator overloads, and a method called `Pointer()` that gives us direct access to the matrix's elements for making OpenGL calls.

By encapsulating the state required for engine uniforms in a class, different instances of the class can be passed to different draw methods. This is less error prone than setting some subset of global states between draw calls, which was required with GLSL built-in uniforms. An engine can even take this one step further and define separate scene state and object state classes. For example, the view and projection matrices from our `State` class may be part of the scene state, and the model matrix would be part of the object state. With this separation, an engine can then pass the scene state to all objects, which then issue draw commands using the scene state and their own state. For conciseness, we only use one state class in this chapter.

6.3 Implementation

To implement this framework, we build on the shader program and uniform abstractions discussed in the previous chapter. In particular, we require a `Uniform` class that encapsulates a 4×4 matrix uniform (`mat4`). We also need a `Shader-Program` class that represents an OpenGL shader program and contains the program's uniforms in a `std::map`, like the following:

```
std::map<std::string, Uniform *>   m_uniformMap;
```

```
class IEngineUniform
{
    public:

        virtual ~IEngineUniform() {}
        virtual void Set(const State& state) = 0;
};

class IEngineUniformFactory
{
    public:

        virtual ~IEngineUniformFactory() {}
        virtual IEngineUniform *Create(Uniform *uniform) = 0;
};
```

Listing 6.3. Interfaces used for setting and creating engine uniforms.

Our framework uses a list of engine uniforms' names to determine what engine uniforms a program uses. A program then keeps a separate list of engine uniforms, which are set using a `State` object before every draw call. We start our implementation by defining two new interfaces: `IEngineUniform` and `IEngine-UniformFactory`, shown in Listing 6.3.

Each engine uniform is required to implement both classes: `IEngineUniform` is used to set the uniform before a draw call, and `IEngineUniformFactory` is used to create a new instance of the engine uniform when the shader program identifies an engine uniform.

Implementing these two interfaces is usually very easy. Let's consider implementing them for an engine uniform named `u_ModelViewMatrix`, which is the model-view matrix similar to the built-in uniform `gl_ModelViewMatrix`. The implementation for `IEngineUniform` simply keeps the actual uniform as a member and sets its value using `State::GetModelView()`, as shown in Listing 6.4. The implementation for `IEngineUniformFactory` simply creates a new `ModelViewMatrixUniform` object, as shown in Listing 6.5.

The relationship among these interfaces and classes is shown in Figure 6.1. Implementing additional engine uniforms, such as a model-view-projection matrix, is usually just as straightforward as this case. In isolation, classes for engine uniforms are not all that exciting. They need to be integrated into our `ShaderProgram` class to become useful. In order to identify engine uniforms, a

```
class ModelViewMatrixUniform : public IEngineUniform
{
    public:

        ModelViewMatrixUniform(Uniform *uniform) : m_uniform(uniform)
        {
        }

        virtual void Set(const State& state)
        {
            m_uniform->SetValue(state.GetModelView());
        }

    private:

        Uniform     *m_uniform;
};
```

Listing 6.4. Implementing an engine uniform for the model-view matrix.

program needs access to a map from uniform name to engine uniform factory. The program can use this map to see which of its uniforms are engine uniforms.

Each program could have a (potentially different) copy of this map, which would allow different programs access to different engine uniforms. In practice, I have not found this flexibility useful. Instead, storing the map as a private static member and providing a public static InitializeEngineUniforms() method, as in Listing 6.6, is generally sufficient. The initialization method should be called once during application startup so each program has thread-safe, read-only

```
class ModelViewMatrixFactory : public IEngineUniformFactory
{
    public:

        virtual IEngineUniform *Create(Uniform *uniform)
        {
            return (new ModelViewMatrixUniform(uniform));
        }
};
```

Listing 6.5. A factory for the model-view matrix engine uniform.

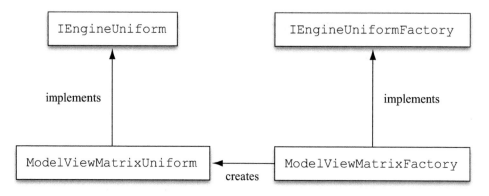

Figure 6.1. The relationship among engine uniform interfaces and the model-view matrix implementation.

```
class ShaderProgram : public ICleanableObserver
{
    public:

        static void InitializeEngineUniforms()
        {
            m_engineUniformFactories["u_modelViewMatrix"] =
                new ModelViewMatrixFactory();
            // ... Add factories for all engine uniforms.
        }

        static void DestroyEngineUniforms()
        {
            std::map<std::string, IEngineUniformFactory *>::iterator i;
            for (i = m_engineUniformFactories.begin();
                i != m_engineUniformFactories.end(); ++i)
            {
                delete i->second;
            }

            m_engineUniformFactories.clear();
        }

        // ... Other public methods.
```

```
    private:

        static std::map<std::string, IEngineUniformFactory *>
            m_engineUniformFactories;
};
```

Listing 6.6. Initialization and destruction for engine uniform factories.

access to the map afterward. It is also a best practice to free the memory for the factories on application shutdown, which can be done by calling `Destroy-EngineUniforms()`, also shown in Listing 6.6.

There are only two steps left to our implementation: identify and store a program's engine uniforms and set the engine uniforms before a draw call. As stated earlier, each program keeps a map from a uniform's name to its uniform implementation, `m_uniformMap`. This map is populated after the program is linked. To implement engine attributes, a program also needs to keep a list of engine uniforms:

```
std::vector<IEngineUniform *>   m_engineUniformList;
```

The relationship among a program, its various pointers to uniforms, and the uniform factories is shown in Figure 6.2. Using the uniform map and factory

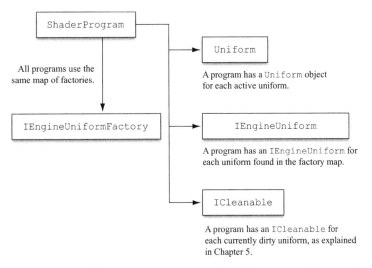

Figure 6.2. The relationship among a program, its uniforms, and the engine uniform factories.

```
std::map<std::string, Uniform *>::iterator     i;

for (i = m_uniformMap.begin(); i != m_uniformMap.end(); ++i)
{
    std::map<std::string, IEngineUniformFactory *>::iterator j =
        m_engineUniformFactories.find(i->first);

    if (j != m_engineUniformFactories.end())
    {
        m_engineUniformList.push_back(j->second->Create(i->second));
    }
}
```

Listing 6.7. Identifying and creating engine uniforms for a program.

map, the list of engine uniforms can be populated by creating an engine uniform for each uniform that has a factory for the uniform's name, as shown in Listing 6.7. In our implementation, engine uniforms are not reference counted, so the program's destructor should delete each engine uniform.

Now that a program has a list of its engine uniforms, it is easy to set them before a draw call. Assuming we are using the delayed technique for setting uniforms introduced in the previous chapter, a shader program has a `Clean()` method that is called before each OpenGL draw call to set each dirty uniform. To set engine uniforms, this method is modified to take a `State` object (which presumably is also passed to the draw method that calls this) and then call the `Set()` method for each automatic uniform, as shown in Listing 6.8.

```
void Clean(const State& state)
{
    std::vector<IEngineUniform *>::iterator     i;

    for (i = m_engineUniformList.begin();
            i != m_engineUniformList.end(); ++i)
    {
        (*i)->Set(state);
    }
}
```

```
    std::for_each(m_dirtyUniforms.begin(), m_dirtyUniforms.end(),
        std::mem_fun(&ICleanable::Clean));
    m_dirtyUniforms.clear();
}
```

Listing 6.8. A modified `ShaderProgram::Clean()` method that automatically sets engine uniforms.

6.4 Beyond GLSL Built-in Uniforms

Thus far, we've focused on using our engine uniform framework to replace the functionality of the old GLSL built-in uniforms. Our framework extends far beyond this, however. Its true usefulness is shown when engine uniforms based on higher-level engine or application-specific data are added. For example, a planet-based game may have engine uniforms for the sun position or the player's altitude. Since engine uniforms are defined using interfaces, new engine uniforms can be added with very little impact.

These high-level engine uniforms can include things like the current time or frame number, which are useful in animation. For example, texture animation may be implemented by translating texture coordinates in a fragment shader based on a uniform containing the current time. To use the uniform, the shader author doesn't have to do any additional work other than define the uniform.

If the `ShaderProgram::InitializeEngineUniforms()` method is coded carefully, applications that have access to headers for `IEngineUniform` and `IEngineUniformFactory`, but not necessarily access to all of the engine's source code, can also add engine attributes.

Engine uniforms can even go beyond setting uniform values. In Insight3D,[1] we use engine uniforms to provide shaders access to the scene's depth and silhouette textures. In these cases, the implementation of the `IEngineUniform::Set()` method binds a texture to a texture unit instead of actually setting a uniform's value (the value for the `sampler2D` uniform is set just once to the texture unit index in the engine uniform's constructor). It is also common for engines to provide engine uniforms for noise textures.

6.5 Implementation Tips

To reduce the amount of code required for new engine uniforms, it is possible to use C++ templates to let the compiler write a factory for each engine uniform for

---

[1] See http://www.insight3d.com/.

```
template <typename T>
class EngineUniformFactory : public IEngineUniformFactory
{
    public:

        virtual IEngineUniform *Create(Uniform *uniform)
        {
            return (new T(uniform));
        }
};
```

Listing 6.9. C++ templates can reduce the amount of handwritten factories.

you. For example, using the factory in Listing 6.8, the ModelViewMatrixFactory class would be replaced with the EngineUniformFactory<ModelViewMatrixUniform> class using the template shown in Listing 6.9.

If you are up for parsing GLSL code, you could also eliminate the need for shader authors to declare engine uniforms by carefully searching the shader's source for them. This task is nontrivial considering preprocessor transformations, multiline comments, strings, and compiler optimizations that could eliminate uniforms altogether.

Finally, a careful implementation of the State class can improve performance. Specifically, derived state, like the model-view-projection matrix, can be cached and only recomputed if one of its dependents change.

6.6 Concluding Remarks

Similar concepts to engine uniforms are in widespread use. For example, OpenSceneGraph[2] has preset uniforms, such as osg_FrameTime and osg_DeltaFrameTime, which are automatically updated once per frame. Likewise, RenderMonkey[3] contains predefined variables for values such as transformation matrices and mouse parameters. RenderMonkey allows the names for these variables to be customized to work with different engines.

Acknowledgements

Thanks to Kevin Ring and Sylvain Dupont from Analytical Graphics, Inc., and Christophe Riccio from Imagination Technologies for reviewing this chapter.

---

[2] See http://www.openscenegraph.org/projects/osg.
[3] See http://developer.amd.com/gpu/rendermonkey/Pages/default.aspx.

7

A Spatial and Temporal Coherence Framework for Real-Time Graphics

Michał Drobot

Reality Pump Game Development Studios

With recent advancements in real-time graphics, we have seen a vast improvement in pixel rendering quality and frame buffer resolution. However, those complex shading operations are becoming a bottleneck for current-generation consoles in terms of processing power and bandwidth. We would like to build upon the observation that under certain circumstances, shading results are temporally or spatially coherent. By utilizing that information, we can reuse pixels in time and space, which effectively leads to performance gains.

This chapter presents a simple, yet powerful, framework for spatiotemporal acceleration of visual data computation. We exploit spatial coherence for geometry-aware upsampling and filtering. Moreover, our framework combines motion-aware filtering over time for higher accuracy and smoothing, where required. Both steps are adjusted dynamically, leading to a robust solution that deals sufficiently with high-frequency changes. Higher performance is achieved due to smaller sample counts per frame, and usage of temporal filtering allows convergence to maximum quality for near-static pixels.

Our method has been fully production proven and implemented in a multi-platform engine, allowing us to achieve production quality in many rendering effects that were thought to be impractical for consoles. An example comparison of screen-space ambient occlusion (SSAO) implementations is shown in Figure 7.1. Moreover, a case study is presented, giving insight to the framework usage and performance with some complex rendering stages like screen-space ambient occlusion, shadowing, etc. Furthermore, problems of functionality, performance, and aesthetics are discussed, considering the limited memory and computational power of current-generation consoles.

(a)

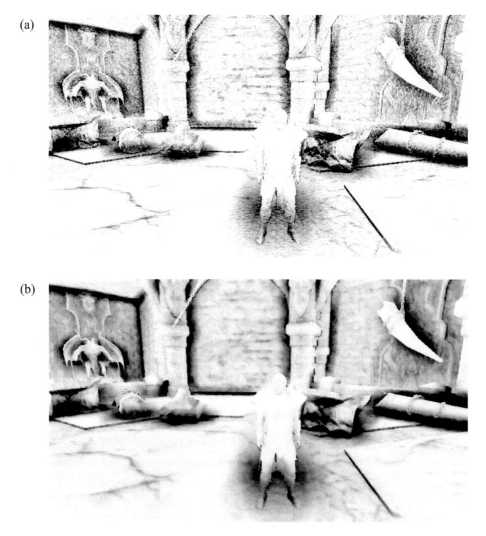

(b)

Figure 7.1. (a) A conventional four-tap SSAO pass. (b) A four-tap SSAO pass using the spatiotemporal framework.

7.1 Introduction

The most recent generation of game consoles has brought some dramatic improvements in graphics rendering quality. Several new techniques were introduced, like deferred lighting, penumbral soft shadows, screen-space ambient occlusion, and even global illumination approximations. Renderers have touched the limit of current-generation home console processing power and bandwidth. However, expectations are still rising. Therefore, we should focus more on the overlooked subject of computation and bandwidth compression.

Most pixel-intensive computations, such as shadows, motion blur, depth of field, and global illumination, exhibit high spatial and temporal coherency. With ever-increasing resolution requirements, it becomes attractive to utilize those similarities between pixels [Nehab et al. 2007]. This concept is not new, as it is the basis for motion picture compression.

If we take a direct stream from our rendering engine and compress it to a level perceptually comparable with the original, we can achieve a compression ratio of at least 10:1. What that means is that our rendering engine is calculating huge amounts of perceptually redundant data. We would like to build upon that.

Video compressors work in two stages. First, the previous frames are analyzed, resulting in a motion vector field that is spatially compressed. The previous frame is morphed into the next one using the motion vectors. Differences between the generated frame and the actual one are computed and encoded again with compression. Because differences are generally small and movement is highly stable in time, compression ratios tend to be high. Only keyframes (i.e., the first frame after a camera cut) require full information.

We can use the same concept in computer-generated graphics. It seems attractive since we don't need the analysis stage, and the motion vector field is easily available. However, computation dependent on the final shaded pixels is not feasible for current rasterization hardware. Current pixel-processing pipelines work on a per-triangle basis, which makes it difficult to compute per-pixel differences or even decide whether the pixel values have changed during the last frame (as opposed to ray tracing, where this approach is extensively used because of the per-pixel nature of the rendering). We would like to state the problem in a different way.

Most rendering stages' performance to quality ratio are controlled by the number of samples used per shaded pixel. Ideally, we would like to reuse as much data as possible from neighboring pixels in time and space to reduce the sampling rate required for an optimal solution. Knowing the general behavior of a stage, we can easily adopt the compression concept. Using a motion vector field, we can fetch samples over time, and due to the low-frequency behavior, we can utilize spatial coherency for geometry-aware upsampling. However, there are several pitfalls to this approach due to the interactive nature of most applications, particularly video games.

This chapter presents a robust framework that takes advantage of spatiotemporal coherency in visual data, and it describes ways to overcome the associated problems. During our research, we sought the best possible solution that met our demands of being robust, functional, and fast since we were aiming for Xbox 360- and PlayStation 3-class hardware. Our scenario involved rendering large outdoor scenes with cascaded shadow maps and screen-space ambient occlusion for additional lighting detail. Moreover, we extensively used advanced material shaders combined with multisampling as well as a complex postprocessing system. Several applications of the framework were developed for various rendering stages. The discussion of our final implementation covers several variations, performance gains, and future ideas.

7.2 The Spatiotemporal Framework

Our spatiotemporal framework is built from two basic algorithms, bilateral upsampling and real-time reprojection caching. (Bilateral filtering is another useful processing stage that we discuss.) Together, depending on parameters and application specifics, they provide high-quality optimizations for many complex rendering stages, with a particular focus on low-frequency data computation.

Bilateral Upsampling

We can assume that many complex shader operations are low-frequency in nature. Visual data like ambient occlusion, global illumination, and soft shadows tend to be slowly varying and, therefore, well behaved under upsampling operations. Normally, we use bilinear upsampling, which averages the four nearest samples to a point being shaded. Samples are weighted by a spatial distance function. This type of filtering is implemented in hardware, is extremely efficient, and yields good quality. However, a bilinear filter does not respect depth discontinuities, and this creates leaks near geometry edges. Those artifacts tend to be disturbing due to the high-frequency changes near object silhouettes. The solution is to steer the weights by a function of geometric similarity obtained from a high-resolution geometry buffer and coarse samples [Kopf et al. 2007]. During interpolation, we would like to choose certain samples that have a similar surface orientation and/or a small difference in depth, effectively preserving geometry edges. To summarize, we weight each coarse sample by bilinear, normal-similarity, and depth-similarity weights.

Sometimes, we can simplify bilateral upsampling to account for only depth discontinuities when normal data for coarse samples is not available. This solu-

```
for (int i = 0; i < 4; i++)
{
    normalWeights[i] = dot(normalsLow[i], normalHi);
    normalWeights[i] = pow(vNormalWeights[i], contrastCoef);
}

for (int i = 0; i < 4; i++)
{
    float depthDiff = depthHi - depthLow[i];
    depthWeights[i] = 1.0 / (0.0001F + abs(depthDiff));
}

for (int i = 0; i < 4; i++)
{
    float sampleWeight = normalWeights[i] * depthWeights[i] *
        bilinearWeights[texelNo][i];
    totalWeight += sampleWeight;
    upsampledResult += sampleLow[i] * fWeight;
}

upsampledResult /= totalWeight;
```

Listing 7.1. Pseudocode for bilateral upsampling.

tion is less accurate, but it gives plausible results in most situations when dealing with low-frequency data.

Listing 7.1 shows pseudocode for bilateral upsampling. Bilateral weights are precomputed for a 2×2 coarse-resolution tile, as shown in Figure 7.2. Depending on the position of the pixel being shaded (shown in red in Figure 7.2), the correct weights for coarse-resolution samples are chosen from the table.

Reprojection Caching

Another optimization concept is to reuse data over time [Nehab et al. 2007]. During each frame, we would like to sample previous frames for additional data, and thus, we need a history buffer, or cache, that stores the data from previous frames. With each new pixel being shaded in the current frame, we check whether additional data is available in the history buffer and how relevant it is. Then,

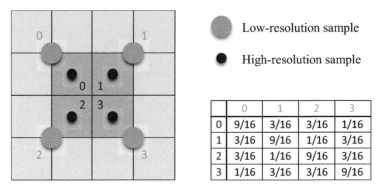

Figure 7.2. Bilinear filtering with a weight table.

we decide if we want to use that data, reject it, or perform some additional computation. Figure 7.3 presents a schematic overview of the method.

For each pixel being shaded in the current frame, we need to find a corresponding pixel in the history buffer. In order to find correct cache coordinates, we need to obtain the pixel displacement between frames. We know that pixel movement is a result of object and camera transformations, so the solution is to reproject the current pixel coordinates using a motion vector. Coordinates must be treated with special care. Results must be accurate, or we will have flawed history values due to repeated invalid cache resampling. Therefore, we perform computation in full precision, and we consider any possible offsets involved when working with render targets, such as the half-pixel offset in DirectX 9.

Coordinates of static objects can change only due to camera movement, and the calculation of pixel motion vectors is therefore straightforward. We find the

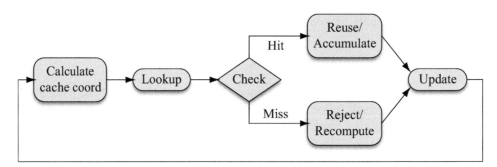

Figure 7.3. Schematic diagram of a simple reprojection cache.

pixel position in camera space and project it back to screen space using the previous frame's projection matrix. This calculation is fast and can be done in one fullscreen pass, but it does not handle moving geometry.

Dynamic objects require calculation of per-pixel velocities by comparing the current camera-space position to the last one on a per-object basis, taking skinning and transformations into consideration. In engines calculating a frame-to-frame motion field (i.e., for per-object motion blur [Lengyel 2010]), we can reuse the data if it is pixel-correct. However, when that information is not available, or the performance cost of an additional geometry pass for velocity calculation is too high, we can resort to camera reprojection. This, of course, leads to false cache coordinates for an object in motion, but depending on the application, that might be acceptable. Moreover, there are several situations where pixels may not have a history. Therefore, we need a mechanism to check for those situations.

Cache misses occur when there is no history for the pixel under consideration. First, the obvious reason for this is the case when we are sampling outside the history buffer. That often occurs near screen edges due to camera movement or objects moving into the view frustum. We simply check whether the cache coordinates are out of bounds and count it as a miss.

The second reason for a cache miss is the case when a point **p** visible at time t was occluded at time $t-1$ by another point **q**. We can assume that it is impossible for the depths of **q** and **p** to match at time $t-1$. We know the expected depth of **p** at time $t-1$ through reprojection, and we compare that with the cached depth at **q**. If the difference is bigger than an epsilon, then we consider it a cache miss. The depth at **q** is reconstructed using bilinear filtering, when possible, to account for possible false hits at depth discontinuities, as illustrated in Figure 7.4.

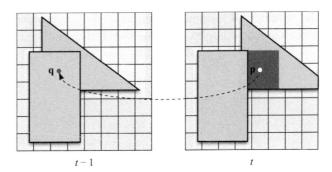

$t-1$ t

Figure 7.4. Possible cache-miss situation. The red area lacks history data due to occlusion in the previous frame. Simple depth comparison between projected **p** and **q** from $t-1$ is sufficient to confirm a miss.

If there is no cache miss, then we sample the history buffer. In general, pixels do not map to individual cache samples, so some form of resampling is needed. Since the history buffer is coherent, we can generally treat it as a normal texture buffer and leverage hardware support for linear and point filtering, where the proper method should be chosen for the type of data being cached. Low-frequency data can be sampled using the nearest-neighbor method without significant loss of detail. On the other hand, using point filtering may lead to discontinuities in the cache samples, as shown in Figure 7.5. Linear filtering correctly handles these artifacts, but repeated bilinear resampling over time leads to data smoothing. With each iteration, the pixel neighborhood influencing the result grows, and high-frequency details may be lost. Last but not least, a solution that guarantees high quality is based on a higher-resolution history buffer and nearest-neighbor sampling at a subpixel level. Nevertheless, we cannot use it on consoles because of the additional memory requirements.

Motion, change in surface parameters, and repeated resampling inevitably degrade the quality of the cached data, so we need a way to refresh it. We would like to efficiently minimize the shading error by setting the refresh rate proportional to the time difference between frames, and the update scheme should be dependent on cached data.

If our data requires explicit refreshes, then we have to find a way to guarantee a full cache update every n frames. That can easily be done by updating one of n parts of the buffer every frame in sequence. A simple tile-based approach or jittering could be used, but without additional processing like bilateral upsampling or filtering, pattern artifacts may occur.

The reprojection cache seems to be more effective with accumulation functions. In computer graphics, many results are based on stochastic processes that combine randomly distributed function samples, and the quality is often based on the number of computed samples. With reprojection caching, we can easily amortize that complex process over time, gaining in performance and accuracy.

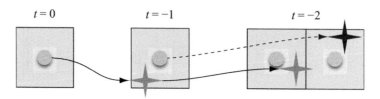

Figure 7.5. Resampling artifacts arising in point filtering. A discontinuity occurs when the correct motion flow (yellow star) does not match the approximating nearest-neighbor pixel (red star).

This method is best suited for rendering stages based on multisampling and low-frequency or slowly varying data.

When processing frames, we accumulate valid samples over time. This leads to an exponential history buffer that contains data that has been integrated over several frames back in time. With each new set of accumulated values, the buffer is automatically updated, and we control the convergence through a dynamic factor and exponential function. Variance is related to the number of frames contributing to the current result. Controlling that number lets us decide whether response time or quality is preferred.

We update the exponential history buffer using the equation

$$h(t) = h(t-1)w + r(1-w),$$

where $h(t)$ represents a value in the history buffer at time t, w is the convergence weight, and r is a newly computed value. We would like to steer the convergence

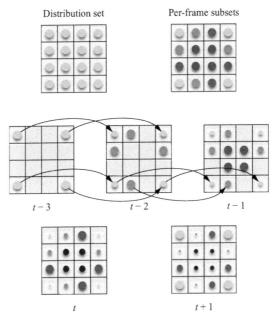

Figure 7.6. We prepare the expected sample distribution set. The sample set is divided into N subsets (four in this case), one for each consecutive frame. With each new frame, missing subsets are sampled from previous frames, by exponential history buffer look-up. Older samples lose weight with each iteration, so the sampling pattern must repeat after N frames to guarantee continuity.

weight according to changes in the scene, and this could be done based on the per-pixel velocity. This would provide a valid solution for temporal ghosting artifacts and high-quality convergence for near-static elements.

Special care must be taken when new samples are calculated. We do not want to have stale data in the cache, so each new frame must bring additional information to the buffer. When rendering with multisampling, we want to use a different set of samples with each new frame. However, sample sets should repeat after N frames, where N is the number of frames being cached (see Figure 7.6). This improves temporal stability. With all these improvements, we obtain a highly efficient reprojection caching scheme. Listing 7.2 presents a simplified solution used during our SSAO step.

```
float4 ReproCache(vertexOutput IN)
{
    float4 ActiveFrame = tex2D(gSampler0PT, IN.UV.xy);
    float freshData = ActiveFrame.x;
    float3 WSpacePos = WorldSpaceFast(IN.WorldEye, ActiveFrame.w);
    float4 LastClipSpacePos = mul(float4(WorldSpacePos.xyz, 1.0),
        IN_CameraMatrixPrev);
    float lastDepth = LastClipSpacePos.z;
    LastClipSpacePos = mul(float4(LastClipSpacePos.xyz, 1.0),
        IN_ProjectionMatrixPrev);
    LastClipSpacePos /= LastClipSpacePos.w;
    LastClipSpacePos.xy *= float2(0.5, -0.5);
    LastClipSpacePos.xy += float2(0.5, 0.5);

    float4 reproCache = tex2D(gSampler1PT, LastClipSpacePos.xy);

    float reproData = reproCache.x;
    float missRate = abs(lastDepth - reproCache.w) - IN_MissThreshold;

    missRate = clamp(missRate, 0.0, 100.0);
    missRate = 1.0 / (1.0 + missRate * IN_MissSlope);

    if (LastClipSpacePos.x < 0.0 || LastClipSpacePos.x > 1.0 ||
        LastClipSpacePos.y < 0.0 || LastClipSpacePos.y > 1.0)
            missRate = 0.0;

    missRate = saturate(missRate * IN_ConvergenceTime);
    freshData += (reproData - freshData.xy) * missRate;
```

```
    float4 out = freshData;
    out.a = ActiveFrame.w;
    return (out);
}
```

Listing 7.2. Simplified reprojection cache.

Bilateral Filtering

Bilateral filtering is conceptually similar to bilateral upsampling. We perform Gaussian filtering with weights influenced by a geometric similarity function [Tomasi and Manduchi 1998]. We can treat it as an edge-aware smoothing filter or a high-order reconstruction filter utilizing spatial coherence. Bilateral filtering proves to be extremely efficient for content-aware data smoothing. Moreover, with only insignificant artifacts, a bilateral filter can be separated into two directions, leading to $O(n)$ running time. We use it for any kind of slowly-varying data, such as ambient occlusion or shadows, that needs to be aware of scene geometry. Moreover, we use it to compensate for undersampled pixels. When a pixel lacks samples, lacks history data, or has missed the cache, it is reconstructed from spatial coherency data. That solution leads to more plausible results compared to relying on temporal data only. Listing 7.3 shows a separable, depth-aware bilateral filter that uses hardware linear filtering.

```
float Bilateral3D5x5(sampler2D inSampler, float2 texelSize,
        float2 UV, float2 Dir)
{
    const float centerWeight = 0.402619947;
    const float4 tapOffsets = float4(-3.5, -1.5, 1.5, 3.5);
    const float4 tapWeights = float4(0.054488685, 0.244201342,
        0.244201342, 0.054488685);

    const float E = 1.0;
    const float diffAmp = IN_BilateralFilterAmp;

    float2      color;
    float4      pSamples, nSamples;
    float4      diffIp, diffIn;
    float4      pTaps[2];

    float2 offSize = Dir * texelSize;
```

```
pTaps[0] = UV.xyxy + tapOffsets.xxyy * offSize.xyxy;
pTaps[1] = UV.xyxy + tapOffsets.zzww * offSize.xyxy;

color = tex2D(inSampler, UV.xy).ra;

// r - contains data to be filtered
// a - geometry depth
pTaps[0].xy = tex2D(inSampler, pTaps[0].xy).ra;
pTaps[0].zw = tex2D(inSampler, pTaps[0].zw).ra;
pTaps[1].xy = tex2D(inSampler, pTaps[1].xy).ra;
pTaps[1].zw = tex2D(inSampler, pTaps[1].zw).ra;

float4 centralD = color.y;

diffIp = (1.0 / (E + diffAmp * abs(centralD - float4(pTaps[0].y,
    pTaps[0].w, pTaps[1].y, pTaps[1].w)))) * tapWeights;
float Wp = 1.0 / (dot(diffIp, 1) + centerWeight);

color.r *= centerWeight;
color.r = Wp * (dot(diffIp, float4(pTaps[0].x, pTaps[0].z,
    pTaps[1].x, pTaps[1].z)) + color.r);

return (color.r);
}
```

Listing 7.3. Directional bilateral filter working with depth data.

Spatiotemporal Coherency

We would like to combine the described techniques to take advantage of the spatiotemporal coherency in the data. Our default framework works in several steps:

1. Depending on the data, caching is performed at lower resolution.
2. We operate with the history buffer (HB) and the current buffer (CB).
3. The CB is computed with a small set of current samples.
4. Samples from the HB are accumulated in the CB by means of reprojection caching.
5. A per-pixel convergence factor is saved for further processing.

6. The CB is bilaterally filtered with a higher smoothing rate for pixels with a lower convergence rate to compensate for smaller numbers of samples or cache misses.
7. The CB is bilaterally upsampled to the original resolution for further use.
8. The CB is swapped with the HB.

The buffer format and processing steps differ among specific applications.

7.3 Applications

Our engine is composed of several complex pixel-processing stages that include screen-space ambient occlusion, screen-space soft shadows, subsurface scattering for skin shading, volumetric effects, and a post-processing pipeline with depth of field and motion blur. We use the spatiotemporal framework to accelerate most of those stages in order to get the engine running at production-quality speeds on current-generation consoles.

Screen-Space Ambient Occlusion

Ambient occlusion AO is computed by integrating the visibility function over a hemisphere H with respect to a projected solid angle, as follows:

$$AO = \frac{1}{\pi} \int_H V_{p,\omega} (\mathbf{N} \cdot \boldsymbol{\omega}) d\boldsymbol{\omega},$$

where $\mathbf{N}$ is the surface normal and $V_{p,\omega}$ is the visibility function at $\mathbf{p}$ (such that $V_{p,\omega} = 0$ when occluded in the direction $\boldsymbol{\omega}$, and $V_{p,\omega} = 1$ otherwise). It can be efficiently computed in screen space by multiple occlusion checks that sample the depth buffer around the point being shaded. However, it is extremely taxing on the GPU due to the high sample count and large kernels that trash the texture cache. On current-generation consoles, it seems impractical to use more than eight samples. In our case, we could not even afford that many because, at the time, we had only two milliseconds left in our frame time budget.

After applying the spatiotemporal framework, we could get away with only four samples per frame, and we achieved even higher quality than before due to amortization over time. We computed the SSAO at half resolution and used bilateral upsampling during the final frame combination pass. For each frame, we changed the SSAO kernel sampling pattern, and we took care to generate a uniformly distributed pattern in order to minimize frame-to-frame inconsistencies.

Due to memory constraints on consoles, we decided to rely only on depth information, leaving the surface normal vectors available only for SSAO computation. Furthermore, since we used only camera-based motion blur, we lacked per-pixel motion vectors, so an additional pass for motion field computation was out of the question. During caching, we resorted to camera reprojection only. Our cache-miss detection algorithm compensated for that by calculating a running convergence based on the distance between a history sample and the predicted valid position. That policy tended to give good results, especially considering the additional processing steps involved. After reprojection, ambient occlusion data was bilaterally filtered, taking convergence into consideration when available (PC only). Pixels with high temporal confidence retained high-frequency details, while others were reconstructed spatially depending on the convergence factor. It is worth noticing that we were switching history buffers after bilateral filtering. Therefore, we were filtering over time, which enables us to use small kernels without significant quality loss. The complete solution required only one millisecond of GPU time and enabled us to use SSAO in real time on the Xbox 360. Figure 7.7 shows final results compared to the default algorithm.

Figure 7.7. The left column shows SSAO without using spatial coherency. The right column shows our final Xbox 360 implementation.

Soft Shadows

Our shadowing solution works in a deferred manner. We use the spatiotemporal framework for sun shadows only since those are computationally expensive and visible all the time. First, we draw sun shadows to an offscreen low-resolution buffer. While shadow testing against a cascaded shadow map, we use a custom percentage closer filter. For each frame, we use a different sample from a well-distributed sample set in order to leverage temporal coherence [Scherzer et al. 2007]. Reprojection caching accumulates the samples over time in a manner similar to our SSAO solution. Then the shadow buffer is bilaterally filtered in screen space and bilaterally upsampled for the final composition pass. Figures 7.8 and 7.9 show our final results for the Xbox 360 implementation.

Figure 7.8. Leveraging the spatiotemporal coherency of shadows (bottom) enables a soft, filtered, look free of undersampling artifacts, without raising the shadow map resolution of the original scene (top).

Figure 7.9. The spatiotemporal framework efficiently handles shadow acne and other flickering artifacts (right) that appear in the original scene (left).

Shadows and Ambient Occlusion Combined

Since our shadowing pipeline is similar to the one used during screen-space ambient occlusion, we integrate both into one pass in our most efficient implementation running on consoles. Our history buffer is half the resolution of the back buffer, and it is stored in RG16F format. The green channel stores the minimum depth of the four underlying pixels in the Z-buffer. The red channel contains shadowing and ambient occlusion information. The fractional part of the 16-bit floating-point value is used for occlusion because it requires more variety, and the integer part holds the shadowing factor. Functions for packing and unpacking these values are shown in Listing 7.4.

Every step of the spatiotemporal framework runs in parallel on both the shadowing and ambient occlusion values using the packing and unpacking functions. The last step of the framework is bilateral upsampling combined with the main deferred shading pass. Figure 7.10 shows an overview of the pipeline. The performance gained by using our technique on the Xbox 360 is shown in Table 7.1.

```
#define PACK_RANGE    31.0
#define MIN_FLT       0.01

float PackOccShadow(float Occ, float Shadow)
{
    return (floor(saturate(Occ) * PACK_RANGE) +
        clamp(Shadow, MIN_FLT, 1.0 - MIN_FLT));
}

float2 UnpackOccShadow(float OccShadow)
{
    return (float2((floor(OccShadow)) / PACK_RANGE, frac(OccShadow)));
}
```

Listing 7.4. Code for shadow and occlusion data packing and unpacking.

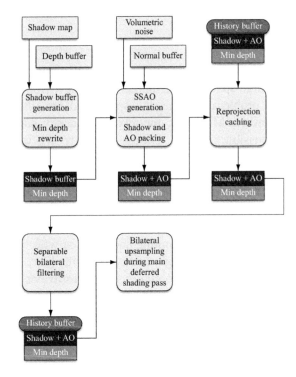

Figure 7.10. Schematic diagram of our spatiotemporal framework used with SSAO and shadows.

Stage	ST Framework	Reference
Shadows	0.7 ms	3.9 ms
SSAO generation	1.1 ms	3.8 ms
Reprojection caching	0.35 ms	–
Bilateral filtering	0.42 ms (0.2 ms per pass)	–
Bilateral upsampling	0.7 ms	0.7 ms
Total	3.27 ms	8.4 ms

Table 7.1. Performance comparison of various stages and a reference solution in which shadowing is performed in full resolution with 2×2 jittered PCF, and SSAO uses 12 taps and upsampling. The spatiotemporal (ST) framework is 2.5 times faster than the reference solution and still yields better image quality.

Postprocessing

Several postprocessing effects, such as depth of field and motion blur, tend to have high spatial and temporal coherency. Both can be expressed as a multisampling problem in time and space and are, therefore, perfectly suited for our framework. Moreover, the mixed frequency nature of both effects tends to hide any possible artifacts. During our tests, we were able to perform production-ready postprocessing twice as fast as with a normal non-cached approach.

Additionally, blurring is an excellent candidate for use with the spatiotemporal framework. Normally, when dealing with extremely large blur kernels, hierarchical downsampling with filtering must be used in order to reach reasonable performance with enough stability in high-frequency detail. Using importance sampling for downsampling and blurring with the spatiotemporal framework, we are able to perform high-quality Gaussian blur, using radii reaching 128 pixels in a 720p frame, with no significant performance penalty (less than 0.2 ms on the Xbox 360). The final quality is shown in Figure 7.11.

First, we sample nine points with linear filtering and importance sampling in a single downscaling pass to 1/64 of the screen size. Stability is sustained by the reprojection caching, with different subsets of samples used during each frame. The resulting image is blurred, cached, and upsampled. Bilateral filtering is used when needed by the application (e.g., for depth-of-field simulation where geometry awareness is required).

Figure 7.11. The bottom image shows the result of applying a large-kernel (128-pixel) Gaussian blur used for volumetric water effects to the scene shown in the top image. This process is efficient and stable on the Xbox 360 using the spatiotemporal coherency framework.

7.4 Future Work

There are several interesting concepts that use the spatiotemporal coherency, and we performed experiments that produced surprisingly good results. However, due to project deadlines, additional memory requirements, and lack of testing, those concepts were not implemented in the final iteration of the engine. We would like to present our findings here and improve upon them in the future.

Antialiasing

The spatiotemporal framework is also easily extended to full-scene antialiasing (FSAA) at a reasonable performance and memory cost [Yang et. Al 2009]. With deferred renderers, we normally have to render the G-buffer and perform lighting computation at a higher resolution. In general, FSAA buffers tend to be twice as big as the original frame buffer in both the horizontal and vertical directions. When enough processing power and memory are available, higher-resolution antialiasing schemes are preferred.

The last stage of antialiasing is the downsampling process, which generates stable, artifact-free, edge-smoothed images. Each pixel of the final frame buffer is an average of its subsamples in the FSAA buffer. Therefore, we can easily reconstruct the valid value by looking back in time for subsamples. In our experiment, we wanted to achieve 4X FSAA. We rendered each frame with a subpixel offset, which can be achieved by manipulating the projection matrix. We assumed that four consecutive frames hold the different subsamples that would normally be available in 4X FSAA, and we used reprojection to integrate those subsamples over time. When a sample was not valid, due to unocclusion, we rejected it. When misses occurred, we could also perform bilateral filtering with valid samples to leverage spatial coherency.

Our solution proved to be efficient and effective, giving results comparable to 4X FSAA for near-static scenes and giving results of varying quality during high-frequency motion. However, pixels in motion were subject to motion blur, which effectively masked any artifacts produced by our antialiasing solution. In general, the method definitely proved to be better than 2X FSAA and slightly worse than 4X FSAA since some high-frequency detail was lost due to repeated resampling. Furthermore, the computational cost was insignificant compared to standard FSAA, not to mention that it has lower memory requirements (only one additional full-resolution buffer for caching). We would like to improve upon resampling schemes to avoid additional blurring.

High-Quality Spatiotemporal Reconstruction

We would like to present another concept to which the spatiotemporal framework can be applied. It is similar to the one used in antialiasing. Suppose we want to draw a full-resolution frame. During each frame, we draw a $1/n$-resolution buffer, called the *refresh buffer*, with a different pixel offset. We change the pattern for each frame in order to cover the full frame of information in n frames. The final image is computed from the refresh buffer and a high-resolution history buffer. When the pixel being processed is not available in the history or refresh buffer, we resort to bilateral upsampling from coarse samples. See Figure 7.12 for an overview of the algorithm. This solution speeds up frame computation by a factor of n, producing a properly resampled high-resolution image, with the worst-case per-pixel resolution being $1/n$ of the original. Resolution loss would be mostly visible near screen boundaries and near fast-moving objects. However, those artifacts may be easily masked by additional processing, like motion blur. We found that setting $n = 4$ generally leads to an acceptable solution in terms of quality and performance. However, a strict rejection and bilateral upsampling policy must be

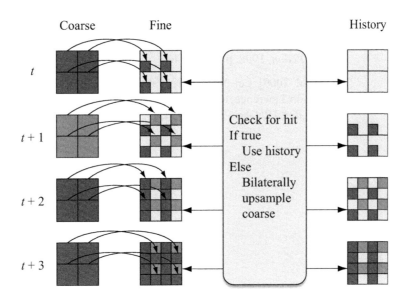

Figure 7.12. Schematic diagram of spatiotemporal reconstruction.

used to avoid instability artifacts, which tend to be disturbing when dealing with high-frequency details. We found it useful with the whole light accumulation buffer, allowing us to perform lighting four times faster with similar quality. Still, several instability issues occurred that we would like to solve.

References

[Kopf et al. 2007] Johannes Kopf, Michael F. Cohen, Dani Liscinski, and Matt Uyttendaele. "Joint Bilateral Upsampling." *ACM Transactions on Graphics* 26:3 (July 2007).

[Lengyel 2010] Eric Lengyel. "Motion Blur and the Velocity-Depth-Gradient Buffer." *Game Engine Gems 1*, edited by Eric Lengyel. Sudbury, MA: Jones and Bartlett, 2010.

[Nehab et al. 2007] Diego Nehab, Padro V. Sander, Jason Lawrence, Natalya Tararchuk, and John R. Isidoro. "Accelerating Real-Time Shading with Reverse Reprojection Caching." *Graphics Hardware*, 2007.

[Scherzer et al. 2007] Daniel Scherzer, Stefan Jeschke, and Michael Wimmer. "Pixel-Correct Shadow Maps with Temporal Reprojection and Shadow Test Confidence." *Rendering Techniques*, 2007, pp. 45–50.

[Tomasi and Manduchi 1998] Carlo Tomasi, and Roberto Manduchi. "Bilateral Filtering for Gray and Color Images." *Proceedings of International Conference on Computer Vision*, 1998, pp. 839–846.

[Yang et al. 2009] Lei Yang, Diego Nehab, Pedro V. Sander, Pitchaya Sitthi-Amorn, Jason Lawrence, and Hugues Hoppe. "Amortized Supersampling." *Proceedings of SIGGRAPH Asia 2009*, ACM.

8

Implementing a Fast DDOF Solver

Holger Grün

Advanced Micro Devices, Inc.

This gem presents a fast and memory-efficient solver for the diffusion depth-of-field (DDOF) technique introduced by Michael Kass et al. [2006] from Pixar. DDOF is a very high-quality depth-of-field effect that is used in motion pictures. It essentially works by simulating a heat diffusion equation to compute depth-of-field blurs. This leads to a situation that requires solving systems of linear equations.

Kass et al. present an algorithm for solving these systems of equations, but the memory footprint of their solver can be prohibitively large at high display resolutions. The solver described in this gem greatly reduces the memory footprint and running time of the original DDOF solver by skipping the construction of intermediate matrices and the execution of associated rendering passes. In contrast to the solver presented by Shishkovtsov and Rege [2010], which uses DirectX 11, compute shaders, and a technique called *parallel cyclic reduction* [Zhang et al. 2010], this gem utilizes a solver that runs only in pixel shaders and can be implemented on DirectX 9 and DirectX 10 class hardware as well.

8.1 Introduction

DDOF essentially relies on solving systems of linear equations described by tridiagonal systems like the following:

$$\begin{pmatrix} b_1 & c_1 & & & 0 \\ a_2 & b_2 & c_2 & & \\ & a_3 & b_3 & c_3 & \\ & & \ddots & \ddots & \ddots \\ 0 & & & a_n & b_n \end{pmatrix} \begin{pmatrix} y_1 \\ y_2 \\ y_3 \\ \vdots \\ y_n \end{pmatrix} = \begin{pmatrix} x_1 \\ x_2 \\ x_3 \\ \vdots \\ x_n \end{pmatrix}. \tag{8.1}$$

It needs to solve such a system of equations for each row and each column of the input image. Because there are always only three coefficients per row of the matrix, the system for all rows and columns of an input image can be packed into a four-channel texture that has the same resolution as the input image.

The x_i represent the pixels of one row or column of the input image, and the y_i represent the pixels of one row or column of the yet unknown image that is diffused by the heat equation. The values a_i, b_i, and c_i are derived from the circle of confusion (CoC) in a neighborhood of pixels in the input image [Riguer et al. 2004]. Kass et al. take a single time step from the initial condition $y_i = x_i$ and arrive at the equation

$$(y_i + x_i) = \beta_i (y_{i+1} - y_i) - \beta_{i-1}(y_i - y_{i-1}). \tag{8.2}$$

Here, each β_i represents the heat conductivity of the medium at the position i. Further on, β_0 and β_n are set to zero for an image row of resolution n so that the row is surrounded by insulators. If one now solves for x_i, then the resulting equation reveals a tridiagonal structure:

$$x_i = -\beta_{i-1}y_{i-1} + (\beta_{i-1} + \beta_i + 1)y_i - \beta_i y_{i+1}. \tag{8.3}$$

It is further shown that β_i is the square of the diameter of the CoC at a certain pixel position i. Setting up a_i, b_i, and c_i now is a trivial task.

As in Kass et al., the algorithm presented in this gem first calculates the diffusion for all rows of the input image and then uses the resulting horizontally-diffused image as an input for a vertical diffusion pass.

Kass et al. use a technique called *cyclic reduction* (CR) to solve for the y_i. Consider the set of equations

$$
\begin{aligned}
a_{i-1}y_{i-2} &+ b_{i-1}y_{i-1} + c_{i-1}y_i & &= x_{i-1} \\
&\quad a_i y_{i-1} + b_i y_i + c_i y_{i+1} & &= x_i \\
&\quad\quad a_{i+1}y_i + b_{i+1}y_{i+1} + c_{i+1}y_{i+2} &= x_{i+1}.
\end{aligned}
\tag{8.4}
$$

CR relies on the simple fact that a set of three equations containing five unknowns y_{i-2}, y_{i-1}, y_i, y_{i+1}, and y_{i+2}, like those shown in Equation (8.4), can be reduced to one equation by getting rid of y_{i-1} and y_{i+1}. This is done by multiplying the first equation by $-a_i/b_{i-1}$ and multiplying the third equation by $-c_i/b_{i+1}$ to produce the equations

$$-a_i \frac{a_{i-1}}{b_{i-1}} y_{i-2} - a_i y_{i-1} - a_i \frac{c_{i-1}}{b_{i-1}} y_i = -\frac{a_i}{b_{i-1}} x_{i-1}$$

$$-c_i \frac{a_{i+1}}{b_{i+1}} y_i - c_i y_{i+1} - c_i \frac{c_{i+1}}{b_{i+1}} y_{i+2} = -\frac{c_i}{b_{i+1}} x_{i+1}. \tag{8.5}$$

When these are added to the second line of Equation (8.4), we arrive at the equation

$$a_i' y_{i-2} + b_i' y_i + c_i' y_{i+2} = x_i', \tag{8.6}$$

which no longer references y_{i-1} or y_{i+1}. This means that we can halve the number of unknowns for each row or column by running a shader that writes its output to a new texture that is half as wide or half as high as the input texture. Repeating this process reduces the number of unknowns to a number that is low enough to be directly solved by the shader. The values of a_i', b_i', and c_i' are functions of the original coefficients in Equation (8.4) and need to be stored in a texture. We also need a texture containing the value of each x_i' (which are also functions of the original coefficients and the x_i), and since it is a color image, all operations are, in fact, vector operations.

Figure 8.1 illustrates the process of reducing the set of unknowns for an input row that is seven pixels wide. Each circle, or e_i, stands for an equation with a set of a_i, b_i, c_i, x_i values and their respective input/output textures. Now, at e_0'' we have reached the case where the tridiagonal system for one row has been reduced to a row with just one element. We can assume that this row of size one is still surrounded by insulators, and thus, the equation

$$a_i'' y_{i-2} + b_i'' y_i + c_i'' y_{i+2} = x_i'' \tag{8.7}$$

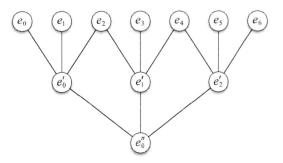

Figure 8.1. Reduction of a set of seven unknowns.

simplifies to the equation

$$b_i'' y_i = x_i'', \qquad (8.8)$$

since the a_i'' and the c_i'' need to be set to zero to implement the necessary insulation. This now allows us to trivially solve for y_i.

For modern GPUs, reducing the system of all image rows to size one does not make use of all the available hardware threads, even for high display resolutions. In order to generate better hardware utilization, the shaders presented in this chapter actually stop at two or three systems of equations and directly solve for the remaining unknown y_i. Each of the two or three results gets computed, and the proper one is chosen based on the output position (SV_POSITION) of the pixel in order to avoid dynamic flow control. The shaders need to support solving for the three remaining unknown values as well because input resolutions that are not a power of two are never reduced to a point where two unknown values are left.

If we reduce the rows of an input image and a texture containing the a_i, b_i and c_i, then the resulting y_i now represent one column of the resulting horizontally diffused image. The y_i values are written to a texture that is used to create a full-resolution result. In order to achieve this, the y_i values need to be substituted back up the resolution chain to solve for all the other unknown y_i. This means running a number of solving passes that blow the results texture up until a full-resolution image with all y_i values is reached. Each such pass doubles the number of known y_i.

If we assume that a particular value y_i kept during reduction is on an odd index i, then we just have to use the value of y_i that is available one level higher. What needs to be computed are the values of y_i for even indices. Figure 8.2 illustrates this back-substitution process and shows which values get computed and which just get copied.

In fact, the first line from Equation (8.4) can be used again because y_i and y_{i-2} are already available from the last solving pass. We can now just compute y_{i-1} as follows:

$$y_{i-1} = \frac{x_{i-1} - a_{i-1} y_{i-2} - c_{i-1} y_i}{b_{i-1}}. \qquad (8.9)$$

Running all solving passes results in the diffused image y.

For an in-depth discussion of DDOF, please consult Kass et al. [2006], as such a discussion is beyond the scope of this gem.

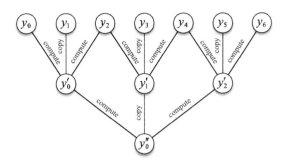

Figure 8.2. Back-substituting the known y_i.

8.2 Modifying the Basic CR Solver

As mentioned at the beginning of this chapter, one of the big problems with the solvers proposed so far is memory consumption. So the main goal of this gem is to reduce the memory footprint of the solver by as much as possible. Experiments have shown that these changes also increase the speed of the solver significantly as a side effect. The following changes to the original CR solver algorithm achieve this goal:

1. Never construct the highest-resolution or second highest-resolution (see the second change below) textures containing the a_i, b_i, and c_i. Instead, compute the a_i, b_i, and c_i on the fly when doing the first reduction and the last solving pass. This saves the memory of one fullscreen buffer and two half-sized buffers that need to have a 32-bit floating-point format. It also makes a construction pass for the highest-resolution a_i, b_i, and c_i redundant.

2. Perform an initial four-to-one reduction instead of a two-to-one reduction to reduce the memory footprint even more. Instead of getting rid of the two unknown variables y_{i-1} and y_{i+1}, the solver gets rid of y_{i-2}, y_{i-1}, y_{i+1}, and y_{i+2}, leaving only y_{i-3}, y_i, and y_{i+3}. It is possible to derive formulas that describe the four-to-one reduction and to write a shader computing these. Instead, a more hands-on approach is used. Listing 8.1 shows a shader model 4 code fragment implementing a four-to-one horizontal reduction. This shader performs the four-to-one reduction in three phases (see comments in the shader). Phase 1 gathers and computes all the data that is necessary to compute the four-to-one reduction. Phase 2 performs a series of two-to-one reductions on the data from Phase 1. Phase 3 performs a final two-to-one reduction, completing the four-to-one reduction. A shader performing vertical reduction can be easily derived from this code fragment.

```
struct ReduceO
{
    float4 abc : SV_TARGET0;
    float4 x  : SV_TARGET1;
};


ReduceO InitialReduceHorz4(PS_INPUT input)
{
    ReduceO   output;

    float fPosX = floor(input.Pos.x) * 4.0 + 3.0;
    int3 i3LP = int3(fPosX, input.Pos.y, 0);


    // Phase 1: Gather and compute all data necessary
    // for the four-to-one reduction.

    // Compute the CoC values in the support needed for
    // the four-to-one reduction.
    float fCoC_4 = computeCoC(i3LP, int2(-4, 0));
    float fCoC_3 = computeCoC(i3LP, int2(-3, 0));
    float fCoC_2 = computeCoC(i3LP, int2(-2, 0));
    float fCoC_1 = computeCoC(i3LP, int2(-1, 0));
    float fCoC0 = computeCoC(i3LP, int2(0, 0));
    float fCoC1 = computeCoC(i3LP, int2(1, 0));
    float fCoC2 = computeCoC(i3LP, int2(2, 0));
    float fCoC3 = computeCoC(i3LP, int2(3, 0));
    float fCoC4 = computeCoC(i3LP, int2(4, 0));


    // Ensure insulation at the image borders by setting
    // the CoC to 0 outside the image.
    fCoC_4 = (fPosX - 4.0 >= 0.0) ? fCoC_4 : 0.0;
    fCoC_3 = (fPosX - 3.0 >= 0.0) ? fCoC_3 : 0.0;
    fCoC4  = (fPosX + 4.0) < g_vImageSize.x ? fCoC4 : 0.0;
    fCoC3  = (fPosX + 3.0) < g_vImageSize.x ? fCoC3 : 0.0;
    fCoC2  = (fPosX + 2.0) < g_vImageSize.x ? fCoC2 : 0.0;
    fCoC1  = (fPosX + 1.0) < g_vImageSize.x ? fCoC1 : 0.0;


    // Use the minimum CoC as the real CoC as described in Kass et al.
    float fRealCoC_4 = min(fCoC_4, fCoC_3);
    float fRealCoC_3 = min(fCoC_3, fCoC_2);
```

```
    float fRealCoC_2 = min(fCoC_2, fCoC_1);
    float fRealCoC_1 = min(fCoC_1, fCoC0);
    float fRealCoC0  = min(fCoC0, fCoC1);
    float fRealCoC1  = min(fCoC1, fCoC2);
    float fRealCoC2  = min(fCoC2, fCoC3);
    float fRealCoC3  = min(fCoC3, fCoC4);

    // Compute beta values interpreting the CoCs as the diameter.
    float bt_4 = fRealCoC_4 * fRealCoC_4;
    float bt_3 = fRealCoC_3 * fRealCoC_3;
    float bt_2 = fRealCoC_2 * fRealCoC_2;
    float bt_1 = fRealCoC_1 * fRealCoC_1;
    float bt0  = fRealCoC0 * fRealCoC0;
    float bt1  = fRealCoC1 * fRealCoC1;
    float bt2  = fRealCoC2 * fRealCoC2;
    float bt3  = fRealCoC3 * fRealCoC3;

    // Now compute the a, b, c and load the x in the support
    // region of the four-to-one reduction.
    float3 abc_3 = float3(-bt_4, 1.0 + bt_3 + bt_4, -bt_3);
    float3 x_3 = txX.Load(i3LP, int2(-3, 0)).xyz;

    float3 abc_2 = float3(-bt_3 ,1.0 + bt_2 + bt_3, -bt_2);
    float3 x_2 = txX.Load(i3LP, int2(-2, 0)).xyz;

    float3 abc_1 = float3(-bt_2, 1.0 + bt_1 + bt_2, -bt_1);
    float3 x_1 = txX.Load(i3LP, int2(-1, 0)).xyz;

    float3 abc0 = float3(-bt_1, 1.0 + bt0 + bt_1, -bt0);
    float3 x0 = txX.Loadi3LP, int2(0, 0)).xyz;

    float3 abc1 = float3(-bt0, 1.0 + bt1 + bt0, -bt1);
    float3 x1 = txX.Load(i3LP, int2(1, 0)).xyz;

    float3 abc2 = float3(-bt1, 1.0 + bt2 + bt1, -bt2);
    float3 x2 = txX.Load(i3LP, int2(2, 0)).xyz;

    float3 abc3 = float3(-bt2, 1.0 + bt3 + bt2, -bt3);
    float3 x3 = txX.Load(i3LP, int2(3, 0)).xyz;

    // Phase 2: Reduce all the data by doing all two-to-one
```

```
    // reductions to get to the next reduction level.
    float a_1 = -abc_2.x / abc_3.y;
    float g_1 = -abc_2.z / abc_1.y;
    float a0 = -abc0.x / abc_1.y;
    float g0 = -abc0.z / abc1.y;
    float a1 = -abc2.x / abc1.y;
    float g1 = -abc2.z / abc3.y;

    float3 abc_p = float3(a_1 * abc_3.x,
        abc_2.y + a_1 * abc_3.z + g_1 * abc_1.x, g_1 * abc_1.z);
    float3 x_p = float3(x_2 + a_1 * x_3 + g_1 * x_1);

    float3 abc_c = float3(a0 * abc_1.x,
        abc0.y + a0 * abc_1.z + g0 * abc1.x, g0 * abc1.z);
    float3 x_c = float3( x0 + a0 * x_1 + g0 * x1);

    float3 abc_n = float3(a1 * abc1.x,
        abc2.y + a1 * abc1.z + g1 * abc3.x, g1 * abc3.z);
    float3 x_n = float3(x2 + a1 * x1 + g1 * x3);

    // Phase 3: Do the final two-to-one reduction to complete
    // the four-to-one reduction.
    float a = -abc_c.x / abc_p.y;
    float g = -abc_c.z / abc_n.y;

    float3 res0 = float3(a * abc_p.x,
        abc_c.y + a * abc_p.z + g * abc_n.x, g * abc_n.z);
    float3 res1 = float3(x_c + a * x_p + g * x_n);

    output.abc = float4(res0, 0.0);
    output.x = float4(res1, 0.0);
    return (output);
}
```

Listing 8.1. Horizontal four-to-one reduction.

3. Perform a final one-to-four solving pass to deal with the initial four-to-one reduction pass. Again, a very hands-on approach for solving the problem at hand is used, and it also has three phases. Since an initial four-to-one reduction shader was used, we don't have all the data available to perform the

needed one-to-four solving pass. Phase 1 of the shader therefore starts to reconstruct the missing data from the unchanged and full-resolution input data in the same fashion that was used in Listing 8.1. Phase 2 uses this data to perform several one-to-two solving steps to produce the missing y_i values of the intermediate pass that we skip. Phase 3 finally uses all that data to produce the final result. Listing 8.2 shows a shader model 4 code fragment implementing the corresponding algorithm for that final solver stage. Again, only the code for the horizontal version of the algorithm is shown.

```
float4 FinalSolveHorz4(PS_INPUT input) : SV_TARGET
{
    // First reconstruct the level 1 x, abc.
    float fPosX = floor(input.Pos.x * 0.25) * 4.0 + 3.0;
    int3 i3LoadPos = int3(fPosX, input.Pos.y, 0);

    // Phase 1: Gather data to reconstruct intermediate data
    // lost when skipping the first two-to-one reduction step
    // of the original solver.
    float fCoC_5 = computeCoC(i3LoadPos, int2(-5, 0));
    float fCoC_4 = computeCoC(i3LoadPos, int2(-4, 0));
    float fCoC_3 = computeCoC(i3LoadPos, int2(-3, 0));
    float fCoC_2 = computeCoC(i3LoadPos, int2(-2, 0));
    float fCoC_1 = computeCoC(i3LoadPos, int2(-1, 0));
    float fCoC0  = computeCoC(i3LoadPos, int2(0, 0));
    float fCoC1  = computeCoC(i3LoadPos, int2(1, 0));
    float fCoC2  = computeCoC(i3LoadPos, int2(2, 0));
    float fCoC3  = computeCoC(i3LoadPos, int2(3, 0));
    float fCoC4  = computeCoC(i3LoadPos, int2(4, 0));

    fCoC_5 = (fPosX - 5.0 >= 0.0) ? fCoC_5 : 0.0;
    fCoC_4 = (fPosX - 4.0 >= 0.0) ? fCoC_4 : 0.0;
    fCoC_3 = (fPosX - 3.0 >= 0.0) ? fCoC_3 : 0.0;
    fCoC4  = (fPosX + 4.0 < g_vImageSize.x) ? fCoC4: 0.0;
    fCoC3  = (fPosX + 3.0 < g_vImageSize.x) ? fCoC3 : 0.0;
    fCoC2  = (fPosX + 2.0 < g_vImageSize.x) ? fCoC2 : 0.0;
    fCoC1  = (fPosX + 1.0 < g_vImageSize.x) ? fCoC1 : 0.0;

    float fRealCoC_5 = min(fCoC_5, fCoC_4);
    float fRealCoC_4 = min(fCoC_4, fCoC_3);
    float fRealCoC_3 = min(fCoC_3, fCoC_2);
```

```
float fRealCoC_2 = min(fCoC_2, fCoC_1);
float fRealCoC_1 = min(fCoC_1, fCoC0);
float fRealCoC0  = min(fCoC0, fCoC1);
float fRealCoC1  = min(fCoC1, fCoC2);
float fRealCoC2  = min(fCoC2, fCoC3);
float fRealCoC3  = min(fCoC3, fCoC4);

float b_5 = fRealCoC_5 * fRealCoC_5;
float b_4 = fRealCoC_4 * fRealCoC_4;
float b_3 = fRealCoC_3 * fRealCoC_3;
float b_2 = fRealCoC_2 * fRealCoC_2;
float b_1 = fRealCoC_1 * fRealCoC_1;
float b0  = fRealCoC0 * fRealCoC0;
float b1  = fRealCoC1 * fRealCoC1;
float b2  = fRealCoC2 * fRealCoC2;
float b3  = fRealCoC3 * fRealCoC3;

float3 abc_4 = float3(-b_5, 1.0 + b_4 + b_5, -b_4);
float3 x_4 = txX.Load(i3LoadPos, int2(-4, 0)).xyz;

float3 abc_3 = float3(-b_4, 1.0 + b_3 + b_4, -b_3);
float3 x_3 = txX.Load(i3LoadPos, int2(-3, 0)).xyz;

float3 abc_2 = float3(-b_3, 1.0 + b_2 + b_3, -b_2);
float3 x_2 = txX.Load(i3LoadPos, int2(-2, 0)).xyz;

float3 abc_1 = float3(-b_2, 1.0 + b_1 + b_2, -b_1);
float3 x_1 = xX.Load(i3LP, int2(-1, 0)).xyz;

float3 abc0 = float3(-b_1, 1.0 + b0 + b_1, -b0);
float3 x0 = txX.Load(i3LP, int2(0, 0)).xyz;

float3 abc1 = float3(-b0, 1.0 + b1 + b0, -b1);
float3 x1 = txX.Load(i3LP, int2(1, 0)).xyz;

float3 abc2 = float3(-b1, 1.0 + b2 + b1, -b2);
float3 x2 = txX.Load(i3LP, int2(2, 0)).xyz;

float3 abc3 = float3(-b2, 1.0 + b3 + b2, -b3);
float3 x3 = txX.Load(i3LP, int2(3, 0)).xyz;
```

```
float a_2 = -abc_3.x / abc_4.y;
float g_2 = -abc_3.z / abc_2.y;


float a_1 = -abc_2.x / abc_3.y;
float g_1 = -abc_2.z / abc_1.y;


float a0 = -abc0.x / abc_1.y;
float g0 = -abc0.z / abc1.y;


float a1 = -abc2.x / abc1.y;
float g1 = -abc2.z / abc3.y;


float3 l1_abc_pp = float3(a_2 * abc_4.x,
    abc_3.y + a_2 * abc_4.z + g_2 * abc_2.x, g_2 * abc_2.z);
float3 l1_x_pp = float3(x_3 + a_2 * x_4 + g_2 * x_2 );


float3 l1_abc_p = float3(a_1 * abc_3.x,
    abc_2.y + a_1 * abc_3.z + g_1 * abc_1.x, g_1 * abc_1.z);
float3 l1_x_p = float3(x_2 + a_1 * x_3 + g_1 * x_1);


float3 l1_abc_c = float3(a0 * abc_1.x,
    abc0.y + a0 * abc_1.z + g0 * abc1.x, g0 * abc1.z);
float3 l1_x_c = float3(x0 + a0 * x_1 + g0 * x1);


float3 l1_abc_n = float3(a1 * abc1.x,
    abc2.y + a1 * abc1.z + g1 * abc3.x, g1 * abc3.z);
float3 l1_x_n = float3(x2 + a1 * x1 + g1 * x3);


// Phase 2: Now solve for thethe intermediate-level
// data we need to compute to go up to full resolution.
int3 i3l2_LoadPosC = int3(input.Pos.x * 0.25, input.Pos.y, 0);
float3 l2_y0  = txYn.Load(i3l2_LoadPosC).xyz;
float3 l2_y1  = txYn.Load(i3l2_LoadPosC, int2(1, 0)).xyz;
float3 l2_y_1 = txYn.Load(i3l2_LoadPosC, int2(-1, 0)).xyz;
float3 l2_y_2 = txYn.Load(i3l2_LoadPosC, int2(-2, 0)).xyz;
float3 l1_y_c = l2_y0;
float3 l1_y_p = (l1_x_p - l1_abc_p.x * l2_y_1
                     - l1_abc_p.z * l2_y0) / l1_abc_p.y;
float3 l1_y_pp = l2_y_1;
float3 l1_y_n = (l1_x_n - l1_abc_n.x * l2_y0
                     - l1_abc_n.z * l2_y1) / l1_abc_n.y;
```

```
    // Phase 3: Now use the intermediate solutions to solve
    // for the full result.
    float3 fRes3 = l2_y0;
    float3 fRes2 = (x_1 - abc_1.x * l1_y_p
                        - abc_1.z * l1_y_c ) / abc_1.y;      // y_1
    float3 fRes1 = l1_y_p;                                   // y_2
    float3 fRes0 = (x_3 - abc_3.x * l1_y_pp
                        - abc_3.z * l1_y_p ) / abc_3.y;      // y_3

    float3 f3Res[4] = {fRes0, fRes1, fRes2, fRes3};
    return (float4(f3Res[uint(input.Pos.x) & 3], 0.0));
}
```

Listing 8.2. Final stage of the solver.

4. Stop at two or three unknowns instead of reducing it all down to just one unknown. Given that the number of hardware threads in a modern GPU is in the thousands, this actually makes sense because it keeps a lot more threads of a modern GPU busy compared to going down to just one unknown. Cramer's rule is used to solve the resulting 2×2 or 3×3 equation systems.

5. Optionally pack the evolving y_i and the a_i, b_i, and c_i into just one four-channel `uint32` texture to further save memory and to gain speed since the number of texture operations is cut down by a factor of two. This packing uses Shader Model 5 instructions (see Listing 8.3) and relies on the assumption that the x_i values can be represented as 16-bit floating-point values. It further assumes that one doesn't need the full mantissa of the 32-bit floating-point values for storing a_i, b_i, and c_i, and it steals the six lowest mantissa bits of each one to store a 16-bit x_i channel.

```
// Pack six floats into a uint4 variable. This steals six mantissa bits
// from the three. 32-bit FP values that hold abc to store x.
uint4 pack(float3 abc, float3 x)
{
    uint z = f32tof16(x.z);
    return (uint4(((asuint(abc.x) & 0xFFFFFFC0) | (z & 0x3F)),
            ((asuint(abc.y) & 0xFFFFFFC0) | ((z >> 6) & 0x3F)),
            ((asuint(abc.z) & 0xFFFFFFC0) | ((z >> 12) & 0x3F)),
            (f32tof16(x.x) + (f32tof16(x.y) << 16))));
}
```

```
struct ABC_X
{
    float3      abc;
    float3      x;
};

ABC_X unpack(uint4 d)
{
    ABC_X       res;

    res.abc = asfloat(d.xyz & 0xFFFFFFC0);
    res.x.xy = float2(f16tof32(d.w), f16tof32(d.w >> 16));
    res.x.z = f16tof32(((d.x & 0x3F) + ((d.y & 0x3F) << 6) +
              ((d.z & 0x3F) << 12)));
    return (res);
}
```

Listing 8.3. Packing/unpacking all solver variables into/from one `rgab32_uint` value.

8.3 Results

Table 8.1 shows how various implementations of the DDOF solver perform at various resolutions and how much memory each solver consumes. These performance numbers (run on a system with an AMD HD 5870 GPU with 1 GB of video memory) show that the improved solver presented in this gem outperforms the traditional solvers in terms of running time and also in terms of memory requirements.

In the settings used in these tests, the packing shown in Listing 8.3 does not show any obvious differences (see Figure 8.3). Small differences are revealed in Figure 8.4, which shows the amplified absolute difference between the images in Figure 8.3. If these differences stay small enough, then packing should be used in DirectX 11 rendering paths in games that implement this gem.

Resolution	Solver	Running Time (ms)	Memory (~MB)
1280×1024	Standard solver	2.46	90
1280×1024	Standard solver + Packing	1.97	70
1280×1024	Four-to-one reduction	1.92	50
1280×1024	Four-to-one reduction + packing	1.87	40
1600×1200	Standard solver	3.66	132
1600×1200	Standard solver + packing	2.93	102
1600×1200	Four-to-one reduction	2.87	73
1600×1200	Four-to-one reduction + packing	2.75	58
1920×1200	Standard solver	4.31	158
1920×1200	Standard solver + packing	3.43	122
1920×1200	Four-to-one reduction	3.36	88
1920×1200	Four-to-one reduction + packing	3.23	70
2560×1600	Standard solver	7.48	281
2560×1600	Standard solver + packing	5.97	219
2560×1600	Four-to-one reduction	5.80	156
2560×1600	Four-to-one reduction + packing	5.59	125

Table 8.1. Comparison of solver efficiency.

(a)   (b)

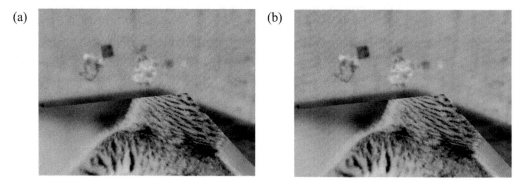

Figure 8.3. A comparison between images for which (a) packing was not used and (b) the packing shown in Listing 8.3 was used.

Figure 8.4. Absolute difference between the images in Figure 8.3, multiplied by 255 and inverted.

References

[Kass et al. 2006] Michael Kass, Aaron Lefohn, and John Owens. "Interactive Depth of Field Using Simulated Diffusion on a GPU." Technical report. Pixar Animation Studios, 2006. Available at http://www.idav.ucdavis.edu/publications/print_pub? pub_id=898.

[Shishkovtsov and Rege 2010] Oles Shishkovtsov and Ashu Rege. "DX11 Effects in Metro 2033: The Last Refuge." Game Developers Conference 2010. Available at http://developer.download.nvidia.com/presentations/2010/gdc/metro.pdf.

[Riguer et al. 2004] Guennadi Riguer, Natalya Tatarchuk, and John Isidoro. "Real-Time Depth of Field Simulation." *ShaderX2*, edited by Wolfgang Engel. Plano, TX: Wordware, 2004. Available at http://ati.amd.com/developer/shaderx/shaderx2_ real-timedepthoffieldsimulation.pdf.

[Zhang et al. 2010] Yao Zhang, Jonathan Cohen, and John D. Owens, "Fast Tridiagonal Solvers on the GPU." *Proceedings of the 15th ACM SIGPLAN Symposium on Principles and Practice of Parallel Programming*. 2010. Available at http://www. idav.ucdavis.edu/func/return_pdf?pub_id=978.

9

Automatic Dynamic Stereoscopic 3D

Jason Hughes
Steel Penny Games

Stereoscopic rendering is not new. In film, it has been used for years to make experiences more immersive and exciting, and indeed, the first theatrical showing of a 3D film was in 1922 [Zone 2007]. A cinematographer can design scenes for film that take advantage of the 3D projection without unduly taxing the audience, intentionally avoiding the various inherent issues that exist with a limited field projection of stereoscopic content. Further, an editor can trim unwanted frames or even modify the content to mask objectionable content within a frame.

Interactive stereoscopic 3D (S3D), however, is fairly new territory. Those same problems that filmmakers consciously avoid must be handled by clever programmers or designers. Sometimes, this means changing some camera angles, or adjusting a few parameters in certain areas of an environment. Sometimes this is not an option, or it may be infeasible to change the content of a title for S3D purposes. It's likely that some visual issues that arise from player control are better dealt with rather than avoided, such as window violations. While some research is available to guide an implementation, practical examples that handle dynamic environments well are hard to come by. This is an exciting time for game programmers: some experimentation and creativity is actually required on our parts until best practices become standard. Here follows the results of our work with Bluepoint Games in porting the excellent *Shadow of the Colossus* and *ICO* to PlayStation 3 in full HD and S3D.

9.1 General Problems in S3D

All stereoscopic content, whether film or games, must be especially mindful of the audience's experience. Some problems that must be avoided in any S3D experience are described in this section.

Figure 9.1. A window violation occurs when one eye sees part of an object but the other does not.

Window Violations

A window violation occurs when an object clips against the left or right edges of the screen while not exactly at the focal plane distance. While this happens in front of and behind the focal plane, it is more disturbing when in negative parallax (closer to the viewer). This very irritating, unnatural feeling is the result of the brain failing to merge two images of an object that it perceives partially but has no visual occlusion to account for the missing information. Figure 9.1 demonstrates a window violation.

Convergence Fatigue

Unlike when viewing 2D images, with S3D, the audience must use eye muscles to refocus on objects as they change in depth. It is uncommon in natural settings for people to experience radical changes in focal depth, certainly not for extended periods of time. This muscular fatigue may not be noticed immediately, but once the viewer's eyes are stressed, they will be hard-pressed to enjoy content further.

It is worth noting that the eye's angular adjustment effort decreases as objects grow more distant and increases as they approach the viewer. With this in mind, it is generally preferable to keep objects of interest at a distance rather than constantly poking out of the screen. This reduces eye strain, but still yields a nice 3D effect.

The film industry has already learned that the audience becomes fatigued if rapid shot changes force a refocusing effort to view the scene. It is unclear whether this fatigue is due to the eye muscles being unable to cope with continued rapid convergence changes or whether the brain is more efficient at tracking objects spatially over time but perhaps must work harder to identify objects initially and then tires quickly when constantly reparsing the scene from different viewpoints. In any case, giving those shot changes relatively similar convergence points from shot to shot helps reduce the stress placed on the audience.

Accommodation/Convergence Deviation

Another problem with S3D is when the brain interprets the focal distance to be significantly closer or farther than the location of the TV screen. This happens when tension placed on the muscles *inside the eye* that controls the focal point of the eyes differs significantly from the expected tension placed on muscles *outside the eye* that controls angular adjustment of the eyes. Figure 9.2 demonstrates a situation where accommodation and convergence are different. When the brain notices these values being in conflict with daily experience, the illusion of depth begins to break down. Related factors seem to involve the content being displayed, the size of the screen, the distance that the viewer sits from the screen, and the spacing between the viewer's eyes. It's complicated, but in short, don't push the effect too much, or it becomes less convincing. For a deeper examination of visual fatigue and how people perceive different content, see the report by Mikšícek [2006].

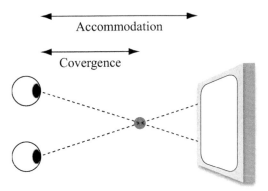

Figure 9.2. Accommodation is the physical distance between the viewer and the display surface. Convergence is the apparent distance between the observed object and the viewer.

9.2 Problems in S3D Unique to Games

In addition to the issues above, S3D rendering engines need to take various other issues into account in order to present a convincing image to each eye at interactive rates. The film industry has an easier time with these problems than games do because the artist designing the scene can ensure optimal viewing conditions for each frame of the movie. They can also afford to devote many minutes worth of CPU cycles to each frame, whereas games get only a fraction of a second to achieve acceptable results. Worse, games typically give the control to the player and often do not have the luxury of predictably good viewing conditions in S3D for every frame. A dynamic solution that is reactive to the current rendered scene would be ideal.

Keystone Distortions

It should be noted that simply rendering with two monoscopic projection matrices that tilt inward around a forward axis creates keystone distortions. Figure 9.3 shows an exaggerated example of how incorrectly set up viewing matrices appear. This kind of distortion is similar to holding two sheets of paper slightly rotated inward, then overlapped in 3D—the intersection is a line rather than a plane, and it warps the image incorrectly for stereoscopic viewing. For the brain

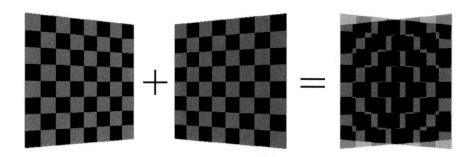

Figure 9.3. Keystone distortion, also known as the tombstone effect, occurs when a rectangle is warped into a trapezoidal shape, stretching the image. This can be seen when pointing a projector at a wall at any nonperpendicular angle. This image shows how two monoscopic projections tilted inward produce a final image that the brain cannot merge as a single checkerboard square.

to merge or fuse two images correctly requires that the focal plane be aligned in a coplanar fashion between the two images. Instead, the correct projection matrices are asymmetrical, off-axis projection matrices. These contain shearing so that the corners of the focal plane exactly match between the left and right stereo images but have a different point of origin. We don't reproduce the math here because it can be found elsewhere with better descriptions of the derivation [Bourke 1999, Schertenleib 2010, Jones et al. 2001].

2D Image-Based Effects

A staple of modern monoscopic games are the image-based effects, commonly called "post effects." These cheap adjustments to rendered images make up for the lack of supersampling hardware, low-resolution rendering buffers, inability for game hardware to render high-quality motion blurs, poor-quality lighting and shadowing with image-space occlusion mapping, blooms and glows, tone mapping, color gamut alterations, etc. It is a major component of a good-quality rendering engine but, unfortunately, is no longer as valuable in S3D. Many of these tricks do not really work well when the left/right images deviate significantly because the alterations are from each projection's fixed perspective. Blurring around the edges of objects, for example, in an antialiasing pass causes those edges to register more poorly with the viewer. Depending on the effect, it can be quite jarring or strange to see pixels that appear to be floating or smeared in space because they do not seem to fit. The result can be somewhat akin to window violations in the middle of the screen and, in large enough numbers, can be very distracting. When considering that post effects take twice the processing power to modify both images, a suggestion is to dial back these effects during S3D rendering or remove them entirely since the hardware is already being pushed twice as hard just to render the scene for both eyes separately.

HUD and Subtitles

The most difficult problem to solve for S3D and games is the heads-up display (HUD). User interfaces (UIs) generally have some issues with current-generation LCD monitors that do not change colors rapidly enough to prevent ghosting between eyes (but this technological problem will probably improve in the next few years to become less of an issue), which makes interesting 3D UIs difficult to enjoy with high-contrast colors.

A systemic problem, though, is where to draw subtitles and UI elements in the game. If they are drawn at the focal plane, then there is no ghosting on LCD monitors since both left and right images display the UI pixels in the same place.

However, any object in 3D that draws in front of the focal plane at that location acts like a window violation because the UI is (typically) rendered last and drawn over that object. The sort order feels very wrong, as if a chunk of the object was carved out and a billboard was placed in space right at the focal plane. And since the parts of the near object that are occluded differ, the occluded edges appear like window violations over the UI. All in all, the experience is very disruptive.

The two other alternatives are to always render the UI in the world, or to move the UI distance based on the nearest pixel. In short, neither works well. Rendering a UI in world space can look good if the UI is composed of 3D objects, but it can be occluded by things in the world coming between the eye and the UI. This can be a problem for gameplay and may irritate users. The second alternative, moving the UI dynamically, creates tremendous eye strain for the player because the UI is constantly flying forward to stay in front of other objects. This is very distracting, and it causes tremendous ghosting artifacts on current LCD monitors.

The best way to handle subtitles and UIs is to remove them as much as possible from gameplay or ensure that the action where they are present exists entirely in positive parallax (farther away than the monitor).

9.3 Dynamic Controls

For purposes of this discussion, the two parameters that can be varied are *parallax separation* and *focal distance*. The parallax separation, sometimes referred to as distortion, is the degree of separation between the left and right stereo cameras, where a value of zero presents the viewer with a 2D image, and a value of one is a maximally stereo-separated image (i.e., the outside frustum planes are parallel with the view vector). Obviously, the higher this parameter is set, the more separated the renderings are, and the more powerful the 3D effect.

The focal distance is the distance between the rendered eye point and the monitor surface where left and right images converge perfectly. Anything rendered at this depth appears to the viewer to be exactly on the plane of the monitor. A convenient method for checking the focal plane in an S3D LCD using shutter glasses is to remove the glasses—an object at focal distance appears completely solid on screen, while objects in front or behind the focal plane have increasingly wide double-images as their depth deviates from the focal plane.

These are not truly independent variables, however. To the viewer, the strength of the effect depends on several factors, most importantly, where the nearest objects are in the scene. Understanding the focal distance intuitively is important. Visualize a large box that surrounds the entire rendered scene. Mentally place this box around your monitor. By decreasing the focal distance, we slide the box farther away from the viewer, which causes more objects in the scene to appear behind the focal plane (the monitor) and fewer to appear in front. By increasing the focal distance, we slide the box toward the viewer and bring more objects in front of the focal plane (the monitor).

9.4 A Simple Dynamic S3D Camera

The first method with which we experimented was a rudimentary system that recognized the importance of the player's character. It has flaws, but it helped us discover how important it is to understand the concept of the "comfortable viewing zone" surrounding the focal plane, and it's easy to look at without inducing fatigue.

Camera Target Tracking

The most obvious method to try was to set the focal distance to the view-space distance to the actor that the camera is tracking. This makes sense because most of the time, the player is watching the character that the camera is watching. Keeping that actor at a comfortable viewing distance (at the plane of the TV) reduces convergence fatigue—even when making "jump cuts" to different cameras, the player does not need to refocus. It was readily apparent that this is not sufficient by itself. In *ICO*, the actor can sometimes be far from the camera, and other times be very close. If the parallax separation is held constant, the degree of stereo deviation that objects in the foreground and background undergo can be extreme and tiring to watch, particularly if something else comes into frame near the camera.

Dynamic Parallax Separation

The next step was to control the parallax separation parameter based on the focal distance. Based on our experimentation, we found that the usable range of parallax separation values fell between 0.010 and 0.350, depending on the content of the scene and the focal distance. Although this was an improvement, it was also quickly recognized that more effort would be necessary to make a truly excellent S3D experience. However, it was clear that a dynamic parallax separation was necessary, and it was implemented this way first to better understand the problem. The following formula was implemented in both *ICO* and *Shadow of the Colossus* initially, with some adjustments to match in-game units:

```
ParallaxSeparation = min(max((FocalDistance - 300.0F) /
                         4800.0F, 0.0F) + 0.05F, 0.35F);
```

In plain English, the parallax separation varies linearly between 0.050 and 0.350 as the focal distance goes from 300 cm to 4800 cm. This was based on the observation that the character, when close to the camera, did not need much parallax separation to make the backgrounds feel 3D. However, in a vast and open room, when the character was far from the camera and no walls were near the camera, the parallax separation value needed to be cranked up to show much 3D detail at large focal distances. The problem with this setup is that anytime a torch or rock or wall comes into the frame from the side, the parallax separation is so great that the object seems to protrude incredibly far out of the TV, and the viewer almost certainly sees window violations. Without taking into account objects pushed heavily toward the viewer, there would inevitably be uncomfortable areas in each game. See Figure 9.4 for an example of this kind of artifact.

Figure 9.4. The simple algorithm always keeps the character in focus. Generally, this is pleasing, except that objects nearer to the camera appear disturbingly close to the viewer. Worse yet, extreme window violations are possible, such as with the tree on the right. (*Image courtesy of Sony Computer Entertainment, Inc.*)

In the end, we realized that the relationship between focal distance and parallax separation is not simple because for any given bothersome scene, there were two ideal settings for these parameters. Nearly any scene could look good by either significantly reducing the strength of the effect by lowering the parallax separation value or by moving the focal plane to the nearest object and reducing the parallax separation value less.

9.5 Content-Adaptive Feedback

Shadow of the Colossus had more issues with S3D because the camera was free to roam with the player, whereas in *ICO*, the camera was provided specific paths based on the player's position in the world. This demanded a better solution that was aware of exactly what the player was seeing and how he would perceive the depth information presented to him. Figure 9.5 summarizes our observations.

This new algorithm is based on several observations about how people react to S3D in interactive environments, especially taking into account the areas where the effect was too strong or too weak to give the correct feeling of grandeur or subtlety without causing the player stress. What we noticed was that objects in the distance were generally acceptable with any parallax separation value, as long as their stereoscopic deviation "felt reasonable" given the distance the player would actually cover to reach them in the game. For example, overemphasizing the parallax separation could be dizzying at times, especially when standing on the edge of a short cliff that the player could easily jump down—a heavy-

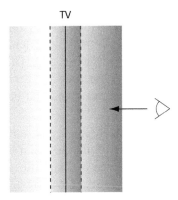

Figure 9.5. The focal plane, where the TV is, has a range indicated in blue where comfortable viewing occurs with minimal stress. The red areas show where discomfort occurs. Notice the tolerance for extremely near objects is very low, whereas distant objects are acceptable.

handed use of S3D incorrectly indicated to players that certain areas were perilously high.

The other observation was that objects that come between the focal plane and the camera with high stereoscopic deviation tend to upset players, especially when they reach the left or right edges of the screen (window violations). Other researchers have suggested variable-width black bars on the sides of the screen to solve the window violation problem [Gunnewiek and Vandewalle 2010], but this is not an ideal solution. While this works for film, it is not ideal for games. Doing this causes a loss of effective resolution for the player, still does not address the issue that objects can feel uncomfortably close at times, and requires tracking all objects that might be drawn to determine which is closest to the camera and visible. This is not an operation most games typically do, and it is not a reasonable operation for world geometry that may be extremely large and sprawling within a single mesh. The only good way to determine the closest object is to do so by examining the depth buffer, at which point we recommend a better solution anyway.

Our approach is to categorize all pixels into three zones: *comfortable*, *far*, and *close*. Then, we use this information to adjust the focal plane nearer to the viewer as needed to force more of the pixels that are currently close into a category corresponding to farther away. This is accomplished by capturing the depth buffer from the renderer just before applying post effects or UI displays, then measuring the distance for each pixel and adding to a zone counter based on its categorization. Figure 9.6 shows an example of a typical scene with this algorithm selecting the parameters. On the PlayStation 3, this can be done quickly on SPUs, and it can be done on GPUs for other platforms.

To simplify the construction of this comfortable zone, we define the half-width of the comfortable zone to extend halfway between the focal distance and the closest pixel drawn (last frame). This is clamped to the focal distance in case the closest pixel is farther away. Since the comfortable zone is symmetrical around the focal plane, the transition between the comfortable zone and far zone is trivial to compute.

Once categorized, we know the ratios of close pixels to comfortable pixels to far pixels. We assume all the pixels in the near zone are going to cause some discomfort if they are given any significant stereoscopic deviation. Given the pixel distribution, we have to react to it by changing S3D parameters to render it better next frame. (The one-frame delay in setting S3D parameters is unfortunate, but game logic can be instrumented to identify hard cuts of the camera a priori and force the S3D parameters back to a nearly 2D state, which are otherwise distressing glitches.)

Figure 9.6. The adaptive dynamic algorithm remains focused on the character in most cases, adjusting the parallax separation based on the ratio of pixels in each category. This screenshot has mostly pixels in the far category. Consequently, the algorithm increases the parallax separation to increase the sense of depth. Notice that the UI elements are at the focal distance as well, but look good to the player because everything else is at or behind them in depth. (*Image courtesy of Sony Computer Entertainment, Inc.*)

Similar to the first algorithm presented, the parallax separation value is adjusted based on the focal distance. However, two important changes allow for content-adaptive behavior: the focal distance is reduced quickly when the near zone has a lot of pixels in it, and the parallax separation is computed based on a normalized weighting of the pixel depth distribution. This additional control dimension is crucial because it allows us to tune the strength of the stereoscopic effect based on the ratio of near and far pixels. Figure 9.7 shows how the adaptive algorithm handles near pixels to avoid window violations.

There are two situations worth tuning for: focal plane is near the camera, and focal plane is relatively far from the camera. In each case, we want to specify how near and far pixels affect the parallax separation, so we use the focal distance to smoothly interpolate between two different weightings. (Weighting the distribution is necessary because a few pixels in the near zone are very, very important to react strongly to when the focal plane is far away, whereas this is not

Figure 9.7. Note by the slight blurry appearance in this superimposed S3D image, the character is not at the focal distance. Indeed, it is set to the rear hooves of the horse because the pixels in the ground plane are in the near category. Since near pixels are heavily weighted, they cause the focal plane to move closer to the camera. The character being slightly beyond the focal plane is not typically noticed by the player. Notice how much parallax separation affects objects in the distance; even the front hooves are noticeably separated. (*Image courtesy of Sony Computer Entertainment, Inc.*)

so important when the focal plane is very close to the viewer. This nonuniform response to pixels at certain depths is crucial for good results. See Figure 9.8 for details.) The resultant weighting is a three-vector (near, comfortable, far) that is multiplied componentwise against the pixel distribution three-vector and then renormalized. Finally, take the dot product between the weighted vector and a three-vector of parallax separation values, each element of which corresponds to the strength of the S3D effect at the current focal distance if all pixels were to fall exclusively inside that zone. This gives context to the overall stereoscopic rendering based on how a scene would look if the focal plane is near or far and based on how many pixels are too close for comfort, how many are very far from the player, or any combination thereof. Figure 9.9 shows a situation where window violations are avoided with the adaptive dynamic algorithm.

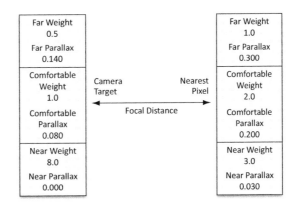

Figure 9.8. This shows actual values in use for two shipped games. As the focal plane floats between these endpoints, the values shown interpolate smoothly to prevent jarring S3D pops.

Figure 9.9. In this case, a large rock is very close to the camera. Setting the focal distance farther away than the rock would increase the pixels that would fall in the near category, causing the focal distance to shorten significantly. This self-correcting feedback is what allows for extremely dynamic environments to be handled comfortably. It effectively stops window violations in most cases, and it prevents uncomfortable protrusions toward the viewer. (*Image courtesy of Sony Computer Entertainment, Inc.*)

Another small but critical detail is that a single frame of objectionable content is not especially noticeable, but some circumstances could exist where the focal distance and parallax separation could bounce around almost randomly if objects came into and out of frame rapidly, such as birds passing by the camera or a horse that comes into frame at certain parts of an animation. The best way to handle these situations is to allow "deadening" of the 3D effect quickly, which tends not to bother the player but makes increases to the effect more gradual. In this way, the chance of objectionable content is minimized immediately, and the increase of the stereoscopic effect is subtle rather than harsh. Our specific implementation allows for the focal distance and parallax separation to decrease instantaneously, whereas they may only increase in proportion to their current value, and we have seen good results from this.

One problem that we ran into while implementing this algorithm was low-valued oscillation. Assuming the units for the near and far focal distances match your game, the oscillation should be observed at less than one percent for focal distance and parallax separation. At this low level, no visual artifacts are apparent. However, if the near and far focal distances do not coincide with reasonable camera-to-focal-object distances, or if the weights on each category are made more extreme, stability cannot be certain. This is because the depth of each rendered frame influences the depth of the next rendered frame. Heavy weights and poorly tuned distances can cause extreme jumps in the focal distance, causing an oscillation. If there is a visible artifact that appears to be an oscillation, it is an indication that some of the control parameters are incorrect.

References

[Zone 2007] Ray Zone. *Stereoscopic Cinema & the Origins of 3-D Film.* University Press of Kentucky, 2007.

[Mikšícek 2006] František Mikšícek. "Causes of Visual Fatigue and its Improvements in Stereoscopy." Technical Report No. DCSE/TR-2006-04, University of West Bohemia in Pilsen, 2006.

[Bourke 1999] Paul Bourke. "Calculating Stereo Pairs." 1999. Available at http://local. wasp.uwa.edu.au/~pbourke/miscellaneous/stereographics/stereorender/.

[Schertenleib 2010] Sébastien Schertenleib. "PlayStation 3 Making Stereoscopic 3D Games." Sony Computer Entertainment Europe, 2010. Available at http://www. technology.scee.net/files/presentations/Stereoscopic_3D/PS3_Making_ Stereoscopic_3D_Games.pdf.

[Jones et al. 2001] Graham Jones, Delman Lee, Nicolas Holliman, and David Ezra. "Controlling Perceived Depth in Stereoscopic Images." Technical Report, Sharp Laboratories of Europe Ltd., 2001.

[Gunnewiek and Vandewalle 2010] René Klein Gunnewiek and Patrick Vandewalle. "How to Display 3D Content Realistically." Technical Report, Philips Research Laboratories, VPQM, 2010.

10

Practical Stereo Rendering

Matthew Johnson
Advanced Micro Devices, Inc.

This chapter discusses practical stereo rendering techniques for modern game engines. New graphics cards by AMD and Nvidia enable application developers to utilize stereoscopic technology in their game engines. Stereo features are enabled by middleware, driver extensions, or the 3D API itself. In addition, new consumer-grade stereoscopic displays are coming to the market, fueled by the excitement over 3D movies such as *Avatar* and *How to Train Your Dragon*.

10.1 Introduction to Stereo 3D

In the real world, people use a variety of methods to perceive depth, including object size, shadows, object occlusion, and other cues. Additionally, having two eyes allows a person to perceive depth by generating a pair of images that are subsequently merged into one image by the human brain. This is called *binocular vision*.

The eyes have several mechanisms to focus and merge a stereoscopic pair into one image:

- *Binocular disparity*. The horizontal displacement between the eyes (called the *interaxial* or *interpupillary distance*) introduces a shift between the images viewed by the eyes. This can be observed, for example, by focusing on an object and alternately closing the left and right eye—the focused object is shifted left and right.
- *Convergence*. Convergence arises through the ability to rotate the eyes inward to help focus on an object in an effort to merge the stereo pair. This often causes discomfort for the person, especially if the object is very close. The opposite of this is *divergence*, but human eyes are only capable of slightly diverging.

■ *Accommodation.* Accommodation is the ability of the eye to focus on an object by changing the curvature of the lens. Accommodation is often simulated today even without stereo. For example, game engines that utilize depth-of-field algorithms often apply a postprocess blur to objects that are deemed out of focus.

The stereo algorithm described in this article takes advantage of binocular disparity to achieve the desired stereo effect.

10.2 Overview of Stereo Displays

There are several types of new stereo displays coming to the market. In the past, these monitors and projectors typically used proprietary display formats to encode stereoscopic data, requiring special software and hardware to drive them. Newer standards, such as Display Port 1.2 and HDMI 1.4, define several required stereo formats that must be supported by qualifying hardware. The availability of common specifications simplifies the implementation of stereo for middleware and application vendors.

The principle challenge in stereo displays is ensuring that the correct image is transmitted to the correct eye. Anaglyph glasses are a relatively low-cost solution to this problem. These glasses are designed to filter certain colors so that each eye can receive only one set of colors. If any of those colors "bleed" into the other eye, a ghosting artifact can occur. In addition, it is difficult to get a full range of colors across the spectrum. Despite this fact, anaglyph glasses are constantly improving with certain technologies or color combinations that diminish these shortcomings.

Because of the disadvantages of anaglyph glasses, newer stereo displays are often bundled with liquid crystal shutter glasses. These glasses work by alternating between one eye and the other (by applying a voltage to one of the lenses to darken it) at a fast refresh rate. Shutter glasses have the advantage of supporting the full color range.

The main disadvantage with shuttering is the flickering that is often observed at lower refresh rates. This is becoming less of an issue as higher refresh rate displays become available. For an HDMI 1.4-compliant stereo display, the high-definition mode 1280×720 (720p) is supported up to 120 Hz (or 60 Hz per eye), while the 1920×1080 mode (1080p) is supported at 48 Hz (24 Hz per eye). A 24-Hz refresh rate is considered the baseline in television and interactive media.

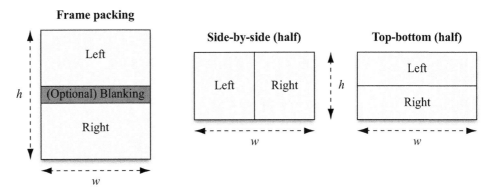

Figure 10.1. Common stereo display formats. Frame packing, side-by-side, and top-bottom are all supported by HDMI 1.4.

Several standards exist for storing the stereo pair content to be delivered to the target display. Frame packing is the most promising format, enabling applications to use full-resolution back buffers for both the left eye and right eye.

A few of the common stereo display formats are shown in Figure 10.1. Some formats, such as side-by-side (half) and top-bottom (half), can be generated with the same display bandwidth and resolution by using half the horizontal or vertical resolution per eye.

10.3 Introduction to Rendering Stereo

The goal of the game engine is to generate the left-eye and right-eye images and render them to a stereo-capable display surface. The displacement between each projected point caused by rendering to the left eye and right eye is called *parallax*. The displacement in the horizontal direction is called *horizontal parallax*. If the left and right eyes are rotated towards the focus (look-at) point, *vertical parallax* can occur. Since convergence causes discomfort in the eye, this method (known as "toe-in") is avoided.

The difference between the three horizontal parallax modes is shown in Figure 10.2. Objects with positive parallax appear behind the screen, while objects with negative parallax appear in front of the screen. When a point is projected to the same position on a plane for the left eye and right eye, it is known as zero parallax. The goal for positioning the camera is to have the scene or object of focus centered at zero parallax, with the front of the object at negative parallax and the back of the object at positive parallax, while ensuring that, at most, the

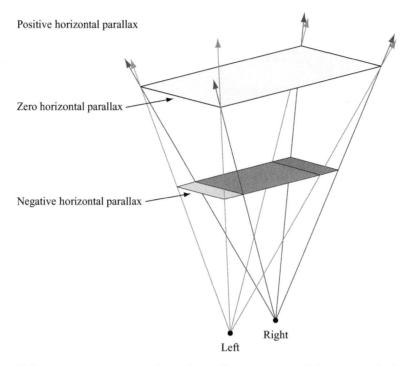

Positive horizontal parallax

Zero horizontal parallax

Negative horizontal parallax

Right

Left

Figure 10.2. A stereo camera configuration utilizing two parallel asymmetric frustums. At zero horizontal parallax, the left eye and right eye overlap completely.

negative parallax does not exceed the interaxial distance. Another alternative is to avoid negative parallax completely and set the near plane distance equal to the zero-parallax distance.

In real life, eye convergence introduces vertical parallax, but this can also cause eye discomfort. To avoid vertical parallax, the frustums should not be rotated to the same look-at point. Using parallel symmetric frustums avoids vertical parallax, but at the cost of introducing excessive negative parallax. Therefore, the preferred way of rendering stereoscopic scenes is utilizing two parallel asymmetric frustums.

10.4 The Mathematics of Stereo Views and Projection

The first step in rendering a scene for stereo is to calculate the desired horizontal parallax. In this chapter, we assume a left-handed coordinate system (positive z is "behind" the screen) in camera space.

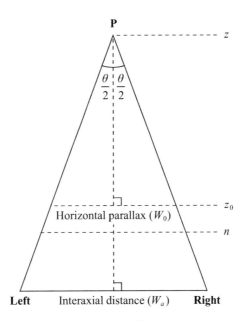

Figure 10.3. The relationship between camera distances and horizontal parallax, based on the projected view-space point **P**.

The general rule of thumb is for the interaxial width W_a between the eyes to be equal to 1/30th of the distance to the horizontal zero-parallax plane, where $\theta \approx 1.9°$ (the angle between the focus point and each eye). This relationship can be visualized in Figure 10.3. The manufacturers of shutter glasses generally prefer a maximum parallax of $\theta = 1.5°$, which comes out to roughly 1/38th the distance to zero parallax.

As an example, suppose that the bounding sphere of a scene in view space has center **C** and radius 45.0 and that we need to calculate the interaxial distance necessary to achieve zero parallax at the center of the scene. In view space, the camera location is $(0,0,0)$. Therefore, the distance from the camera to the center of the sphere is C_z, and thus $z = C_z$. For a maximum parallax of $\theta = 1.5°$, we have the relationship

$$\tan \frac{1.5°}{2} = \frac{W_a}{2z}.$$

So $W_a \approx 0.0262z$. Thus, one can find the desired interaxial distance given a horizontal parallax distance of zero.

Setting the near plane n of the viewing frustum can be done based on the desired acceptable negative parallax. The maximum negative parallax should not exceed the interaxial distance. A relationship between these distances can be established by using similar triangles and limits.

As another example, we calculate the parallax at $z = z_0$, $z = z_0/2$, $z = \infty$, and $z = 0$. By definition, the parallax at z_0 is zero. The parallax can be solved for the other values by similar triangles:

$$\frac{W_0}{W_a} = \frac{z - z_0}{z},$$

$$W_0 = W_a \left(1 - \frac{z_0}{z}\right).$$

For $z = z_0/2$,

$$W_0 = W_a \left(1 - \frac{z_0}{z}\right) = -W_a.$$

For $z = \infty$,

$$W_0 = \lim_{z \to \infty} W_a \left(1 - \frac{z_0}{z}\right) = W_a.$$

For $z = 0$,

$$W_0 = \lim_{z \to 0} W_a \left(1 - \frac{z_0}{z}\right) = -\infty.$$

As z approaches ∞, the horizontal parallax approaches the interaxial distance, which is perceived by the eyes as a far-away object. As z approaches zero, the horizontal parallax approaches $-\infty$. For this reason, introducing excessive negative parallax in the scene adds the risk that the eyes will not be able to converge the two images. To prevent this, set the near plane to $z_0/2$, which results in a negative horizontal parallax equal to the interaxial distance.

In most applications, the camera is designed for monoscopic viewing and is modified for left and right eyes only when stereo rendering is enabled. Given a camera at the monoscopic (center) position, it is straightforward to calculate the left and right plane offsets at the near plane based on the camera offset. Figure 10.4 shows the offset for the left camera. Note that shifting the camera horizontally results in an equal shift at the zero-parallax line.

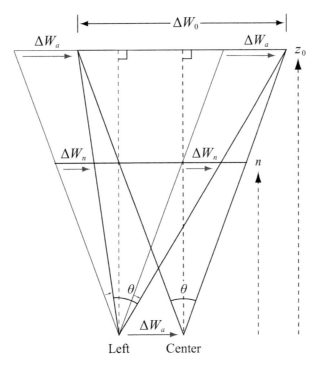

Figure 10.4. Offsetting the left and right frustum planes to form an asymmetric matrix with zero parallax. (For clarity, the right eye is not shown.)

To set up the left and right cameras, offset the left and right cameras horizontally by half the interaxial distance. For projection, offset the left and right frustum planes for each eye in order to setup the asymmetric frustum.

Given a horizontal camera offset of ΔW_a, we can calculate the ΔW_n offset for the left and right projection planes by using similar triangles, as follows:

$$\frac{\Delta W_n}{\Delta W_a} = \frac{n}{z_0}.$$

Therefore,

$$\Delta W_n = \frac{n}{z_0} \Delta W_a.$$

Listing 10.1 outlines the pseudocode for modifying the view frustum and camera transformation for stereoscopic viewing.

```cpp
// Camera structure (left-handed space).
struct Camera
{
    float3 up;          // camera up direction (normalized)
    float3 dir;         // camera look-at direction (normalized)
    float3 pos;         // camera position

    float aspect;       // aspect ratio (w/h)
    float fov;          // horizontal field of view (radians)

    float Wa;           // camera interaxial distance (stereo)
    float z0;           // camera distance to zero parallax

    float zn;           // camera frustum near plane (screen projection)
    float zf;           // camera frustum far plane
    float l;            // camera frustum left plane
    float r;            // camera frustum right plane
    float t;            // camera frustum top plane
    float b;            // camera frustum bottom plane
};

// Create a view-projection matrix from camera.
matrix getViewProjMatrix(const Camera& camera)
{
    matrix     viewMatrix;
    matrix     projMatrix;
    matrix     viewProjMatrix;

    // Build look-at view transformation matrix.
    getCameraMatrix(&viewMatrix, &camera.pos, &camera.dir, &camera.up);

    // Build off-center projection transformation matrix.
    getPerspectiveMatrix(&projMatrix, camera.l, camera.r, camera.b,
        camera.t, camera.zn, camera.zf);

    // Multiply view matrix by projection matrix.
    matrixMultiply(&viewProjMatrix, &viewMatrix, &projMatrix);
    return (viewProjMatrix);
}
```

```
// Creating center/left/right view-projection matrices.
void buildViewProjectionMatrices()
{
    // Get current monoscopic camera in scene.
    Camera camera = getCurrentCamera();

    // Create right vector from normalized up and direction vector.
    float3 right = Cross(&camera.up, &camera.dir);

    // Calculate horizontal camera offsets and frustum plane offsets.
    float3 cameraOffset = 0.5F * camera.Wa * right;
    float planeOffset = 0.5F * camera.Wa * camera.zn / camera.z0;

    // Create left eye camera from center camera.
    Camera cameraLeft = camera;
    cameraLeft.pos -= cameraOffset;
    cameraLeft.l += planeOffset;
    cameraLeft.r += planeOffset;

    // Create right eye camera from center camera.
    Camera cameraRight = camera;
    cameraRight.pos += 0.5F * camera.Wa * right;
    cameraRight.l -= planeOffset;
    cameraRight.r -= planeOffset;

    // Store camera view-projection matrices.
    g_viewProjMatrix = getViewProjMatrix(camera);
    g_viewProjMatrixLeft = getViewProjMatrix(cameraLeft);
    g_viewProjMatrixRight = getViewProjMatrix(cameraRight);
}
```

Listing 10.1. Pseudocode for camera structure (left-handed space).

10.5 Using Geometry Shader to Render Stereo Pairs

On some stereo display formats, the left eye and right eye may be packed in the same back buffer, and we can use viewports to control which eye to render to. In Direct3D 10 and OpenGL 4.1, the geometry shader can be used to generate geometry as well as redirect output to a specific viewport. Instead of rendering stereo in multiple passes, one can use the geometry shader to generate geometry for each eye. This rendering technique may improve performance, especially if the geometry shader is being used anyway.

One method could be to use geometry amplification to do this. This requires that we perform the view-projection transformations in the geometry shader, which may not be as efficient as instancing if the geometry shader is the bottleneck. Another way is to use geometry instancing and write a pass-through geometry shader. Sample HLSL code for Direct3D 10 is shown in Listing 10.2. To render with instancing, set the geometry instanced count to two, as follows:

```
pD3DDevice->DrawIndexedInstanced(numIndices, 2, 0, 0, 0);
```

```
matrix WorldViewProjLeft;
matrix WorldViewProjRight;

struct PS_INPUT
{
    float4      Pos : SV_POSITION;
    float4      Color : COLOR0;
    uint        InstanceId : INSTANCE;
};

struct GS_OUTPUT
{
    float4      Pos : SV_POSITION;
    float4      Color : COLOR0;
    uint        InstanceId : INSTANCE;
    uint        Viewport : SV_ViewportArrayIndex;
};

PS_INPUT VS(float4 Pos : POSITION, float4 Color : COLOR,
        uint InstanceId : SV_InstanceID)
{
    PS_INPUT    input;

    if (InstanceId == 0)
    {
        input.Pos = mul(input.Pos, WorldViewProjLeft);
    }
    else
```

```
    {
        input.Pos = mul(input.Pos, WorldViewProjRight);
    }

    input.Color = Color;
    input.InstanceId = InstanceId;

    return (input);
}

[maxvertexcount(3)]
void GSStereo(triangle PS_INPUT In[3],
        inout TriangleStream<GS_OUTPUT> TriStream)
{
    GS_OUTPUT       output;

    for (int v = 0; v < 3; v++)
    {
        output.Viewport = In[v].InstanceId;
        output.InstanceId = In[v].InstanceId;
        output.Color = In[v].Color;
        output.Pos = In[v].Pos;

        TriStream.Append(output);
    }
}
```

Listing 10.2. This HLSL code renders a stereo pair using the geometry shader with instancing.

References

[AMD 2009] AMD. "AMD Advances 3D Entertainment: Demonstrates Blu-Ray Stereoscopic 3D Playback at 2010 International CES". December 7, 2009. Available at http:// www.amd.com/us/press-releases/Pages/amd-3d-2009dec7.aspx.

[Lengyel 2004] Eric Lengyel. *Mathematics for 3D Game Programming & Computer Graphics*. Hingham, MA: Charles River Media, 2004.

[McAllister 2006] David F. McAllister. "Display Technology: Stereo & 3D Display Technologies." *Encyclopedia of Imaging Science and Technology*, Edited by Joseph P. Hornak, Wiley, 2006.

[HDMI 2010] HDMI Licensing, LLC. "HDMI Licensing, LLC Makes 3D Portion of HDMI Specification Version 1.4 Available for Public Download." February 3, 2010. Available at http://www.hdmi.org/press/press_release.aspx?prid=119.

[National Instruments 2010] National Instruments. "3D Video: One of Seven New Features to Test in HDMI 1.4." 2010. Available at http://zone.ni.com/devzone/cda/tut/p/id/11077.

[Ramm 1997] Andy Ramm. "Stereoscopic Imaging." *Dr. Dobbs Journal*. September 1, 1997. Available at http://www.drdobbs.com/184410279.

[Bourke 1999] Paul Bourke. "Calculating Stereo Pairs." July 1999. Available at http://local.wasp.uwa.edu.au/~pbourke/miscellaneous/stereographics/stereorender/.

11

Making 3D Stereoscopic Games

Sébastien Schertenleib
Sony Computer Entertainment Europe

11.1 Introduction

With the large variety of 3D content being made available (sports events, movies, TV, photos, games, etc.), stereoscopic 3D is gaining momentum. With the support for 3D content on the PC and game consoles such as the PlayStation 3 and Nintendo 3DS, it is likely that it will become even more widespread. In this chapter, we present some topics that need to be considered when creating or converting a game to stereoscopic 3D. We also present some optimization techniques that are targeted to improving both the run-time performance and visual fidelity.

11.2 How Stereoscopic 3D Works

Stereoscopic 3D is produced by creating two separate images (one for each eye) that are then displayed on a 3D screen, as shown in Figure 11.1. Depending on the technology in place, those two images are then separated for the correct eyes through some means. The three major approaches are the following:

- *Active shutter glasses.* The screen alternately displays the left and right images and sends a signal to the LCD screen in the lens for each eye, blocking or transmitting the view as necessary.
- *Passive polarized glasses.* The screen is paired with adjacent right and left images using orthogonal polarizations. The filter on each eye blocks the orthogonally polarized light, allowing each eye to see only the intended image.
- *Parallax barrier.* The screen features a layer of material with some slits placed in front of it, allowing each eye to see a different set of pixels without glasses, but with restricted view angles.

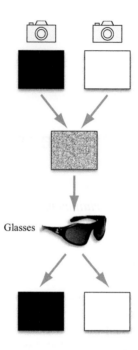

Figure 11.1. Creating a stereoscopic 3D scene.

11.3 How to Set Up the Virtual 3D Cameras

Contrary to other media, video games have the luxury of being able to control the camera properties directly. As we mentioned earlier, stereoscopic 3D requires that we set up two distinct cameras. One possible solution is illustrated in Figure 11.2(a), where we use a simple offset to move the left and right cameras. This approach results in a large portion of each image being visible to only one eye, as shown by the arrows. This tends to generate strong eye strain. Therefore, an alternative approach that is sometimes used with a stereoscopic camcorder is to toe in both cameras by rotating them inward, as shown in Figure 11.2(b). However, the convergence is no longer parallel to the screen, producing a vertical parallax deviation when the camera rotates upward or downward. This is unnatural and uncomfortable for the user. To circumvent these shortcomings, the scheme depicted in Figure 11.2(c) consists of using parallel cameras with an asymmetric projection that minimizes the zone covered by a single image while avoiding vertical parallax. This provides a much more comfortable experience.

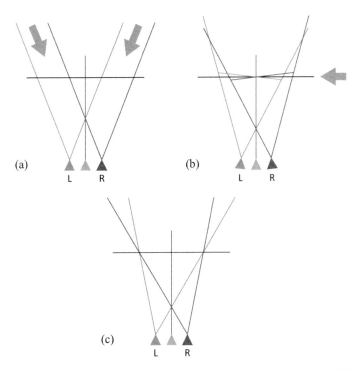

Figure 11.2. (a) A simple offset is applied to the left and right cameras. (b) Both cameras are rotated inward. (c) The cameras have parallel view directions by use asymmetric projections. Configurations (a) and (b) lead to issues that deteriorate the stereoscopic 3D experience. Configuration (c) avoids those shortcomings by using asymmetric projection matrices.

With this model, the usual projection matrix M_{proj} given by

$$M_{proj} = \begin{bmatrix} \dfrac{2n}{r-l} & 0 & \dfrac{r+l}{r-l} & 0 \\ 0 & \dfrac{2n}{t-b} & \dfrac{t+b}{t-b} & 0 \\ 0 & 0 & \dfrac{n+f}{n-f} & \dfrac{2nf}{n-f} \\ 0 & 0 & -1 & 0 \end{bmatrix}$$

changes because it is no longer the case that $r+l=0$ and $t+b=0$. The off-center view frustum that we must use is shown in Figure 11.3.

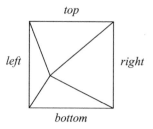

Figure 11.3. The view volume for an asymmetric frustum. The *left* and *right* values represent the minimum and maximum *x* values of the view volume, and the *bottom* and *top* values represent the minimum and maximum *y* values of the view volume, respectively.

Having the ability to alter the camera properties every frame provides a much larger degree of freedom for controlling the 3D scene. For instance, we can adjust the convergence of the cameras to control the depth and size of the objects within the environment. The convergence corresponds to areas of the left and right projected images that superimpose perfectly and therefore have zero parallax, appearing in the plane of the screen. We can also adjust the interaxial distance, which is the separation between both cameras, in order to push back foreground objects. This is very important because it allows us to offer a much more comfortable experience.

11.4 Safe Area

When creating a stereoscopic scene, we need to take into account where objects are located within the 3D space. To make it more comfortable to watch, it is important that we take into account the zone of comfort, shown in Figure 11.4.

The screen acts as a window, and most of the content usually resides inside the screen space, which is the volume of space behind the screen. Content that comes out of the screen is usually small and fast. This also affects the heads-up display (HUD), which can be moved slightly inside the screen space since, otherwise, it might be difficult to focus on the main scene. To reduce eye strain, it is also important to avoid any window violation that occurs when an object touches the edges of the stereo window, resulting in an object being cut off more in one eye than the other. A trivial solution is to alter the frustum culling so that the objects visible to a single eye are properly culled, as illustrated in Figure 11.5. Here, the frustum origin is moved closer to the screen by the distance Δz given by

$$\Delta z = \frac{sd}{s + 2d \tan\left(fov/2\right)},$$

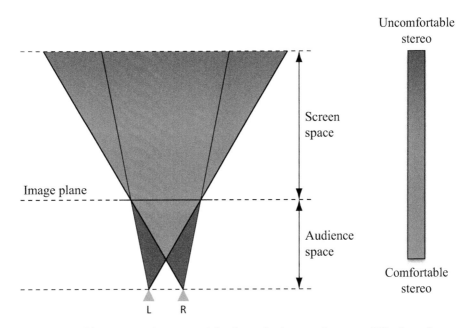

Figure 11.4. Objects very close to and far from the image plane are difficult to focus on and are uncomfortable. Ideally, most of the scene should reside in the safe and comfortable area.

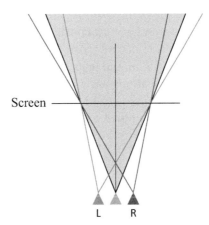

Figure 11.5. Frustum culling for stereoscopic 3D. Instead of combining the view frustum of both cameras, we want to cull any object that would be visible only to one camera, reducing window violations.

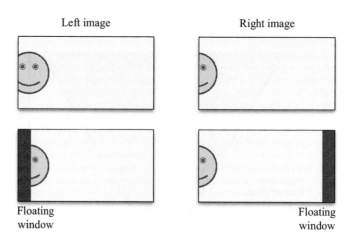

Left image   Right image

Floating   Floating
window   window

Figure 11.6. Without a floating window, some objects might be more visible to one eye, but by using a floating window we can prevent such situations.

where d is the distance to the viewing plane, s is the separation distance between the left and right cameras, and *fov* is the horizontal field-of-view angle.

It is also possible to go one step further and use a dynamic floating window that has a more negative parallax than the closest object, where we avoid that part of an object that becomes more visible to one eye, as shown in Figure 11.6. This creates the illusion of moving the screen surface forward. It is also possible to minimize the difference between the frame and its surface using a graduation or motion blur at the corners of the screen.

We might also have to consider limiting the maximum parallax to prevent any divergence that occurs when the separation of an object for both eyes on the screen is larger than the gap between our eyes (~ 6.4 cm). Thankfully, the HDMI 1.4 specifications allow retrieving the size of the TV, which can be used to calibrate the camera separation. Depending on the screen size, the number of pixels N that are contained within this distance varies as

$$N = \frac{d_{\text{interocular}} \times w_{\text{pixels}}}{w_{\text{screen}}},$$

where $d_{\text{interocular}}$ is the distance between the eyes measured in centimeters, w_{pixels} is the width of the screen in pixels, and w_{screen} is the width of the screen measured in centimeters. For example, for a 46-inch TV and a resolution of 1920×1080 pixels, the number of pixels N for a typical human interocular distance is about 122 pixels.

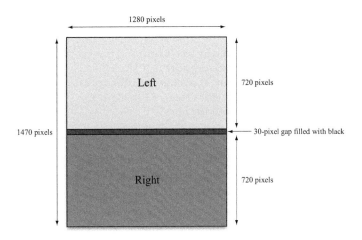

Figure 11.7. Frame packing at 720p.

11.5 Technical Considerations

Creating a stereoscopic 3D scene impacts the run-time performance because there is an additional workload involved. At full resolution, this implies rendering the scene twice. For game engines that are heavily pixel-bound, such as many deferred renderers, this might be critical. Also, the frame buffer and depth buffer need to be larger, especially when using the frame-packing mode exposed by the HDMI 1.4 specification, as shown in Figure 11.7.

To overcome this additional workload some hardware provides an internal scaler that lets us keep the same memory footprint and pixel count as a native monoscopic application with a display mode such as 640×1470. An additional problem is related to swapping the front and back buffers. Monoscopic games can choose to either wait for the next vertical blank or perform an immediate flip to keep a higher frame rate at a cost of possible screen tearing. With stereoscopic games that rely on the frame-packing mode, doing so would generate tearing in one eye only, which is very uncomfortable to watch. As a consequence, it might be a better choice to run at a slower fixed frequency.

11.6 Same Scene, Both Eyes, and How to Optimize

When rendering the scenes, the data needs to be synchronized for both eyes to prevent artifacts. Thankfully, some elements of the scene are not view-dependent and can therefore be shared and computed once. Consider the following typical game loop:

```
while (notdead)
{
    updateSimulation(time);
    renderShadowMaps();
    renderScene(LeftEye, RightEye);
    renderHUD(LeftEye, RightEye);
    vsyncThenFlip();
}
```

Figure 11.8 presents ways to minimize the impact for both the GPU and the CPU by ensuring that view-independent render targets are shared. Some effects that are view-dependent, such as reflections, can sometimes be shared for both views if the surface covered is relatively small, as it often is for mirrors. This leads to artifacts, but they might be acceptable. On some platforms like the PlayStation 3, it is also possible to perform some effects asynchronously on the SPU, such as cascaded shadow maps. In particular, the CPU overhead can also be reduced by caching the relevant rendering states.

It is also possible to use multiple render targets (MRTs) to write to both left and right frame buffers in a single pass. This technique can be used to write to both render targets in a single pass when rendering objects at the screen level or when applying full-screen postprocessing effects, such as color enhancement or crosstalk reduction. This is depicted in Figure 11.9.

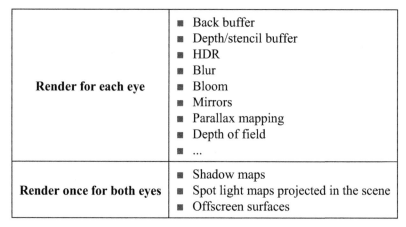

Render for each eye	■ Back buffer ■ Depth/stencil buffer ■ HDR ■ Blur ■ Bloom ■ Mirrors ■ Parallax mapping ■ Depth of field ■ ...
Render once for both eyes	■ Shadow maps ■ Spot light maps projected in the scene ■ Offscreen surfaces

Figure 11.8. Scene management where view-independent render targets are computed once.

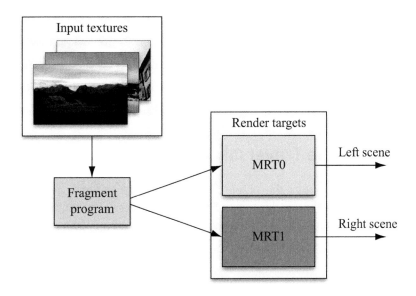

Figure 11.9. Multiple render targets allow us to write to both the left and right frame buffers in a single pass.

Some GPUs flush the rendering pipeline when a new surface is bound as a render target. This might lead to a performance hit if the renderer frequently swaps surfaces for the left and right eyes. A simple solution for avoiding this penalty consists of binding a single surface for both eyes and then moving the viewport between left and right rendering positions, as illustrated in Listing 11.1.

```
setRenderStates();
setLeftEyeProjection();
setLeftEyeViewport();        // surface.x = 0, surface.y = 0
Drawcall();

setRightEyeProjection();
setRightEyeViewport();       // surface.x = 0, surface.y = 720 + 30
Drawcall();

// You can carry on with the same eye for the next object to
// minimize the change of projection matrix and viewport.
setRenderStates();
Drawcall();
```

```
setLeftEyeProjection();
setLeftEyeViewport();
Drawcall();
```

Listing 11.1. This code demonstrates how the images for the left and right eyes can be combined in a single render target by moving the viewport.

11.7 Scene Traversal

To improve the scene traversal for both cameras, it is possible to take into account the similarities between both views. This can be used to improve the scene management at a lower granularity. In fact, if we assume that a point in the relative right-eye viewing position is $\mathbf{P}_{\text{right}} = (x, y, z)$, then the observation that the corresponding point for the relative left-eye viewing position is $\mathbf{P}_{\text{left}} = (x - e, y, z)$, where e is the camera separation, allows us to improve the scene traversal, where a normal vector $\mathbf{N}$ to a polygon in one eye is also valid for the other eye. This could lead to improved hidden surface removal and could also help us discard objects for both eyes using conservative occluders for occlusion queries. This helps minimize the CPU and GPU workload. On the PlayStation 3 platform, a common approach is to perform backface culling using SPU programs. Therefore, it is possible to perform backface culling for both views in a single pass. This function consists of performing a dot product and testing the result. If we compute the view separately, this involves the following operations:

$$2 \times \{3 \times multiplication + 3 \times addition + 1 \times comparison\}$$
$$= 6 \times multiplication + 6 \times addition + 2 \times comparison.$$

This can be improved, and we need to consider the following cases:

- Polygon is front-facing for both views.
- Polygon is back-facing for both views.
- Polygon is front-facing for one view and back-facing for the other view.

Let $\mathbf{N}$ be the normal of a triangle, let $\mathbf{V}_L = (x_L, y, z)$ be the direction from one of the triangle's vertices to the left camera position, and let $\mathbf{V}_R = (x_R, y, z)$ be the direction from the same vertex to the right camera position. The triangle is front-facing for the left camera if $\mathbf{N} \cdot \mathbf{V}_L > 0$, and it is front-facing for the right camera if $\mathbf{N} \cdot \mathbf{V}_R > 0$. Using $x_L = x_R - e$, we need

$$\max\{N_x x_R, N_x(x_R - e)\} > -N_y y - N_z z.$$

This means that in the worst case, it takes

$$4 \times multiplication + 4 \times addition + 1 \times maximum + 1 \times comparison$$

operations. For checking the back-facing triangle for both views, the efficiency of this test is improved by around 33 percent.

11.8 Supporting Both Monoscopic and Stereoscopic Versions

If a game supports both stereoscopic and monoscopic rendering, the stereoscopic version may have to reduce the scene complexity by having more aggressive level of detail (geometry, assets, shaders) and possibly some small objects or post-processing effects disabled. Thankfully, the human visual system does not only use the differences between what it sees in each eye to judge depth, it also uses the similarities to improve the resolution. In other words, reducing the game video resolution in stereoscopic games has less of an impact than with monoscopic games, as long as the texture and antialiasing filters are kept optimal.

In this chapter, we focus on creating 3D scenes by using distinct cameras. It is also possible to reconstruct the 3D scene using a single camera and generate the parallax from the depth map and color separation, which can be performed very efficiently on modern GPUs[1] [Carucci and Schobel 2010]. The stereoscopic image can be created by rendering an image from one view-projection matrix, which can then be projected pixel by pixel to a second view-projection matrix using both the depth map and color information generated by the single view. One of the motivations behind this approach is to keep the impact of stereoscopic rendering as low as possible to avoid compromises in terms of resolution, details, or frame rate.

11.9 Visual Quality

Some of the techniques used in monoscopic games to improve run-time performance, such as view-dependent billboards, might not work very well for close objects, as the left and right quads would face their own cameras at different angles. It is also important to avoid scintillating pixels as much as possible. There-

---

[1] For example, see http://www.trioviz.com/.

fore, keeping good texture filtering and antialiasing ensures a good correlation between both images by reducing the difference of the pixel intensities. The human visual system extracts depth information by interpreting 3D clues from a 2D picture, such as a difference in contrast. This means that large untextured areas lack such information, and a sky filled with many clouds produces a better result than a uniform blue sky, for instance. Moreover, and this is more of an issue with active shutter glasses, a large localized contrast is likely to produce ghosting in the image (crosstalk).

Stereo Coherence

Both images need to be coherent in order to avoid side effects. For instance, if both images have a different contrast, which can happen with postprocessing effects such as tone mapping, then there is a risk of producing the Pulfrich effect. This phenomenon is due to the signal for the darker image being received later by the brain than for the brighter image, and the difference in timing creates a parallax that introduces a depth component. Figure 11.10 illustrates this behavior with a circle rotating at the screen distance and one eye looking through filter (such as sun-glasses). Here, the eye that looks through the filter receives the image with a slight delay, leading to the impression that the circle rotates around the up axis.

Other effects, such as view-dependent reflections, can also alter the stereo coherence.

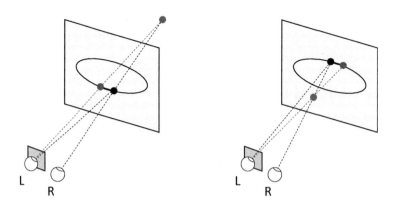

Figure 11.10. The Pulfrich effect.

Fast Action

The brain needs some time to accommodate the stereoscopic viewing in order to register the effect. As a result, very fast-moving objects might be difficult to interpret and can cause discomfort to the user.

3D Slider

Using a 3D slider allows the user to reduce the stereoscopic effect to a comfortable level. The slider controls the camera properties for interaxial distance and convergence. For instance, reducing the interaxial distance makes foreground objects move further away, toward the comfortable area (see Figure 11.4), while reducing the convergence moves the objects closer. The user might adjust it to accommodate the screen size, the distance he is sitting from the screen, or just for personal tastes.

Color Enhancement

With active shutter glasses, the complete image is generally darkened due to the LCD brightness level available on current 3D glasses. It is possible to minimize this problem by implementing a fullscreen postprocessing pass that increases the quality, but it has to be handled with care because the increased contrast can increase the crosstalk between both images.

Crosstalk Reduction

Crosstalk is a side effect where the left and right image channels leak into each other, as shown in Figure 11.11. Some techniques can be applied to reduce this problem by analyzing the color intensity for each image and subtracting them from the frame buffer before sending the picture to the screen. This is done by creating calibration matrices that can be used to correct the picture. The concept consists of fetching the left and right scenes in order to extract the desired and unintended color intensities so that we can counterbalance the expected intensity leakage, as shown in Figure 11.12. This can be implemented using multiple render targets during a fullscreen postprocessing pass. This usually produces good results, but unfortunately, there is a need to have specific matrices tuned to each display, making it difficult to implement on a wide range of devices.

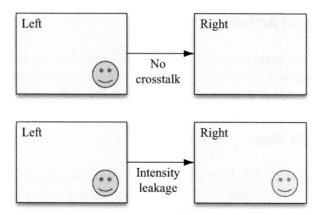

Figure 11.11. Intensity leakage, where a remnant from the left image is visible in the right view.

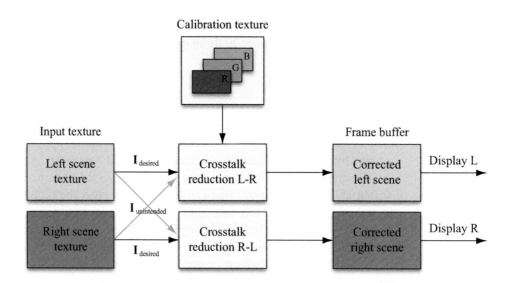

Figure 11.12. Crosstalk reduction using a set of calibration matrices.

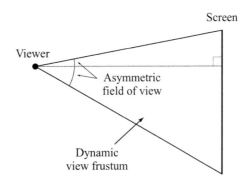

Figure 11.13. Ortho-stereo viewing with head-tracked VR.

11.10 Going One Step Further

Many game devices now offer support for a front-facing camera. These open new possibilities through face tracking, where "ortho-stereo" viewing with head-tracked virtual reality (VR) can be produced, as shown in Figure 11.13. This provides the ability to follow the user so that we can adapt the 3D scene based on his current location. We can also consider retargeting stereoscopic 3D by adopting a nonlinear disparity mapping based on visual importance of scene elements [Lang et al. 2010]. In games, elements such as the main character could be provided to the system to improve the overall experience. The idea is to improve the 3D experience by ensuring that areas where the user is likely to focus are in the comfortable area and also to avoid divergence on far away objects. For instance, it is possible to fix the depth for some objects or to apply depth of field for objects outside the convergence zone.

11.11 Conclusion

Converting from a monoscopic game to stereoscopic 3D requires less than twice the amount of processing but requires some optimization for the additional rendering overhead. However, the runtime performance is only one component within a stereoscopic 3D game, and more effort is needed to ensure the experience is comfortable to watch. A direct port of a monoscopic game might not create the best experience, and ideally, a stereoscopic version would be conceived in the early stages of development.

References

[Carucci and Schobel 2010] Francesco Carucci and Jens Schobel. "AAA Stereo-3D in CryEngine." *Game Developers Conference Europe*, 2010.

[Lang et al. 2010] Manuel Lang, Alexander Hornung, Oliver Wang, Steven Poulakos, Aljoscha Smolic, and Markus Gross. "Nonlinear Disparity Mapping for Stereoscopic 3D." *ACM Transactions on Graphics* 29:4 (July 2010).

12

A Generic Multiview Rendering Engine Architecture

M. Adil Yalçın
Tolga Çapın
Department of Computer Engineering, Bilkent University

12.1 Introduction

Conventional monitors render a single image, which is generally observed by the two eyes simultaneously. Yet, the eyes observe the world from slightly different positions and form different images. This separation between the eyes provides an important depth cue in the real world. Multiview rendering aims to exploit this fundamental feature of our vision system for enhanced 3D rendering.

Technologies that allow us to send different images to the two eyes have been around for years, but it is only now that they can reach the consumer level with higher usability [Bowman et al. 2004]. The existing technologies vary among the different types of 3D displays, and they include shutter glasses, binocular head-mounted displays, and the more recent and popular autostereoscopic displays that require no special glasses.

Recent techniques for multiview rendering differ in terms of visual characteristics, fidelity, and hardware requirements. Notably, multiview rendering engines should be able to support more than two simultaneous views, following recent 3D display technologies that can mix a higher number of simultaneous views than traditional stereo view [Dodgson 2005].

Currently, many available multiview applications are configured for the stereo-view case, and the routines that manage stereo rendering are generally implemented as low-level features targeted toward specific APIs and displays. We present a higher-level multiview rendering engine architecture that is generic, robust, and easily configurable for various 3D display platforms, as illustrated in

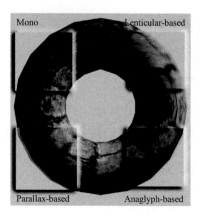

Figure 12.1. A torus rendered for different basic 3Ddisplay platforms.

Figure 12.1. The architecture simplifies the use of multiview components of a rendering engine and it solves the problems of rendering separate views and merging them for the target display. It also includes support for multiview rendering optimization techniques, such as view-specific level-of-detail systems and pipeline modifications. Additional discussions include insights on further extensions, design guidelines, and other possible uses of the presented architecture.

Most of the discussions, terminologies, and guidelines in this chapter follow OpenGL conventions. The implementation of the architecture is included in the OpenREng library,[1] an open-source rendering engine based on modern desktop and mobile OpenGL specifications with multiview rendering support, and the sample applications are built using this library.

12.2 Analyzing Multiview Displays

At its core, rendering for 3D displays aims to generate separate images (views) for separate eyes. All types of 3D displays, summarized below, create the illusion of 3D space based on multiplexing the different images for each eye, whether it be temporally, spatially, or in color. It is a challenge to support all types of displays transparently to application implementations in a way that hides the details of the low-level resource management required for multiview rendering.

Displays that require wearing special eyeglasses can be subdivided into the following categories (as also shown in Figure 12.2):

---

[1] See http://openreng.sourceforge.net/.

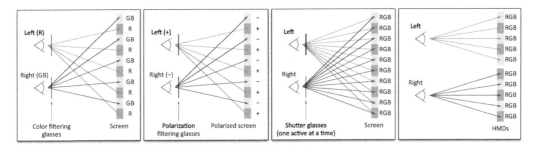

Figure 12.2. Stereo rendering techniques that require wearing glasses. From left to right, anaglyph glasses, polarized glasses, shutter glasses, and head-mounted displays.

■ *Anaglyph glasses*. These are based on multiplexing color channels. The two views are filtered with different colors and then superimposed to achieve the final image.

■ *Head-mounted displays (HMDs)*. These are based on displaying both views synchronously to separate display surfaces, typically as miniaturized LCD, organic light-emitting diode (OLED), or CRT displays.

■ *Shutter glasses*. These are based on temporal multiplexing of the two views. These glasses work by alternatively closing the left or right eye in sync with the refresh rate of the display, and the display alternately displays a different view for each eye.

■ *Polarized glasses*. With passive and active variants, these glasses are based on presenting and superimposing the two views onto the same screen. The viewer wears a special type of eyeglasses that contain filters in different orientations.

While these displays require special hardware, another type of 3D display, called an *autostereoscopic* display, creates the 3D effect without any special eyeglasses. Autostereoscopic displays operate by emitting a different image toward each eye of the viewer to create the binocular effect. This is achieved by aligning an optical element on the surface of the screen (normally an LCD or OLED) to redirect light rays for each eye. A composite image that superimposes the two views is rendered by the display subpixels, but only the correct view is directed to the corresponding eye.

There are two common types of optical filter, a lenticular sheet and a parallax barrier. A lenticular sheet consists of small lenses having a special shape that refract the light in different directions. A parallax barrier is essentially a mask with

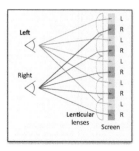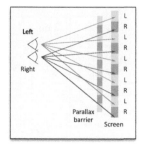

Figure 12.3. Autostereoscopic displays: a lenticular sheet (left) and a parallax barrier (right).

openings that allow light to pass through only in certain directions. These two technologies are illustrated in Figure 12.3. In both cases, the intensity of the light rays passing through the filter changes as a function of the viewing angle, as if the light is directionally projected. The pixels for both eyes are combined in a single rendered image, but each eye sees the array of display pixels from a different angle and thus sees only a fraction of the pixels, those precisely conveying the correct left or right view.

The number of views supported by autostereoscopic displays varies. The common case is two-view, which is generally called stereo-view or stereo-rendering. Yet, some autostereoscopic 3D displays can render 4, 8, or 16 or more views simultaneously. This allows the user to move his head side to side and observe the 3D content from a greater number of viewpoints. Another basic variable is the size of the display. Three-dimensional TVs, desktop LCD displays, and even mobile devices with multiview support are becoming popular and accessible to the mass market.

As a result, it is a challenge to build applications that run on these different types of devices and 3D displays in a transparent way. There is a need for a multiview rendering architecture that hides the details of multiplexing and displaying processes for each type of display.

12.3 The Architecture

The architecture we present consists of the following:

- An extensible multiview camera abstraction.
- An extensible multiview compositor abstraction.

- A configurable multiview buffer that holds intermediate view-dependent rendering results.
- Rendering pipeline modifications to support multiview rendering.
- Level-of-detail implementations for multiview rendering.

To be able to support both single-view and multiview rendering seamlessly, the multiview system architecture is integrated into a viewport abstraction over display surfaces. This further allows multiview rendering in multiple viewports on the screen, even with different multiview configurations for each multiview viewport, as shown in Figure 12.4. With this approach, you can add picture-in-picture multiview regions to your screen, or you can show your single-view 2D graphical user interface (GUI) elements over multiview 3D content by rendering it only a single time after the multiview content is merged into the frame buffer. Briefly, with this approach, you have control over where and how you want your multiview content to be rendered. An overview of our architecture is shown in Figure 12.5.

Multiview rendering is enabled by attaching a multiview camera, a multiview buffer, and a multiview compositor to a viewport in a render window. Since most of these components are configurable on their own and provide abstraction over a distinct set of features, the system can be adjusted to fit into many target scenarios. At render time, the multiview rendering pipeline is activated if the attached

Figure 12.4. The same scene rendered to different viewports with different multiview configurations: anaglyph using color-mode (bottom-left), parallax using both off-target and on-target (stenciling), on-target-wiggle, and on-target-separated.

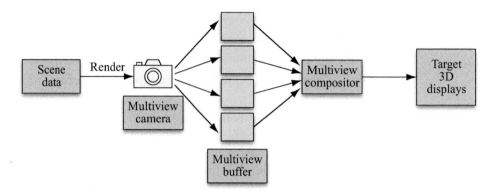

Figure 12.5. Overview of the multiview rendering architecture.

multiview components support the same number of views. As shown in Figure 12.5, the scene is first rendered multiple times after activating a specific camera view and a specific view buffer target. This generates the rendering content to be used by the attached multiview compositor, which outputs its result to the region of the rendered viewport. The last step is performed by the operating system, which swaps the final frame buffer holding the multiview content to be shown on the target displays.

The sample code provided in Listing 12.1 can add full support for anaglyph-based stereo-view rendering using our architecture without any other modifications to the application code.

```
CameraStereoView& camera(CameraStereoView::create(*camNode));

// Set standard parameters.
// (aspect ratio, far-near dist., field-of-view, etc.)
camera.setEyeSeparation(0.3);        // stereo-specific
camera.setFocalDistance(25.0);       // stereo-specific
MVBuffer_Cfg       mvbParams;

mvbParams.viewCount = 2;
mvbParams.type = MVBufferType_Offtarget;
mvbParams.offtarget.colorFormat = ImageFormat_RGBA;
mvbParams.offtarget.sharedDepthStencilTargets = true;
mvbParams.offtarget.sharedFrameBuffer = true;
```

```
// Attach the components to viewport.
RSys.getViewport(0)->mCamera = &camera;
RSys.getViewport(0)->attachMVCompositor(new MVC_Anaglyph);
RSys.getViewport(0)->attachMVBuffer(new MultiViewBuffer, mvbParams);
```

Listing 12.1. Setting up a basic multiview rendering pipeline in an application.

12.4 The Multiview Camera

Multiview rendering requires each of the views to be rendered with view offset parameters, so that the eyes can receive the correct 3D images. To ease the control of cameras in multiview configurations, our architecture provides a multiview camera concept that is integrated into our basic multiview rendering pipeline.

To be able to easily manipulate cameras within a generic multiview architecture, we define a single interface, the *multiview camera*. This interface allows for specialization of different configurations through class inheritance. Implementations can integrate most of the multiview-specific functionality (e.g., activating a specific view) in a generic camera class, and single-view cameras can therefore be treated as simple multiview cameras.

As usual, the camera object is used to identify view and projection transformation matrices that are applied to 3D mesh vertices. The view matrix depends on both the camera position and the multiview camera parameters that offset the view position from a reference camera transformation. The same approach applies to the projection matrix—both basic intrinsic parameters (such as field of view, near and far clip plane distances, and aspect ratio) and multiview configuration parameters (such as focal length) can affect the final result.

In our design, a single multiview camera object aims to reflect a group of perspective-projection cameras. Each view is rendered using perspective projection, which implies that the multiview camera has the properties of a perspective camera and can thus inherit from a perspective camera component, as shown in Listing 12.2. Intuitively, in multiview configurations, one wants to be able to control a single camera object that would internally calculate all view-specific parameters with respect to the base camera position and orientation. The view-specific perspective projection and view transformation are generated using an active view number, and this active view number needs to be automatically managed by the main renderer. Internal camera parameters are specified by extending the multiview camera interface and defining the required parameters that affect the view and projection matrix setup.

Also, each specific multiview camera implementation can define a single bounding volume to be used for one-time object culling over all views. The actual geometry covered inside a multiview volume is likely to be nonconvex, which can make precise intersection tests with the scene bounding volumes computationally less efficient. To set a single, shared culling frustum for a multiview camera, a larger frustum that covers the multiview volume needs to be set up. The corners of a suitable approximation for an extended multiview culling frustum can be composed from the outermost corners of each view-specific frustum.

```cpp
// Base camera interface provides basic multiview operations.
class Camera
{
    public:

        // Get/set aspect ratio, near-far distance, field of view, etc.
        virtual uchar getViewCount() const;      // returns 1 as default
        virtual void setActiveView(uchar viewIndex);  // default no-op
        virtual const Matrix4& getViewMatrix() const;
        const Matrix4& getProjectionMatrix() const;
        const BoundingVolume& getFrustum() const;

    protected:

        mutable Matrix4             mProjMatrix_Cache;
        mutable BoundingVolume      mFrustum_Cache;
        bool                        mProjMatrixMatrix_dirty, mFrustum_dirty;

        virtual void updateProjMatrix() const = 0;
        virtual void updateFrustum() const = 0;
};

class CameraPerspective : public Camera
{
    ...
};

class CameraMultiView : public CameraPerspective
{
    public:
```

```
        void setActiveView(size_t viewIndex);

    protected:

        uchar       mActiveView;
};

class CameraStereoView : public CameraMultiView
{
    public:

        // Set/get focal distance and eye separation.
        const Matrix4& getViewMatrix() const;

    protected:

        float       mFocalDistance, mEyeSeparation;

        void updateProjectionMatrix() const;
};
```

Listing 12.2. Projection and view matrix management for stereo rendering.

The remainder of this section focuses on the special case of stereo cameras and describes the basic parameters that can allow easy and intuitive manipulation of the camera matrices. The concepts described in this section can be extended to multiview configurations that require more views.

As shown in Figure 12.6, different approaches can be applied when creating stereo image pairs. In the figure, d denotes the separation between the eyes. Parallel and oriented frustums use the basic symmetrical perspective projection setup for each view and offset the individual camera positions (and the camera orientations, in the oriented frustum case). The skewed frustum approach modifies the perspective projection matrix instead of updating the camera orientation, so the two image planes are parallel to the zero-parallax plane.

The camera view position offset depends on the right direction vector of the base multiview camera and the d parameter. If the base camera is assumed to be in the middle of the individual view points, the offset distance is simply $d/2$ for each view. To generate skewed frustum projection pairs, assuming that the frustum can be specified with left-right, bottom-top, and near-far values, as is done

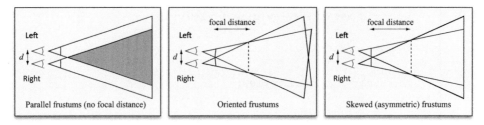

Figure 12.6. Basic approaches for setting up stereo camera projection frustum pairs.

for the OpenGL function `glFrustum()`, only the left and right values need to be modified. The offset Δx for these values can be calculated using the formula

$$\Delta x = \frac{dn}{2f},$$

where n is the distance to the near plane, and f is the focal distance. Note that the projection skew offset Δx is added for the right camera and subtracted for the left camera.

Figure 12.7 shows a simple diagram that can be used to derive the relationship among the angle shift s, the eye separation d, half field-of-view angle θ, and the focal distance f. By using trigonometric relationships and the fact that lines intersect at the point $\mathbf{p}$ in the figure, the following equation can be derived:

$$f = \frac{d\left(1 - \tan^2\theta \tan^2 s\right)}{2\tan s\left(1 + \tan^2\theta\right)}.$$

As expected, a smaller angle shift s results in a larger focal distance, and the eye separation parameter d affects the focal distance linearly.

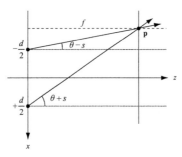

Figure 12.7. Derivation of the relationship between the angle shift s, the eye separation d, and the focal distance f.

12.5 The Multiview Buffer

Multiview rendering requires rendering separate views to distinct targets so that they can be mixed as required. The *multiview buffer* concept in this architecture aims to encapsulate important properties of the render targets that can be used in multiview pipelines.

The instances of multiview buffers are created by the application and attached to the viewports using multiview buffer configuration parameters. These parameters are designed to allow easy high-level configuration of internal resources (such as textures, render buffers, and frame buffers) that will be created. Although the multiview buffer concept is not designed for extensibility through class inheritance, the parameters can provide the required robustness to the application developer and can be extended by updating this single interface when required.

After analyzing possible multiview configurations, we have observed that there are two basic types of multiview buffers:

- On-target buffers for which we render to the viewport's render target directly.
- Off-target buffers that create and manage their own resources as render targets.

An on-target multiview buffer uses the attached viewport's render surface instead of creating any new (offscreen) surfaces. A final compositing phase may not be needed when an on-target multiview buffer is used because the multiview rendering output is stored in a single surface. The rendering pipeline can still be specialized for per-view operations using the multiview compositor attachments. For example, to achieve on-target anaglyph-based rendering, an attached compositor can select per-view color write modes, in turn separating the color channels of each view, or a compositor can select different rendering regions on the same surface. Also, OpenGL quad-buffer stereo mode can be automatically managed as an on-target multiview buffer since no additional surfaces need to be set up other than the operating system window surface, and its usage depends on target surface's native support for left and right view buffering.

An off-target multiview buffer renders to internally managed offscreen surfaces instead of the attached viewport's render surface, and it can thus be configured more flexibly and independently from the target viewport. Offscreen rendering, inherent in off-target multiview buffers, allows rendering the content of each view to different surfaces (such as textures). The application viewport surface is later updated with the composite image generated by the attached multiview compositor. If the composition (merge) step of the multiview display device requires that complex patterns be sampled from each view, as is common in lenticular-based displays, or if the per-view outputs need to be stored in separate resources with different configurations (sizes, component types, etc.) as a multiview optimization step, using an off-target multiview buffers is required.

Some additional aspects of off-target buffer configurations are the following:

- The color channel targets need to be separated for each view.
- The depth and stencil targets can be shared between different views if the view-specific images are rendered sequentially. Clearing depth/stencil buffers after rendering has been completed for each view ensures that each view has its own consistent depth/stencil buffer.
- Specific to OpenGL, a multiview buffer can be assigned a single frame buffer, as opposed to switching frame buffer objects in each view, and the texture attachments may be dynamic. Rendering performance may differ depending on the hardware and the rendering order used.
- For off-target buffers, the sizes of internal render surfaces are based on the attached viewport render surface size since the internal view-specific surfaces are later merged into the viewport render surface.
- The multiview buffers can apply additional level-of-detail settings. Possible approaches are discussed in Section 12.8.

12.6 The Multiview Compositor

The *multiview compositor* component is responsible for merging a given off-target multiview buffer (the render data for specific views) into the target viewport, and it can also be used to define view-specific rendering states. Since the compositing logic is heavily dependent on the target hardware configuration, our architecture supports an extensible multiview compositor design, allowing the programmer to define hardware-specific view-merge routines by inheriting from a base class interface.

The composition phase requires that a multiview buffer provide the rendering results of the different views in separate render targets. Thus, when an on-target multiview buffer is used, there is no need to define a compositing method. Yet, using an off-target multiview buffer and multiview compositor provides a more flexible mechanism, while introducing only slight data, computation, and management overheads.

Since the multiview buffers use GPU textures to store render results, the multiview compositors can process the texture data on the GPU with shaders, as shown in Listings 12.3 and 12.4. Using a shader-driven approach, the view buffers can be upsampled or downsampled in the shaders, using available texture filtering options provided by the GPUs (such as nearest or linear filtering).

```
in vec2     vertexIn;
out vec2    textureCoord;

void main()
{
    textureCoord = vertexIn.xy * 0.5 + 0.5;
    glPosition = vec4(vertexIn.xy, 0.0, 1.0);
}
```

Listing 12.3. A sample vertex shader for a parallax-based multiview rendering composition phase.

```
uniform sampler2D    viewL;
uniform sampler2D    viewR;
varying vec2         textureCoord;

void main()
{
    vec4 colorL = texture2D(viewL, textureCoord);
    vec4 colorR = texture2D(viewR, textureCoord);

    // Creating the stripe pattern for left-right view.
    gl_FragColor = colorR;
    if (mod(gl_FragCoord.x, 2.0) > 0.5) gl_FragColor = ColorL;
}
```

Listing 12.4. A sample fragment shader for a parallax-based multiview rendering composition phase.

12.7 Rendering Management

Given the basic multiview components, rendering an object for a specific view is achieved through the following steps:

- Activate a specific view on a multiview camera and update the projection and view matrices.
- Activate a specific view on a multiview buffer.
- Activate view-specific object materials and geometries, as part of an object level-of-detail (LOD) system (see Section 12.8).

After all objects are rendered to all of the views, the multiview compositor for the viewport can process the view outputs and generate the final multiview image-age.

Once the rendering requirements of an object-view pair are known, there are two options for rendering the complete scene, as shown in Figure 12.8. In the first case, a specific view is activated only once, and all of the visible objects are rendered for that view. This process is continued until all of the views are completed, and such an approach keeps the frame target "hot" to avoid frequent frame buffer swapping. In the second case, each object is activated only once, and it is rendered to all viewports sequentially, this time keeping the object "hot." This approach can reduce vertex buffer or render state switches if view-specific geometry/render state data is not set up. Also, with this approach, the camera should cache projection and view matrix values for each view since the active view is changed very frequently. Depending on the setup of the scene and the number of views, the fastest approach may differ. A mixed approach is also possible, where certain meshes in the scene are processed once into multiple views and the rest are rendered as a view-specific batch.

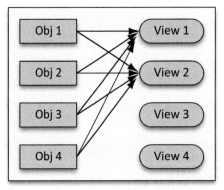 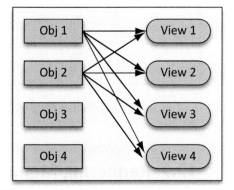

For each view, render all objects in the scene.   For each object in the scene, render to all views.

Figure 12.8. Rendering order considerations for multiview pipelines.

Shared Data between Different Views

It is possible to make use of the coherence between different views during rendering as follows:

- Most importantly, the same scene data is used to render the 3D scene (while this can be extended by using a multiview object LOD system). As a result, animations modifying the scene data need to be only applied once.
- Object or light culling can be applied once per frame using a single shared frustum for multiview camera objects, containing the frustums of view-specific internal cameras.

In summary, only the effective drawing time of a specific viewport is affected when multiview rendering is activated. Other than an increase in the number of draw calls, the multiview composition step also costs render time, especially in low-end configurations such as mobile devices, and it should be optimally implemented. For example, on-target multiview buffers can be preferred to avoid an additional compositing phase if per-view compositing logic can be applied at render time by regular 3D pipelines. With such an approach, color write modes in the graphics pipeline can be used to set up regular anaglyph rendering, or stencil testing can be used to create per-view compositing patterns.

12.8 Rendering Optimizations

The aim of the multiview rendering optimizations discussed in this section is to provide some basic building blocks that can help programmers reduce total rendering time without sacrificing the perceived quality of the final result. Since, according to binocular suppression theory, one of the two eyes can suppress the other eye, the non-dominant eye can receive a lower quality rendering while not reducing the effective quality of the multiview rendering of a scene. A recent study [Bulbul et al. 2010] introduced a hypothesis claiming that "if the intensity contrast of the optimized rendering for a view is lower than its original rendering, then the optimized rendering provides the same percept as if it were not optimized." This and similar guidelines can be studied, implemented and tested to be able to optimize a multiview rendering pipeline and thus to render more complex scenes in real time.

Multiview Level-of-Detail for Objects

In addition to the object distance parameter, engines supporting multiview architectures can introduce a new detail parameter, the active view number, and use this parameter to select the vertex data and the rendering state of an object. Level-of-detail can then be implemented in one or more of the following ways:

- *Level-of-detail for object geometry.* Simplified mesh geometries can be used for some of the views, reducing vertex processing time for complex meshes.

- *Level-of-detail for object materials.* Different views can apply different rendering techniques to a single object. The changes may involve reducing the number of rendering passes or the shading instruction complexity, such as switching between per-pixel and per-vertex shading or using different surface models for different views.

- *Level-of-detail for shaders.* Shaders can execute different code paths based on the active view number, allowing more algorithmic approaches to reducing shader complexities for higher numbers of views. This can also ease development of view-specific shaders and allow the creation of simplified routines for optimized views. Shader level-of-detail is a research area in itself, and various automated techniques have also been proposed [Pellacini 2005]. This can equally be achieved by an application-specific solution, where different shaders are supported for the different views.

Multiview Level-of-Detail for Multiview Buffers

Using the customizability of off-target multiview buffers, it is possible to create buffers of different sizes for different views. One basic approach is to reduce the resolution of the target by half, practically reducing the pixel shading cost. The low-resolution buffers can be upsampled (magnified) during the composition phase, using nearest or linear sampling.

Since the low-resolution buffer on the same viewport will be blurry, it is expected that the sharper view will dominate the depth perception and preserve the sharpness and quality of the overall perceived image [Stelmach et al. 2000]. Thus, it is advisable to use buffer level-of-detail options whenever the pixel shading is time consuming.

We should also note that one of the two eyes of a person can be more dominant than the other. Thus, if the dominant eye observes a higher-quality view, the user experiences a better view. It is not possible to know this information in advance, so user-specific tests would need to be performed and the system adjusted for each user. An approach that avoids dominant-eye maladjustment is to switch the low- and high-resolution buffer pairs after each frame [Stelmach et al. 2000].

Other Optimization Approaches

As surveyed by Bulbul et al. [2008], *graphics-pipeline-based* and *image-based* optimization solutions have also been proposed. Graphics-pipeline-based optimizations make use of the coherence between views, or they are based on approximate rendering where fragment colors in all neighboring views can be approximated from a central view when possible. In image-based optimizations, one view is reconstructed from the other view by exploiting the similarity between the two. In these techniques, the rendering time of the second image depends on only the image resolution, instead of the scene complexity, therefore saving rendering computations for one view. Approaches have been proposed that are based on warping, depth buffers, and view interpolation [Bulbul et al. 2008].

12.9 Discussion

Multiview Scene Setup

Our architecture supports customization and extensible parameterization, but does not further provide guidelines on how to set the multiview camera parameters and scene in order to achieve maximum viewing comfort. In the first volume of *Game Engine Gems*, Hast [2010] describes the plano-stereoscopic view mechanisms, common stereo techniques such as anaglyph, temporal multiplexing (shutter glasses), and polarized displays and discusses their pros and cons. Some key points are that contradicting depth cues should be avoided and that special care needs to be directed at skyboxes and skydomes, billboards and impostors, GUIs, cursors, menus in virtual 3D space, frame rate, view synchronization, and scene-to-scene camera setup consistency (such as focal distance). Viewers may have different eye separation distances and display sizes, and the distance of the viewer to the display can differ among different platforms. It should be kept in mind that creating the right 3D feeling is a process that requires a scalable technical infrastructure (as presented in this chapter) and an analysis of the target platforms, the virtual scene, and animations.

Enabling/Disabling Multiview Rendering at Run Time

It is important to allow the user to select single-view rendering if the hardware supports it; some users [Hast 2010] may not be able to accommodate the multiview content easily and may prefer single-view rendering because it can produce a higher-quality image. The architecture natively supports switching between single-view and multiview configurations through run-time attachment and detachment of multiview components (camera, buffer, and compositors as required) to a specific viewport on the render target. Viewport rendering logic that easily adapts itself to follow a single-view or multiview rendering pipeline is possible, and the implementation within OpenREng provides a sample solution.

Postprocessing Pipelines

Post-processing pipelines are commonly used, and their adaptation to multiview rendering can present a challenge. Most of the post-processing filters use spatial information about a fragment to calculate the output. The spatial information is partly lost when different views are merged into a single image. Thus, applying the same post-processing logic to the single composited image may not produce the expected output. If spatial data is not used, such as in color filters, the post-processing can natively interact with the results in separate views. However, filters like high dynamic range and bloom may interact with spatial data and special care may need to be taken [Hast 2010]. In our architecture, the post-processing logic can be integrated into multiview compositor logic (shaders) to provide another rendering pass optimization.

Integration with Other Stereo-Rendering APIs

As discussed in Section 12.5, our architecture can benefit from OpenGL quad-buffer stereo mode support directly. Yet there are other proprietary APIs that manage the stereo rendering at the driver level. As an example, Nvidia's 3DVision API only supports DirectX implementations. Basically, the multiview rendering is handled by the graphics driver when the application follows specific requirements. Since such APIs offer their own abstractions and optimizations for stereo rendering, it may not be possible to wrap their APIs over our architecture.

3D Video Playback

To be able to playback 3D video over our architecture, it is possible to send the decoded 3D video data for separate views to their corresponding multiview buffer color render targets and specify the composition by defining your own multiview compositors. It is also possible to skip the multiview buffer interface and perform the composition work directly using the decoded video data inside the multiview compositor merge routines.

Using Multiview Pipeline for Other Rendering Techniques

Our multiview rendering architecture can be extended to support soft shadow techniques that use multi-lights to generate multiple depth results from different locations. Yang et al. [2009] show an example of the multi-light approach for soft shadow rendering.

Acknowledgements

This project has been supported by 3DPHONE, a project funded by the European Union EC 7th Framework Programme.

References

[Bowman et al. 2004] Doug A. Bowman, Ernst Kruijff, Joseph J. LaViola, and Ivan Poupyrev. *3D User Interfaces: Theory and Practice.* Reading, MA: Addison-Wesley, 2004.

[Bulbul et al. 2010] Abdullah Bulbul, Zeynep Cipiloglu, and Tolga Çapın. "A Perceptual Approach for Stereoscopic Rendering Optimization." *Computers & Graphics* 34:2 (April 2010), pp. 145–157.

[Dodgson 2005] Neil A. Dodgson. "Autostereoscopic 3D Displays." *Computer* 38:8 (August 2005), pp. 31–36.

[Hast 2010] Anders Hast. "3D Stereoscopic Rendering: An Overview of Implementation Issues." *Game Engine Gems 1*, edited by Eric Lengyel. Sudbury, MA: Jones and Bartlett, 2010.

[Pellacini 2005] Fabio Pellacini. "User-Configurable Automatic Shader Simplification." *ACM Transactions on Graphics* 24:3 (July 2005), pp. 445–452.

[Stelmach et al. 2000] L. Stelmach, Wa James Tam, D. Meegan, and A. Vincent. "Stereo Image Quality: Effects of Mixed Spatio-Temporal Resolution." *IEEE Transactions on Circuits and Systems for Video Technology* 10:2 (March 2000), pp. 188–193.

[Yang et al. 2009] Baoguang Yang, Jieqing Feng, Gaël Guennebaud, and Xinguo Liu. "Packet-Based Hierarchal Soft Shadow Mapping." *Computer Graphics Forum* 28:4 (June–July 2009), pp. 1121–1130.

13

3D in a Web Browser

Rémi Arnaud

Screampoint Inc.

13.1 A Brief History

The idea 3D graphics in a web browser is not a new concept. Its prototype implementations can be traced to almost two decades ago, almost as old as the concept of the world wide web (WWW) itself, as it was first introduced in a paper presented at the first WWW Conference organized by Robert Cailliau in 1994. The virtual reality markup language (VRML, pronounced *vermal*, renamed to virtual reality modeling language in 1995) was presented by Dave Raggett in a paper submitted to the first WWW conference and discussed at the VRML birds of a feather (BOF) established by Tim Berners-Lee, where Mark Pesce [1] presented the Labyrinth demo he developed with Tony Parisi and Peter Kennard [2]. This demonstration is one of the first, if not the first, of 3D graphics for the web.

The first version of VRML was published in November 1994 by Gavin Bell, Tony Parisi, and Mark Pesce. It very closely resembles the API and file format of the Open Inventor software, originally developed by Paul Strauss and Rikk Carey at Silicon Graphics, Inc. (SGI) [3]. The current and functionally complete version is VRML97 (ISO/IEC 14772-1:1997). SGI dedicated engineering and public relations resources to promote the CosmoPlayer and ran a web site at vrml.sgi.com on which was hosted a string of regular short performances of a character called Floops who was a VRML character in a VRML world. VRML has since been superseded by X3D (ISO/IEC 19775-1) [4], an XML encoding of VRML.

Despite its ISO standardization status, VRML/X3D has not had the same success as HTML. HTML is definitely the standard for publishing content on the web, but VRML/X3D has failed to garner the same level of adoption for 3D content publishing. HTML has evolved from static content to dynamic content (a.k.a. Web 2.0) and has fueled the economy with billions of dollars in businesses that are still growing despite the internet bubble bursting circa 2000. Currently, there are over two dozen web browsers available for virtually all platforms, including desktop and laptop computers, mobile phones, tablets, and embedded systems. The browser war started in 1995, and Microsoft (with Internet Explorer) won the first round against Netscape to dominate the market by early 2000. The browser wars are not over as Google (Chrome), Mozilla (Firefox), Opera (Opera) and Apple (Safari) are now eroding Microsoft's dominance.

During the same period of time, 3D has grown significantly as a mass-market medium and has generated large revenues for the entertainment industry through games and movies. 3D display systems have materialized in movie theaters and generate additional revenues. So the question remains: Why has VRML/X3D not had the same pervasive path as HTML? The web is filled with tons of opinions as to why this did not work out. (Note: X3D is still being proposed to the W3C HTML Working Group for integration with HTML 5.) Mark Pesce himself offered his opinion in an interview published in 2004, ten years after introducing VRML to the WWW conference [2]:

> John Carmack pronounced VRML dead on arrival. His words carried more weight with the people who really mattered—the core 3D developers—so what should have been the core constituency for VRML, games developers, never materialized. There was never a push to make VRML games-ready or even games-capable because there were no market-driven demands for it. Instead, we saw an endless array of "science experiments."

This comment is of particular interest in the context of this book since its target audience is game developers. According to Mark Pesce, game developers should have been all over VRML and creating content for it. Indeed, content is what makes a medium successful, and games represent a significant amount of 3D interactive content, although not all games require 3D graphics.

Game developers are important because they are recognized for pushing the limits of the technology in order to provide the best possible user experience. Game technology needs to empower artists and designers with tools to express their creativity and enable nonlinear interactive storytelling that can address a good-sized audience and build a business case for 3D on the web. Game developers do not care if a technology is recognized by ISO as a standard. They are more interested in the availability of tools they can take immediate advantage of, and they require full control and adaptability of the technology they use, for the creation of a game is an iterative process where requirements are discovered as the game development progresses.

The main issue game developers seem to have with VMRL/X3D is its core foundation. The Open Inventor scene graph structure (see Figure 13.1) is used to both store the 3D data as well as define its behavior and interaction with the user, dictating a specific run-time design that would constrict game engines.

The objectives of Open Inventor, which have been followed closely in the design of VRML/X3D, are well described in the Siggraph 1992 paper that introduced it:

- *Object representation.* Graphical data should be stored as editable objects and not just as collections of the drawing primitives used to represent them. That is, applications should be able to specify what it is and not have to worry about how to draw it.
- *Interactivity.* An event model for direct interactive programming must be integrated with the representation of graphical objects.
- *Architecture.* Applications should not have to adapt to object representation or interaction policies imposed by the toolkit. Instead, the toolkit mechanisms should be used to implement the desired policies. Such flexibility should also be reflected in the ability to extend the toolkit when necessary.

The problem is that these design goals have not been proven to be universal or able to solve the needs for all representations of and interaction with 3D content. In fact, another scene graph technology was developed at SGI at the same time as Open Inventor in 1991: Iris Performer [5]. Its principal design goals were to allow application developers to more easily obtain maximum performance from 3D graphics workstations featuring multiple CPUs and to support an immediate-mode rendering library. In other words, both Open Inventor and Iris Performer used a scene graph technology, but one was designed for performance and the other for object-oriented user interactivity:

- Inventor defined a file format, which was then repurposed as VRML/X3D. Performer did not have a documented native file format; instead, it offered a loader facility so third-party's modeling tools could provide an importer. MultiGen's OpenFlight was a popular tool and format for this.
- Performer did not offer any default run-time library, but there was sample code and the often-used and often-modified "perfly" sample application, which contributed to Performer's reputation for being difficult to use.

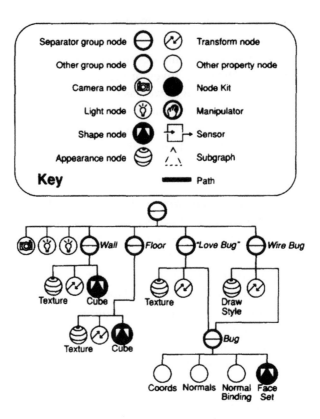

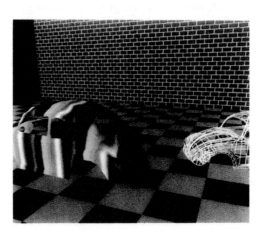

Figure 13.1. An Open Inventor simple scene graph and resultant rendering. (*Image ©
1992 ACM.*)

■ Inventor provided user interaction with the 3D data as well as an interface to
it. Performer did not have much in terms of built-in tools for user interaction,
but instead focused on generating images at fixed frame rates. Performer was
often used in a simulation environment where the interface to the user was an
external system, such as an airplane cockpit.

SGI tried several times to combine both the performance of Performer and
usability of Open Inventor. First, SGI introduced the Cosmo 3D library, which
offered an Open Inventor-style scene graph built on an Iris Performer-style low-
level graphics API. After the first beta release, SGI joined with Intel and IBM to
push OpenGL++ on the OpenGL architecture review board (ARB) [6] as a standard scene graph layer on top of OpenGL that could be used to port Performer or
Inventor (or Optimizer, yet another scene graph for the computer-assisted design
(CAD) market). The ARB was also interested in seeing OpenGL++ become a
standard for the web. This project died when SGI turned their attention to an almost identical project with Microsoft named Fahrenheit. The idea was that SGI
would focus on the high-level API (scene graph) while Microsoft worked on the
low-level Fahrenheit (FLL) API that would eventually replace both OpenGL and
Direct3D. Fahrenheit was killed when it became clear Microsoft was playing SGI
and was instead focused on releasing DirectX 7 in 1999 [7].

Sun Microsystems was also interested in creating a standard scene graph API
that could be universal and bridge desktop applications with web application:
Java3D [8], based on the cross-platform Java development language and run-time
library. The first version was released in December 1998, but the effort was dis-
continued in 2004. It was restarted as a community source project but then "put
on hold" in 2008 [9]. Project Wonderland, a Java virtual world project based on
Java3D, was ported to the Java Monkey Engine (jME) API, a Java game engine
used by NCSoft that produced better visuals and performance and reduced design
constraints.

Today, game engines are closer in design to Iris Performer than to Open Inventor. For instance, game engines provide offline tools that can preprocess the
data into a format that is closer to the internal data structures needed on the target
platform, thus eliminating complex and time-consuming data processing in the
game application itself in order to save precious resources, such as CPU time,
memory, and user patience. Also, game engines often create interactivity and
user interfaces with native programming through scripting languages such as
Lua. This provides maximum flexibility for tuning the user experience without
having to spend too much time in designing the right object model and file
format.

A side effect of this divergence of goals is the creation of a gap in bridging assets between content creation tools and game engines. X3D, as a publishing format, was not designed to carry the information from the modeling tools and content pipeline tools to the game. Other formats were proprietary to each modeling tool. Ten years after VRML was first introduced, Rémi Arnaud and Mark Barnes from Sony Computer Entertainment proposed COLLADA [10, 11], also based on XML, with the goal of providing a standard language for helping game developers to build their content pipeline [12]. Its design goal is to stay away from run-time definitions and, more specifically, from scene graphs. COLLADA has gained significant momentum in many areas (being used by Google Earth, Photoshop, game engine pipelines, modeling tool interchange, CAD, and 3D for the web). More recently, the 3D modeling industry has gone through consolidation, as Autodesk now owns 3DS Max, Maya, and XSI, representing a significant market share in the games industry. Autodesk has made significant progress in providing FBX, a proprietary library that is used for data interchange between Autodesk tools, as well as interfacing with game engines.

What we can say from this short history is that a lot of energy has been expended by many clever people to attempt creating the universal high-level representation of 3D data through various implementations of scene graphs, but all have failed so far. Most efforts aimed to provide 3D on the web (VRML, X3D, Java3D, OpenGL++, etc.) by standardizing a scene graph representation as the universal 3D publishing format, with the thinking that what worked for 2D with HTML should work for 3D as well. What we should learn from these two decades of effort is that it won't happen because there are too many incompatible goals. Even if restricted to one market segment—games—it is not even clear that a universal game engine can be created covering all genres.

13.2 Fast Forward

Fast forwarding to 2010, the landscape has changed significantly. There is now a huge enticement to publish games over the internet, effectively targeting the web as a platform. Revenues from online and mobile games currently generate about a third of all game software revenues globally and are predicted to represent 50 percent of the market by 2015 [13]. Online games generate billions of dollars of revenue in China and other Asian countries. The U.S. market is expected to hit $2 billion by 2012. Social games, such as Zynga's *FarmVille*, are on their way to becoming a mass market phenomenon, with more than 60 million people playing online (compared to half a billion people reported on Facebook).

This new business has evolved very fast, so fast that major game publishers are having trouble adapting from a culture of management of mostly multimillion dollar, multiyear-development AAA titles for game consoles to a very different culture of low-budget games distributed and marketed primarily through social networking. Some prominent game developers predict this current trend will have a profound impact on the overall game industry business model [14]. The video game industry has been through massive changes in the past, including a crash in 1984. This time, the industry experts are not predicting a crash but a massive shift that is poised to push more game publishing onto the web.

The need for superior interactive animated graphical content running in a web page has been growing since the introduction of HTML. In 1997, Macromedia (part of Adobe since April 2005) released Flash 1.0, commonly used to create animations and advertisements, to integrate video into web pages such as YouTube, and more recently, to develop rich internet applications (RIAs). Flash is a growing set of technologies that includes an editing tool, a scripting language closely related to JavaScript called ActionScript, a file format called .swf, and a browser plug-in available for many platforms.

In order to visualize Flash content, a plug-in needs to be installed by the client in the web browser. In order to maximize their market reach, Macromedia provided the plug-in for free and worked out several deals to ensure the plug-in (a.k.a. Flash Player) came preinstalled on all computers. In 2001, 98% of web browsers came preinstalled with the Flash player (mostly because market-dominant Microsoft Internet Explorer included the Flash player), so that users could directly visualize Flash content.

This created enough fertile ground for games to start spreading on the web, and the term "Flash game" quickly became popular. Thousands of such games exist today and can be found on aggregator sites such as addictinggames.com, owned by Nickelodeon. Some games created with the first releases of Flash, such as *Adventure Quest* (launched in 2002, see Figure 13.2), are still being updated and are played by thousands.

Most of the games created with Flash are 2D or fake 3D since Flash does not provide a 3D API. Still, there are several 3D engines, open source and commercial, that have been developed in ActionScript and are used to create 3D games. We look into this in more detail in Section 13.3.

Another technology created about at the same time as Flash, with the goal of enhancing web development, is Java, first released in 1996. Java is a programming language for which a virtual machine (VM) executes the program in a safe environment regardless of the hardware platform and includes a just-in-time (JIT) compiler that provides good performance. Unfortunately, Sun Microsystems did

Figure 13.2. *Adventure Quest* (Flash). See http://aqworlds.com/. (*Image © 2010 Artix Entertainment, LLC.*)

not enjoy a good relationship with Microsoft, since they saw Java as a competitor rather than an enhancement to Microsoft's products [15], and it was usually the case that a plug-in had to be installed. Unlike Flash, Java does offer bindings to 3D hardware acceleration and therefore offers much better performance than Flash. We explore this technology in Section 13.4.

Still, if a plug-in has to be downloaded, why not package a real game engine as a browser plug-in in order to provide the best possible performance and user experience? There is a reduction in the addressable market because the game distribution websites have to agree to support the technology, and the user to download and install the plug-in, as well as agree to the license, but this may be the only viable technology available immediately to enable the 3D experience. Several game engines are available as plug-ins, and these are discussed in Section 13.5.

One way to deal with the plug-in situation is to improve the mechanism by which a browser is extended, and that is what Google Native Client is about. It also introduces a hardware accelerated graphics API secure enough to be included in a web browser. This new technology is explored in Section 13.6.

Recently, HTML5 was pushed to the front of the scene, specifically when Apple CEO Steve Jobs took a public stand about why he doesn't allow Flash on

Apple's mobile platforms [16] and instead pushed for either native applications or HTML5 technologies. The HTML5 suite of standards is not yet published, but some portions are considered stable, such as the Canvas 3D API, which provides graphics hardware acceleration to JavaScript. Canvas 3D is now implemented as the WebGL API [17], a standard developed by the Khronos Group, and is a close adaptation of OpenGL ES (another Khronos Group standard) available for mobile phones (such as iPhone/iPad, Android, etc.). The Khronos Group is also the home of the COLLADA standard, which, coupled with WebGL, provides a standard solution for both content and an API for 3D on the web [18]. At the time of this writing, WebGL has not yet been released to the general public, but we describe it in Section 13.7.

13.3 3D with Flash

The main benefits of using Flash are that it's preinstalled (so it does not require the user to install anything), it is supported by the main outlets that reach customers (e.g., Facebook), it comes with professional creation tools (a Flash editor and Adobe AIR for running and testing locally), and there are tons of developers well-trained in ActionScript. The main issue is its performance, since the 3D rendering is done in ActionScript running on the CPU. Flash does not provide access to 3D hardware acceleration.

There are several available libraries that provide 3D with Flash:

- *Papervision3D*. Carlos Ulloa [19] started this project in December 2005. He had been programming 3D in software for games in the old times of Atari and Amiga, and he decided to reuse the same technology to provide 3D in Flash. By the end of 2006, Papervision3D was open-sourced. The library evolved from Flash 7 through 10 and is now in version 2.1, available at http://code.google.com/p/papervision3d/.
- *Sandy3D*. This project was started by Thomas Pfeiffer in October 2005 and is now open-sourced. It can be downloaded in version 3.1.2 at http://www.Flashsandy.org/download.
- *Away3D*. Away3D started as a reorganization of the Papervision3D code for better performance. Now in version 3.5.0, available at http://away3d.com/downloads, it is its own codebase. There is also a "lite" version that is only 25 kilobytes in size for faster download. One remarkable difference is the availability of a tool called PreFab3D (see Figure 13.3), the "missing link" between designers, modelers, and Flash 3D developers. This tool enables nonprogrammers to optimize the data for 3D in Flash. For example, the tool

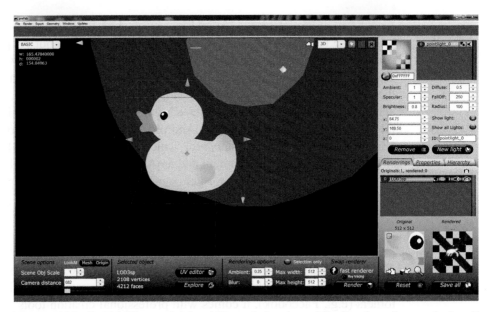

Figure 13.3. PreFab3D, an Adobe AIR tool for Away3D. See http://www.closier.nl/
prefab/. (*Image © Fabrice Closier.*)

can bake lights and shadows into textures using ray tracing, create low polygon models with normal maps, refine UV mapping, and apply real-time shading techniques.

■ *Alternativa3D.* This company develops and licenses a 3D engine for Flash
with support for multiuser real-time connection of clients and players. Their
commercial product, now in version 5.6.0, is aimed at game developers.
They offer a multiplayer tank battle game (see Figure 13.4) as a live demonstration of the capabilities of their engine.

All of these projects have a very active developer community,[1] and they provide tutorials, samples, documentation, and books [20, 21]. They enable loading
of 3D assets using COLLADA and 3DS Max formats. The engine's capabilities
are related to the performance of the virtual machine, which does not provide any
native 3D hardware acceleration. Flash 10 introduced some hardware-accelerated
features that are used to accelerate composition and rendering on a 2D surface.

---

[1] At the time of this writing, Away3D and Alternativa3D are currently the preferred solutions for performance and features because they have a more active development community, but this is not a measure of future performance.

Figure 13.4. TankiOnline, a Flash 3D game using Alternativa3D. See http://alternativa
platform.com/en/. (*Image © 2010 Alternativa LLC.*)

Flash Player 10 also introduced accelerated shaders (e.g., Pixel Bender) that are
exploited as material properties by those libraries. But still there is no 3D acceleration, so the data needs to be organized in a BSP tree in order to accelerate the
CPU polygon sorting, since a hardware Z-buffer is not available.

Listing 13.1 shows what the code that loads and displays a model in ActionScript with Papervision3D looks like.

```
package
{
    import org.papervision3d.Papervision3D;
    import org.papervision3d.cameras.*;
    import org.papervision3d.materials.*;
    import org.papervision3d....

    public class Main extends Sprite
    {
        // The PV3D scene to render
        public var scene : Scene3D;
        private var model : DisplayObject3D;
```

```
    // The PV3D camera
    public var camera : FrustumCamera3D;

    // The PV3D renderer
    public var renderer : BasicRenderEngine;

    // The PV3D viewport
    public var viewport : Viewport3D;

    public function Main() : void
    {
        init();
    }

    private function init() : void
    {
        // create a viewport
        viewport = new Viewport3D();

        // add
        addChild(viewport);

        // create a frustum camera with FOV, near, far
        camera = new FrustumCamera3D(viewport, 4.5, 500, 5000);

        // create a renderer
        renderer = new BasicRenderEngine();

        // create the scene
        scene = new Scene3D();

        // initialize the scene
        initScene();

        // render each frame
        addEventListener(Event.ENTER_FRAME, handleRender);
    }

    private function initScene() : void
    {
        model = new DAE();
```

```
            model.addEventListener(FileLoadEvent.LOAD_COMPLETE,
                OnModelLoaded);

            DAE(model).load("duck_triangulate.dae");
        }

        /**
         * show model once loaded
         */
        private function OnModelLoaded(e : FileLoadEvent) : void
        {
            e.target.removeEventListener(FileLoadEvent.LOAD_COMPLETE,
                OnModelLoaded);
            scene.addChild(model);
        }

        /**
         * Render!
         */
        private function handleRender(event : Event = null) : void
        {
            // orbit the camera
            camera.orbit(_camTarget, _camPitch, _camYaw, _camDist);

            // render
            renderer.renderScene(scene, camera, viewport);
        }
    }
}
```

Listing 13.1. Papervision3D code snippet for loading and displaying a COLLADA model.

Even though the performance is limited to 10,000 to 20,000 triangles at 30 frames per second, this technology provides excellent results for games that do not need to update the entire 3D display every frame, such as games mixing 2D and 3D or games where the point of view changes infrequently, as in golf games (see Figure 13.5). Note that by the time this book is published, Adobe will most likely have introduced a hardware-accelerated 3D API for the next version of Flash [22].

Figure 13.5. Golf game using Papervision3D Flash. See http://www.ogcopen.com/Ogc OpenPlay.php?lang=en. (*Image © 2010 MoreDotsMedia GBR.*)

13.4 3D with Java

Java is a programming language expressly designed for use in the distributed environment of the Internet. It was designed to have the "look and feel" of the C++ language, and it enforces an object-oriented programming model. Java can be used to create complete applications that may run on a single computer or be distributed among servers and clients in a network. It can also be used to build a small application module or applet for use as part of a web page. A lightweight version called Java ME (Micro Edition) was created specifically for mobile and embedded devices.

Java is a compiled language, but it is not compiled into machine code. It is compiled into Java bytecodes that are executed by the Java virtual machine (JVM). Because the JVM is available for many hardware platforms, it should be possible to run the same application on any device without change. Web browsers are such platforms where a Java program can be executed. Java performance is excellent and approaches the performance of languages such as C that are compiled to the machine language of the target processor directly. This performance is due to the use of the JIT compiler technology, which replaces interpretation of Java bytecodes with on-the-fly compilation to machine instruction code, which is then executed.

Java has included support for 3D hardware acceleration for a long time through bindings to OpenGL (JSR-231, a.k.a. JOGL) and to OpenGL ES for Java ME (JSR-239). In addition to binding to OpenGL and OpenGL ES, the Mobile 3D Graphics (M3G) API (a.k.a. JSR-184) was created specifically for gaming on mobile devices, and it includes a higher-level scene description specification and its file format .m3g. M3G is a kind of mini-Java 3D (JSR-912, JSR-189, JSR-926 [8]). Version 2 of the M3G API (JSR-297) has been released to catch up quickly with evolving graphics features on mobile devices, such as shaders. It is interesting that the first technology we looked at, Flash, offers no 3D acceleration and that Java provides a confusing choice of 3D hardware acceleration that may not be available on all platforms. Per our discussion earlier, we know that we should

Figure 13.6. *Poisonville*, a jME *Grand Theft Auto*-like game launched from the web, but running in its own (small) window. See http://poisonville.bigpoint.com/. (*Image © Bigpoint.*)

probably ignore the high-level scene graph technologies that limit the game engine to a predefined behavior, narrowing down our choice to JOGL. This wrapper library provides full access to the APIs in the specifications for OpenGL 1.3–3.0, OpenGL 3.1 and later, OpenGL ES 1.x, and OpenGL ES 2.x, as well as nearly all vendor extensions. It is currently an independent open source project under the BSD license [23]. Note that other similar bindings have also been created for audio and GPU compute APIs—OpenAL (JOAL) and OpenCL (JOCL) are available on the same web page as JOGL.

This is the choice made by the few game engines that have been built with this technology, such as the jME, now in version 3 (jME3), that can run inside the browser as an applet or launched in an external JVM application that can run fullscreen. There are many applications taking advantage of OpenGL 3D acceleration with Java on the web, but the jME is probably the only mature technology available to make games with Java plus JOGL (see Figure 13.6).

Despite its long existence, its advanced technology providing 3D hardware acceleration, and both its desktop and mobile availability, Java has had more success as a language and framework to run on servers than it has on clients, and it has had no real success in the game space. The confusion created by the Sun-Microsoft dispute over the Java technology (which created incompatibilities), the confusion as to which style of 3D graphics API to use, the perceived lack of sup-

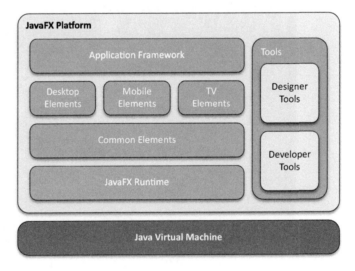

Figure 13.7. JavaFX, a new language running inside a JVM language. (*Image © 2010 Oracle Corporation.*)

port for OpenGL on Windows, and the very poor support by integrated chipsets
[24] may have, until recently, caused a lot of damage to Java's popularity. Java is
also handicapped by its 80 percent installed base (compared to 97 percent for
Flash [25]), vulnerabilities, large download size ($\sim 15\,$MB), and long startup
times.

The main issue seems to be the design of Java itself. It is very structured and
object-oriented but is not well suited for the creation of applications with intense
graphical user interfaces. Web programming requires more dynamic structures
and dynamic typing. Enter JavaFX [26], a completely new platform and language
(see Figure 13.7) that includes a declarative syntax for user interface development. However, there is no clear path stated about when 3D will become part of
this new JVM technology, so JavaFX is not suitable for writing 3D games for the
web for the time being.

13.5 3D with a Game Engine Plug-In

Extending a browser with a game engine plug-in is probably the best way to get
the optimum performance and features. But asking users to download a plug-in
for each game is not viable. So keeping the same philosophy that one plug-in can
run many applications, the game engine plug-in, once installed, can load and run
a specific game as a package containing the content and the game logic in the
scripting language(s) provided by the engine. Programming-wise, it is not much
different than creating a Flash game, but it has far superior performance because
all the functionality necessary for writing a game is provided through high-
performance libraries, including not only 3D hardware acceleration but also
components such as a terrain engine, a collision and physics engine, and an animation system. Also, a major difference in the Flash and Java solutions is that
game engines come with sophisticated editors and content pipelines, which reduce the efforts required of game developers a lot. Good tools are necessary for
cost-effective development.

The web is not (yet?) the place for AAA titles with long development cycles.
Therefore, having rapid prototyping tools becomes essential for developing 3D
games in a short period of time with a small budget. Developing for the web platform also means developing games that run on the low end of the performance
scale, which is a problem for the type of experience associated with AAA productions.

Because of the lack of interest in AAA titles in a browser, the small budget,
and the need to support low-end platforms, the traditional game engines (such as
Unreal and CryEngine) did not see a business for providing engines for the web

and mobile platforms. Therefore, the opportunity was available for new companies with a new business model to provide those engines. Engines such as Unity and ShiVa (see Figure 13.8) are already powering hundreds of games on the web or natively on Windows, Mac OS, Linux, and mobile platforms such as iPhone, iPad, Android, and Palm. The business model is adapted to the market, the plug-in is always free and small in size (Unity is 3.1 MB, and ShiVa is 700 kB), there are no royalties, there is a completely free entry-level or learning edition, and upgrades to the professional full-featured versions cost less than $2,000. So even the most expensive version is well within the budget of any game developer and is in the same range as a modeling tool that is necessary to create the 3D content to be used in the game.

Responding to the game development team needs, these two engines also offer asset management servers and collaboration tools. The motto for these engines dedicated to web and mobile development is *ease-of-use*, *cost effectiveness*, and *performance*. They give a new meaning to cross-platform development, which used to mean developing a game across all the game consoles. The success of ShiVa and Unity is bringing additional competition from the engines that used to serve exclusively the console game developers (Xbox 360, PlayStation 3, and Wii), such as the recently announced Trinigy WebVision engine, but it is not clear whether they will be able to adopt the same low price model or whether they bring enough differentiation to justify a major difference in price. In fact,

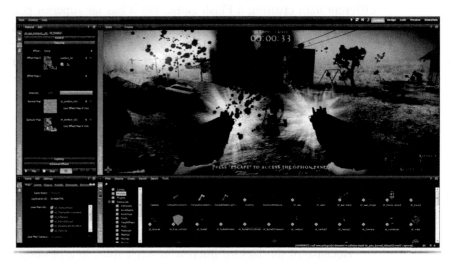

Figure 13.8. ShiVa Editor with one-click play button. See http://www.stonetrip.com/. (*Image © 2010 Stonetrip.*)

it could very well be the other way around, as ShiVa and Unity (both already on Wii) could be made available for cross-console development in the future.

Regarding developing and delivering 3D games for the web, those two technologies today provide by far the best performance and quality compared to the other technologies studied in this chapter. They also provide essential tools that provide ease-of-use, real-time feedback, editing, and tuning that are mandatory for the productivity demanded by short schedules and tight budgets in web development. But they also enable deployment of the game outside of the web as mobile phone and tablet native applications, as well as PC/Mac standalone applications. (However, PC/Mac versions of these engines do not offer the same level of quality as PC-specific engines.) Except for the market-limiting fact that on the web the user has to install a plug-in, the cross-platform aspect of those technologies is sometimes the only choice that makes sense when publishing a game on both the web and mobile devices in order to grow potential revenues without multiplying the development cost by the number of target platforms. As the vendors of those engines are focusing on ease-of-use and are very responsive to game developer needs, the results are quite astonishing in quality in terms of what non-highly-specialized developers can do in a few weeks.

To maximize flexibility and performance, Unity is taking advantage of the JIT compiler technology of .NET (Mono for cross-platform support) and providing support for three scripting languages: JavaScript, C#, and a dialect of Python. ShiVa is using a fast version of Lua. Scripts are compiled to native code and, therefore, run quite fast. Unity scripts can use the underlying .NET libraries, which support databases, regular expressions, XML, file access and networking. ShiVa provides compatibility with JavaScript (bidirectional interaction), PHP, ASP, Java, etc., using XML (send, receive, and simple object access protocol).

Integration with Facebook is also possible, opening the door to the half-billion customers in that social network. This is done though an integration with Flash,[2] taking advantage of the fact that Flash is on 97 percent of the platforms and available as supported content by most websites.

As a tip: don't miss the Unity statistics page (see Figure 13.9) that provides up-to-date information about what hardware platforms are being used. This provides a good indication of the range of performance and types of hardware used to play 3D games in a browser. (Note that these statistics show that most web game players have very limited GPUs and would not be able to run advanced engines from Epic or Crytek.)

---

[2] For example, see http://code.google.com/p/aquiris-u3dobject/.

Figure 13.9. Unity graphics card statistics for Q2 2010. See http://unity3d.com/webplayer/hardware-stats.html. (*Image © 2010 Unity Technologies.*)

13.6 Google Native Client

In order to enable loading and executing a game in a 3D environment, all of the plug-ins have to provide a safe environment in which to execute code and call optimized routines, and they have to provide access to graphics hardware acceleration on the GPU. All the technologies we have looked at so far offer a facility to create machine-executable code from both byte code (JIT and VM) and text parsers. What if there was a way to create a secure environment in which to run applications that are already compiled into the CPU binary format?

This is the goal of Google Native Client, an open-source technology for running native code in web applications, with the goal of maintaining the browser neutrality, operating system portability, and safety that people expect from web applications. Currently focused mainly on supporting a large subset of the x86 instruction set, this project also targets other CPUs, such as those based on ARM technology, and it should be available to integrate with any browser, either natively or as a plug-in.

This can be viewed as a meta plug-in that allows all the existing plug-ins that can obey the security issues (limited access to local drive, no self-modifying code, etc.) to be executed directly. Therefore, ShiVa and Unity will be perceived as a "no plug-in" version once Google Native Client becomes pervasive. It will be ideal to have a single plug-in (Google Native Client) that provides maximum performance and flexibility to all the applications, while providing "no plug-in" usability to the user. Native Client could very well be the ultimate cross-platform technology development, provided that vendors do not find a way to oppose its deployment (Apple may block this the same way it is blocking Flash on their iOS devices).

Unity (and soon ShiVa) can accept C/C++ libraries that can be linked with the game to provide additional optimized functionality that is too CPU-intensive to be done with the scripting language. But this has not been mentioned thus far in this chapter because it is not useful for the development of web games, as those libraries are not linked with the standard web plug-in and are therefore not available to the game running on the web. But with Google Native Client, it will be possible for the game developer to provide his own libraries, compiled with the Google Native Client GCC compiler, and link those dynamically with the game running on the web.

Native Client also offers an audio and graphics API. The 3D hardware-accelerated API is OpenGL ES 2.0. It is really nice that Google did not go ahead and provide a standard API that offers shader programs and a relatively slim layer over the GPU. Kudos to Google for choosing to embrace a standard, OpenGL ES, rather than inventing a new graphics API. Choosing the ES version of the API (as opposed to the desktop version of the API) enables the same graphics code to run on any desktop, mobile, or tablet hardware.

The problem with this choice is that although OpenGL ES 2.0 can be implemented rather easily given an OpenGL 2.0 driver, only 35 percent of Windows XP computers (XP currently represents 60 percent of all computers [27]) have an OpenGL 2.0 driver, and the others have drivers that are not very stable. On mobile platforms, there is a large dominance of OpenGL ES 2.0, so in order to unify the terrain, Google has created ANGLE, an implementation of OpenGL ES 2.0 on top DirectX 9, which is well supported on all the desktops. This includes a shader translator from GLSL ES to HLSL.

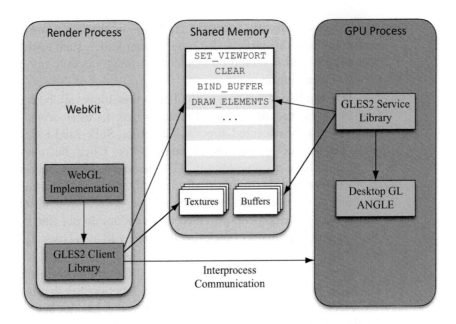

Figure 13.10. Secure access to the GPU as a separate process [28].

Access to the GPU in the Chrome browser is provided through a protection mechanism (see Figure 13.10) that is mandatory in order to provide the level of security that is required by a web browser running plug-in code. The GPU is isolated in its own process. The web browser is using separate render processes that do not talk directly to the GPU, but instead run an OpenGL ES driver that converts each command into a tokens in a command list (or display list) that is then used by the GPU process on the other end. A similar architecture is used in Native Client.

Although Google Native Client technology is not ready for prime time at the time of this writing, it is definitely one to keep an eye on for the development of 3D games for the web in the near future.

13.7 3D with HTML5

The last technology we look at can be categorized at the other end of the spectrum relative to all the technologies we have seen so far. The idea is to provide 3D acceleration through an embedded JavaScript API in web browsers. It is similar to the OpenGL for Java bindings technology we saw earlier, but done by a group of developers representing many browser vendors so that it is embedded within the web browser and provides OpenGL ES 2.0 functionality that can be called directly from JavaScript.

JavaScript is very popular for scripting on the client, although not so long ago, the preference was to write as much as possible of a web app in the server to avoid malicious code running in the browser or annoying pop-ups and other anomalies. The rule of thumb was to turn off JavaScript in the browser by default and use it only for trusted websites. This has changed as security has improved in the browser implementation of JavaScript and with advanced user interfaces that AJAX has brought to the user by leveraging the power of JavaScript to partially update HTML pages without a full reload from the server at each event.

Performance has been the main concern of JavaScript applications (see Figure 13.11). Recently, the JIT technology that was developed and refined by Java was taken advantage of by JavaScript as well as Flash ActionScript (AS3 2006) to improve performance. In fact, JavaScript is now approaching the performance of Java in most cases. The remaining problem with JavaScript is the large discrepancy in performance (by a factor of ten in some cases) among the various implementations that a given hardware configuration may be running. It is already difficult to target the web since there are so many hardware platforms and performance profiles, but when multiplied by the factor of ten in the JavaScript performance spread, this becomes an impossible problem. On the other hand,

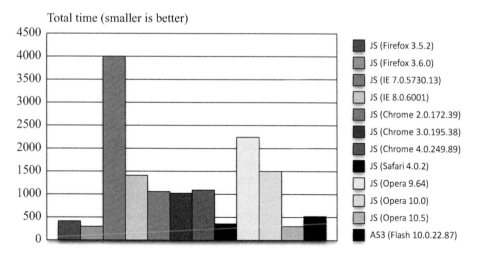

Figure 13.11. AS3 versus JavaScript (JS) performance test (Data from http://www.JacksonDunstan.com/articles/618, March 2010).

even if not the fastest in all cases, Flash provides consistent and good performance, regardless of which browser it is running inside. A given plug-in can impose the browser to update the Flash plug-in to the AS3 version if needed, but there is no such thing for native JavaScript because the entire browser would have to be upgraded.

So equipped with a fast scripting language and a strong desire to add hardware-accelerated 3D for web application developers, Google, Apple, Mozilla, and Opera announced during Siggraph 2009 that they would create the WebGL working group under the intellectual property (IP) protection umbrella of the Khronos Group [17] and joined the many working groups already working on graphics standards, such as OpenGL, OpenCL, and COLLADA. When Vladimir Vukićević (Mozilla) was thinking about where to create the new standard working group he basically had two choices: the W3C, home of all web standards, and Khronos, home of the graphics standards. Since his group was composed of web browser specialists, he thought they should join the standard body, where they could meet with graphic specialists, because expertise in both areas is required to create this new standard. This also enables complementary standards to be used conjointly and solve a bigger piece of the puzzle, such as how COLLADA and WebGL can be used together to bring content to the web [18].

Technically speaking, WebGL is an extension to the HTML canvas element (as defined by the W3C's WHATWG HTML5 specification), being specified and standardized by the Khronos Group. The HTML canvas element represents an element on the page into which images can be rendered using a programmatic interface. The only interface currently standardized by the W3C is the CanvasRenderingContext2D. The Khronos WebGL specification describes another interface, WebGLRenderingContext, which faithfully exposes OpenGL ES 2.0 functionalities. WebGL brings OpenGL ES 2.0 to the web by providing a 3D drawing context to the familiar HTML5 canvas element through JavaScript objects that offer the same level of functionality.

This effort proved very popular, and a public mailing list was established to keep up general communication with the working group, working under strict IP protection, which was quite a new way of functioning for the Khronos Group. Even though the specification has not yet been released, several implementations are already available for the more adventurous web programmers, and at Siggraph 2010, a dozen applications and frameworks were already available for demonstration. This is quite an impressive community involvement effort, indicating the large interest in having 3D acceleration without the need for a plug-in. After the demonstration, "finally" was whispered in the audience, since some have been waiting for this since the first HTML demo was made almost two decades ago.

The inclusion of native 3D rendering capabilities inside web browsers, as witnessed by the interest and participation in the Khronos Group's WebGL working group, aims at simplifying the development of 3D for the web. It does this by eliminating the need to create a 3D web plug-in and requiring a nontrivial user download with manual installation before any 3D content can be viewed by the user.

When creating a 3D game for the web, graphics is fundamental but is only a subset of the full application. Other features are to be provided by various nongraphics technologies that together form the HTML5 set of technologies that web browsers are implementing (see Figure 13.12). In order to bring data into the web browser, the game application will have to either embed the content in the HTML page or use the XMLHttpRequest API to fetch content through the web browser. Unfortunately, the programmer will also have to obey the built-in security rules, which in this case restricts access to content only from the same server from which the HTML page containing the JavaScript code was obtained. A possible workaround then needs to be implemented on the server in order to request external content, possibly through a simple script that relays the request.

One major issue with JavaScript is the fact that the entire source code of the application is downloaded to the browsers. There are utilities to obfuscate the code, which make it impossible to debug as well, but it is not too hard to reverse engineer. This may not be the ideal, however, since game developers are not necessarily happy to expose their source code.

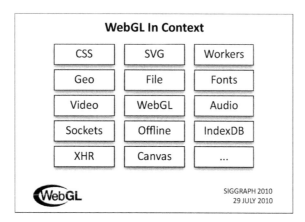

Figure 13.12. Vladimir Vukićević (Mozilla) presents how WebGL sits in the context of the HTML5 suite of standards at the Siggraph 2010 WebGL BOF.

Another issue with HTML5 is that there are a lot of hacks involved for cross-browser compatibility. It is not clear if this issue will be resolved anytime soon, so there is a need for libraries and other tools to isolate the developer from these issues so 3D game development won't be as painful on native HTML5. Flash, Unity, and ShiVa are doing a good job at isolating the developer from those browser compatibility issues.

WebGL is a cutting-edge technology with many things to be discovered before it can safely be used to develop 3D games for the web. The lack of tools for game developers is probably the most problematic point because this makes it impractical and certainly not cost effective. Many WebGL-supporting initiatives are under way (and more are coming along every month), such as GLGE, SpiderGL, CopperLicht, the Seneca College Canvas 3D (C3DL) project, and the Sirikata project. From these efforts and those as yet unforeseen, new and compelling content will be developed.

13.8 Conclusion

The conclusion to this chapter is not the one I had hoped for. But the research is clear: there is no ideal solution to what technology to use today to write 3D games for the web.

Unity and ShiVa are a primary choice based on their performance, success stories, and professional tools. But they still require the installation of a plug-in, which may be a nonstarter because the client paying for the development of the game may dictate that either Flash or no plug-in at all is to be used.

This is why Flash, especially if Adobe decides to add game-related features, will always be viable in the near future. But 3D acceleration is not enough because there are many more features required by the game engine (including physics and artificial intelligence) that would have to be provided and optimized for Flash, and Adobe may have no interest in developing these.

HTML5 definitely has a bright future. 3D has lots of usages besides games that will be covered by WebGL, and the no-plug-in story will be of interest moving forward. But this is still in theory, and it will depend on how Internet Explorer maintains market share and whether they provide WebGL (right now, there is no sign of this support). So that means WebGL will require a plug-in installation on Internet Explorer, which defeats the purpose.

Google Native Client is perhaps the one technology I had not considered at first, but it seems to be the most promising. It will provide optimum performance and will require a plug-in in many cases, but that plug-in will be required by so many applications that it will be very common. It also provides access to OpenGL ES 2.0 and a good level of security from the user's perspective.

Of course, there is a chance that the mobile market is going to be divided (iOS versus Android), and it is possible that Native Client won't be a choice on iOS. It is also possible that HTML5 will not be a valid choice on iOS mobile devices, given that a native-built application will most likely be much faster. Maybe this is Apple's plan, to support standard applications through HTML5 but force advanced applications, such as games, to be provided only as native applications, offering Apple better control of what applications are to be published on iOS platforms.

References

[1]   XMediaLab. "Mark Pesce, Father of Virtual Reality Markup Language." Video interview. November 25, 2005. Available at http://www.youtube.com/watch?v= DtMyTn8nAig.

[2]   3d-test. "VRML: The First Ten Years." Interview with Mark Pesce and Tony Parisi. March 21, 2004. Available at http://www.3d-test.com/interviews/media machines_2.htm.

[3]   Paul S. Strauss and Rikk Carey. "An Object-Oriented 3D Graphics Toolkit." *Computer Graphics* 26:2 (July 1992), pp. 341–349.

[4]   Web3D Consortium. "Open Standards for Real-Time 3D Communication." Available at http://www.web3d.org/.

[5]   John Rohlf and James Helman. "IRIS Performer: A High Performance Multiprocessing Toolkit for Real-Time 3D Graphics." *Proceedings of Siggraph 1994*, ACM Press / ACM SIGGRAPH, Computer Graphics Proceedings, Annual Conference Series, ACM, pp. 381–394.

[6]   OpenGL ARB. "OpenGL++-relevant ARB meeting notes." 1996. Available at http://www.cg.tuwien.ac.at/~wimmer/apis/opengl++_summary.html.

[7]   Randall Hand. "What Led to the Fall of SGI? – Chapter 4." Available at http:// www.vizworld.com/2009/04/what-led-to-the-fall-of-sgi-chapter-4/.

[8]   Oracle. Maintenance of the Java 3D specification. Available at http://www.jcp. org/en/jsr/detail?id=926.

[9]   Sun Microsystems. "ANNOUNCEMENT: Java 3D plans." 2008. Available at http://forums.java.net/jive/thread.jspa?threadID=36022&start=0&tstart=0.

[10]   The Khronos Group. "COLLADA – 3D Asset Exchange Schema." Available at http://khronos.org/collada/.

[11] Rémi Arnaud and Mark Barnes. *COLLADA: Sailing the Gulf of 3D Digital Content Creation*. Wellesley, MA: A K Peters, 2006.

[12] Rémi Arnaud. "The Game Asset Pipeline." *Game Engine Gems 1*, edited by Eric Lengyel. Sudbury, MA: Jones and Bartlett, 2010.

[13] Tim Merel. "Online and mobile games should generate more revenue than console games." GamesBeat, August 10, 2010. Available at http://venturebeat.com/2010/08/10/online-and-mobile-games-should-generate-more-revenue-than-console-games/.

[14] Jason Rubin. "Naughty Dog founder: Triple-A games 'not working'." CVG, August 3, 2010. Available at http://www.computerandvideogames.com/article.php?id=258378.

[15] "Microsoft and Sun Microsystems Enter Broad Cooperation Agreement; Settle Outstanding Litigation." Microsoft News Center, April 2, 2004. Available at http://www.microsoft.com/presspass/press/2004/apr04/04-02SunAgreementPR.mspx.

[16] Steve Jobs. "Here's Why We Don't Allow Flash on the iPhone and iPad." Business Insider, April 29, 2010. Available at http://www.businessinsider.com/steve-jobs-heres-why-we-dont-allow-flash-on-the-iphone-2010-4.

[17] The Khronos Group. "WebGL – OpenGL ES 2.0 for the Web." Available at http://khronos.org/webgl/.

[18] Rita Turkowski. "Enabling the Immersive 3D Web with COLLADA & WebGL." The Khronos Group, June 30, 2010. Available at http://www.khronos.org/collada/presentations/WebGL_Collada_Whitepaper.pdf.

[19] Steve Jenkins. "Five Questions with Carlos Ulloa." *Web Designer*, September 22, 2009. Available at http://www.webdesignermag.co.uk/5-questions/five-questions-with-carlos-ulloa/.

[20] Paul Tondeur and Jeff Winder. *Papervision3D Essentials*. Birmingham, UK: Packt Publishing, 2009.

[21] Richard Olsson and Rob Bateman. *The Essential Guide to 3D in Flash*. New York: friends of ED, 2010.

[22] 3D Radar. "Adobe Flash 11 to Get 3D Functionality." Available at http://3dradar.techradar.com/3d-tech/adobe-flash-11-get-3d-functionality-09-07-2010.

[23] JogAmp.org. The JOGL project hosts the development version of the Java Binding for the OpenGL API (JSR-231). Available at http://jogamp.org/jogl/www/.

[24] Linden Research. "Intel chipsets less than a 945 are NOT compatible." Quote from "Second Life hardware compatibility." Available at http://secondlife.com/support/system-requirements/.

[25]   Stat Owl. "Web Browser Plugin Market Share." Available at http://www.statowl.
        com/plugin_overview.php.

[26]   Oracle. "Better Experiences for End-Users and Developers with JavaFX 1.3.1."
        Available at http://javafx.com/.

[27]   Ars Technica. "Worldwide OS Share Trend." Available at http://arstechnica.com/
        microsoft/news/2010/08/windows-7-overtakes-windows-vista.ars.

[28]   Ken Russell and Vangelis Kokkevis. "WebGL in Chrome." Siggraph 2010
        WebGL BOF, The Khronos Group. Available at http://www.khronos.org/
        developers/library/2010_siggraph_bof_webgl/WebGL-BOF-2-WebGL-in-
        Chrome_SIGGRAPH-Jul29.pdf.

14

2D Magic

Daniel Higgins
Lunchtime Studios, LLC

Magician street performers don't require a big stage, scantily clothed assistants, or saw boxes to amaze an audience. They are extraordinary in their ability to entertain with a coin or a deck of cards. Many modern-day game developers are like street performers and do amazing things on a smaller stage. They often work on mobile platforms and in small groups instead of giant teams. Small budgets and limited time are no excuses for developers producing subpar products, however. Limited resources means that these teams must stretch what little they have in order to produce magic on screen.

The magic described in this chapter is simple, powerful, and requires only a small amount of graphics knowledge, such as how to construct and render a quad [Porter and Duff 1984]. This isn't an introduction to graphics programming, and it isn't about using shaders or the latest API from DirectX or OpenGL. Instead, it's a collection of easy-to-implement graphical tricks for skilled (non-graphics specialist) programmers. In this article, we first explore very basic concepts that are central to graphics programming, such as vertices, colors, opacity, and texture coordinates, as well as creative ways to use them in two dimensions. Next, we examine how to combine these building blocks into a powerful and complex structure that allows for incredible 2D effects (with source code included on the website). Finally, we conclude by conjuring our own magic and demonstrating some simple examples of this structure in action.

14.1 Tools of the Trade

When a magician develops a new trick featuring a prop, he must first examine the prop's properties. Take a coin for example. It's small, round, light, and sturdy. It easily hides in the palm of our hand and is the right size to roll across our fingers. Once we have a good understanding of the coin's properties, we can develop tricks that suit its potential. Likewise in graphics, our props are vertices. Knowing what composes a vertex gives us clues about how to develop tricks that harness its powers.

In its most basic form, a vertex is a point with properties that are used while rendering. It's graphics information for a drawing location somewhere on (or off) the screen. In 2D, it's most commonly used as one of four points that make up a quad (two triangles sharing an edge) and is generally used in the rendering of text or an image. A vertex contains position, color, opacity, and often texture coordinates. Let's examine each property in detail.

14.2 Position

A vertex's position determines where it is rendered on the screen. Location on its own may not seem interesting, but when combined with the other vertices of a quad, we open up a window for scale, motion, and perspective effects.

Scale Effects

Imagine a boat race game where we want to display a countdown prior to the start of a race. Displaying the count involves starting each number small, then increasing its size over the course of one second. We accomplish this by altering the position of each vertex in a quad relative to the others in such a way that the screen space becomes larger. Consider dressing this up with some acceleration modifications that affect how fast it scales. Does it overshoot its destination scale and have to snap back, producing a wobbling jelly-like effect? Perhaps it starts quickly and eases into its destination scale, like a car pulling into a parking spot.

Regardless of what acceleration modifier you choose, don't choose linear acceleration—how boring! Adding unique scaling techniques, such as the smooth step shown in Figure 14.1, adds character to a game. Smooth step is just a simplification of a Hermite spline [Pipenbrinck 1998] but produces elegant results with little CPU usage. Other functions to consider when modifying percentages are sine, cosine, power, square, and cube. Once you determine a scaling acceleration theme, develop a cohesive style by using it throughout the user interface (UI).

Motion Effects

Motion effects, such as sliding or bouncing a quad across the screen, are achieved by applying a physics force to a quad (and eventually each vertex). We are not limited to maintaining a flat border in 2D space. That is, the y coordinate

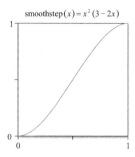

Figure 14.1. The smooth-step function remaps the range [0,1].

for our top-left vertex and the y coordinate for our top-right vertex do not have to be the same. We can just as easily alter the orientation of our objects, spin them, or flip them upside down. We accomplish this by storing an orientation value in the range $[0, 2\pi)$ on the quad. When the orientation changes, the quad updates the position of each of its four vertices to be relative to its new orientation. We could also store the new orientation in a transform matrix that is applied before rendering.

Perspective Effects

A big advantage 3D has over 2D is lighting. In the 2D world, we're often stuck with the lighting rendered into the art during production. We shouldn't bemoan this problem, however—while we cannot change the lighting in the art, we can

Figure 14.2. (a) The original image. (b) A shadow generated by darkening and blurring its pixels. (c) Squashing the vertices to build perspective. (d) The final images rendered together.

enhance it at run time. A perfect example of this is using shadows with our 2D objects. Consider what happens in Figure 14.2. Here we take a rendering of one of our characters, darken it, blur it, and squash its vertices, projecting it down to create a slightly angled shadow. The result is an almost magical, subtle effect that provides depth to our game world visuals.

14.3 Color and Opacity

Tinting and opacity are the primary responsibilities of a vertex's color. Normally, vertices are rendered with a white coloring (no tint), and the color's alpha value is used to indicate the transparency of a quad's corner. That's just the tip of the iceberg, though, as color is one of the most valuable tools for taking our 2D games to the next level, especially for UI design.

Colored UI Model

Optimization isn't just for frame rates; its principles should be applied to a company's development processes. With that in mind, look to vertex coloring as a means for small, budget-conscious teams to reduce art pipelines, promote reuse of textures, and develop consistency in UI design.

Traditionally, UI art is built in third-party tools and rendered verbatim to the screen. When it's time to render, each vertex is rendered white (no visible change) and may use its alpha value as the vertex's transparency. This has the advantage that artists know what the outcome will be, but it does not promote art reuse and requires third-party tools if color changes are needed. What if we want to reuse much of our game's art, save memory, or want our UI to adapt to the player's preferences? We need to look at our UI rendering differently than we have in the past and move to the colored UI model.

The colored UI model involves building reusable pieces of UI in grayscale, focusing on making them as white as possible, with darkened areas to indicate shadow and depth, then coloring these pieces when rendering. This allows for UIs to quickly change their color and opacity based on run-time data or be initialized from data files during load. Consider looking into luminance format textures for more performance improvements.

Black and white is all about predictability. Render a colorful texture to a green-colored quad, and the result is predictably a modulation of source texel and destination color but is likely not the programmer's intended result. If, however, our UI is black and white, we know that the ending colors will all be shades of the original colors we applied through the vertex. The catch with this is that we are changing only the purely white areas of the original texture to our desired color, and all other final display colors are darker depending on the shade of gray in the original texture.

The most common example of white art and vertex coloring is text rendering. When rendering a string that uses a font where each character has been preren-dered in white onto a transparent background, we have the ability to color text dynamically. To accomplish this, we look up a character's texture coordinates, choose the color for each character, and then render it on a quad of that color. Our result? Characters from the same font file are shown on screen but colored differently, with no extra texture memory required. We now extend this reuse to our entire UI.

Imagine building a sports game where each team has a light and dark color. Our designer wants to connect our gamers with the identity of their team through the use of team colors. We could build a uniquely colored UI for each team, but that takes a lot of effort and resources. Instead, artists design a black and white UI, with the knowledge that some UI pieces will use the light team colors and other areas will use the dark team colors. Now we have one art set instead of many, resulting in a savings in memory, art development time, and likely load time as well.

This isn't to say an image of a person somersaulting off a building should be in black and white and tinted when rendered. Use white art and tint it when there is an asset that can be reused if colored differently. Often, buttons, frames, bor-ders, scroll bars, tab controls, and other basic UI components are easily colored to fit into different scenes.

Opacity Separate from Color

Gone are the days when it was acceptable to have UI objects jarringly pop on and off the screen. Modern games fade objects or provide some other type of smooth transition. These transitions (and indeed all transparency) are achieved by modi-fying the alpha component of a vertex's color. A value of one means the vertex is fully opaque. A value of zero means the vertex is completely invisible.

Opacity is closely related to color due to the effect of a color's alpha value on transparency. It's arguable that opacity is unnecessary since alpha *is* transparen-cy. Why store opacity then? The problem is that temporary effects, like fades, pollute the true value of a color's alpha. It's safer to apply opacity during render-ing (`finalAlpha = alpha * opacity`) and preserve the true state of a vertex's color. We can optimize this a bit by not storing opacity per vertex and opt for storing it on a quad instead, since most opacity effects (like fade) don't need to affect vertices individually.

Interpolation

We're not done with colored UIs yet. A white UI and colored vertices give us another advantage, dynamic interpolation. Initialization time isn't the only time we alter color. Game-state changes we want to communicate should also affect color. For example, if we have a button we want the player to press, we can make it pulse back and forth between two colors (and opacities). Interpolating between two color values using a parameter t in the range $[0,1]$ is easy using a simple expression such as the following:

```
RGBAtarget = RGBAstart + (RGBAend - RGBAstart) * t;
```

Day and Night

A good use of vertex coloring in a 2D game is to simulate day and night. While dynamic shadows are very effective in providing time indication, they are tricky to get right. Tinting, on the other hand, is a simple and effective way to add day and night effects, and it provides players with a feeling of time passage without much headache for the programmer. At most, it's a matter of tinkering with color values until design and art come to an agreement on what looks good.

14.4 Texture (UV) Coordinates

Texture coordinates [Dietrich 2000] are the location in a texture that a vertex uses when rendering. Conceptually similar to using (x, y) coordinates on the screen, textures use the coordinates (u, v) (often called UVs) and range over a normalized square $(0,0)$ to $(1,1)$ when rendering a complete texture to a quad. We aren't limited to this narrow range, however. If we use a texture address mode such as "wrap," we have the ability to repeat a texture as we extend beyond the range $[0,1]$.

UI Frames

Using the wrap texture address mode is particularly useful when building reusable UI components. Consider UI objects such as window frames where sizes may be unknown or where dynamic scaling is desirable. If we want to build a frame that supports many sizes, we would construct art for each corner and then create tileable sections for the top middle, bottom middle, left middle, and right middle. In Figure 14.3, we demonstrate how we can render complete corner images having (u, v) values from $(0,0)$ to $(1,1)$, but then repeat thin slices from our middle

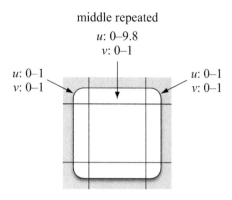

Figure 14.3. Frame where middle area of the top, bottom, left, and right are repeatable.

sections to create a dynamically sized box. We repeat the middle section by increasing its u or v value depending on its direction. For example, if the middle section of a frame should be 15 pixels in x, and the art is only 10 pixels wide, then a u value of 1.5 and a texture-addressing mode of wrap would repeat half of the texture.

Texture Scrolling

While most motion is done using the positions of vertices, UVs have their role, especially with tileable textures. Given a repeatable (tileable) texture such as water or sky and slowly adding to its texture coordinates, known as texture scrolling, we create the appearance of a flowing river or clouds moving across a sky. This is particularly effective when done at different speeds to indicate depth, such as water scrolling slowly in the distance while a river close to the player's perceived location appears to scroll faster; this effect is known as parallax scrolling [Balkan et al. 2003].

Texture Sheets

Why waste textures space? Some graphics APIs require textures to be a power of two, and if not, then it's often better for performance reasons anyway, so what if we have textures that don't fit nicely into a power-of-two size? By building in a simple addressing system, we put multiple objects in one texture (known as a texture sheet) and store their coordinates in an atlas (such as an XML file). In Figure 14.4, we show such a texture sheet from the mobile phone application *atPeace*. The texture sheet is divided into many 32×32-pixel blocks, with each

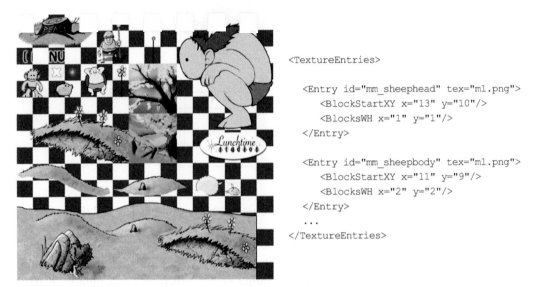

```
<TextureEntries>

    <Entry id="mm_sheephead" tex="m1.png">
        <BlockStartXY x="13" y="10"/>
        <BlocksWH x="1" y="1"/>
    </Entry>

    <Entry id="mm_sheepbody" tex="m1.png">
        <BlockStartXY x="11" y="9"/>
        <BlocksWH x="2" y="2"/>
    </Entry>
    ...
</TextureEntries>
```

Figure 14.4. Texture sheet of 32×32 pixel blocks and an XML database used to locate objects.

individual image's starting block, width, and height stored in the texture atlas. When the final image is made, the background blocks are hidden, and the background is transparent. Anytime an image in the texture is needed, we use an ID and consult our texture atlas. There we find in which texture an image is located and what its texture coordinates are.

Having blocks of 32×32 pixels does waste some texture space, but the size is simple to understand and use. Luckily, there are numerous resources available online to help with generating texture sheets and their corresponding atlases [Ivanov 2006].

Enough of these basic properties! It is time to put everything together and make some magic.

14.5 What a Mesh!

As in life, building blocks can be combined into systems that become both incredible and complex. While vertices aren't exactly atoms or strands of DNA, they are powerful building blocks for graphics programming. They link together and form wonderful tapestries of interesting and powerful structures. One such structure is a mesh. Seen often in 3D games, a mesh allows programmers to manipulate many areas of a texture, resulting in breathtaking effects that far surpass

Figure 14.5. Image with four vertices (left), and image with many vertices formed into a mesh (right).

what can be done when simply rendering as a quad. These structures are formed through the dissection of a simple image, composed of four vertices, and subsequent partitioning into a mesh of many vertices, as shown in Figure 14.5.

The partitioning process creates a series of related quads known as a mesh. Each quad is responsible for rendering a portion of the texture, but it's up to the mesh to manage the quads intelligently. With the mesh holding these quads together, we can now apply a global intelligence across the image to produce a synchronized system of effects, such as flickering skies, flashlight effects, river ripples, trampoline physics, and water illusions.

So why not just build separate texture objects and let them act independently? Having a mesh gives us the ability to add structure, intelligence, and fine-grain control over our texture. Creating the illusion of a treasure chest beneath the water would be difficult without a mesh. If designed for easy vertex modification, a mesh makes seemingly impossible tasks like water easy to implement.

14.6 Mesh Architecture

Accompanying this chapter is a folder on the website full of goodies, including source code. We provide a fully implemented C++ mesh that is cross-platform and ready to be dropped into a new code base and used with minimal effort. Optimization was not the primary goal of this implementation, although recommendations for performance are included later in this chapter. Ease of understanding and extensibility were our focus. With minimal effort, new effects can be added to our mesh that go well beyond the several examples we provide here.

The mesh architecture consists of three important items: vertex, mesh, and modifiers. First, we use a vertex class to hold each vertex's position, color, opacity, and texture coordinates. Next, we store all our vertices inside a mesh class and use the mesh class as a way to coordinate all the vertices, acting as a type of brain for the overall image. Last, our modifiers do the dirty work and alter the mesh in such a way that we conjure magic on the screen.

Mesh

The mesh is a class that contains a 2D array of vertex objects. Every place where there is a line intersection in Figure 14.5, there exists a vertex. Using Cartesian coordinates, a vertex can be quickly retrieved and modified. Our mesh design has a base `Mesh` class and a derived `MeshUV` class. In the base implementation, we do not support texture coordinates. It's useful for times when we want to use colored lighting without a texture, as demonstrated in our double rainbow example later in this chapter.

Space

There is a good argument to be made about whether a mesh should exist in screen (or local) space or in world space. On the side of screen space, if a mesh changes size or position, then we don't need to recompute any of our vertices. On the side of world space, everything with the mesh requires just a bit more work, calculation, and CPU time on a frame-by-frame basis. That is to say, if we're not moving or scaling our mesh, then we're wasting CPU time by going to and from screen space. Our implementation assumes screen space; that is, the mesh considers its top left corner to be $(0,0)$, and vertices see their (x, y) positions as percentage offsets into the mesh. Therefore, only if we change the number of mesh points per row or column do we need to rebuild our mesh from scratch.

State

Another contention point in mesh design is whether to build vertices that contain state. Having state means that a vertex has two sets of properties, original and working. The original properties could be considered the normal, "resting" properties that would make the texture appear as if it had no mesh. The working properties are copies of the originals that have been modified by an effect and, thus, have altered the state of the vertex. The vertex then fights its way back towards its original properties by using mesh-defined physics and other parameters to alter its working set.

One side of the argument contends that without state, vertices are simple objects, and the entire system is easier to understand. The problem with this approach, however, is that combination effects can often get dicey, and we have to recompute modifications on a vertex many more times than we would if there was state.

The argument for having state is quite persuasive. If a vertex contains a destination set of properties (position, color, etc.) and uses mesh parameters to morph from their current states toward the destination states, then we end up with a very consistent and predictable set of movements. It allows easy control of vertex physics and provides the ability to apply a mesh-wide set of physical properties. It's a "fire-and-forget" mentality where a force gets applied to the mesh, and the mesh is responsible for reacting. Our result can be quite interesting considering that we can alter a texture's physics on a mesh-wide basis. Take a texture that appears as rubble and a texture that appears as cloth. Both can now react to the same force in very different ways. A texture of rubble would be rigid and morph little given a physics force, while a cloth texture would have drastic mesh-morphing results. While both have their uses, for simplicity, our version uses the stateless implementation, and we recompute our effects each rendering cycle. In accordance with that, the burden of vertex modification is passed from vertex physics onto modifier objects.

Modifiers

Modifiers are what the mesh is all about. After we split an image into a mesh, it still looks exactly the same to the player. It's when we apply modifier objects to the mesh that we change its appearance with effects that amaze.

Modifier objects are given the mesh each frame update and are allowed to modify the vertex properties prior to rendering. Each mesh holds onto a collection of these modifiers, which are sorted into priorities since some may need to happen prior to others. For example, if deforming a vertex, one may want the deformation to happen before normal colored lighting so that the darkening effect of simulated shadow happens last.

14.7 Mesh Examples

Below are several simple examples of mesh manipulation. These are just scratching the surface of what can be done, and following the patterns of these samples, adding new modifier objects should be quick and easy.

```
float       mPosition[2];       // 2D position (in world space)
float       mWH[2];             // width, height

Mesh        *mMesh;             // points to our parent mesh

uint32      mID;                // unique ID for this modifier
uint32      mDeathTime;         // world time we're scheduled to perish
uint16      mPriority;          // used to sort our modifiers
bool        mDirty;             // if true, recheck our params
```

Listing 14.1. Data in base modifier class.

Modifiers derive from a modifier base class, which consists of the data members shown in Listing 14.1.

Point Light

We often color our vertices. A simple way to achieve this is to develop a light source similar to a 3D point light. A point light contains a position, radius, color, and fall-off function, as shown in Listing 14.2. The fall-off function determines how hard or soft the light's boundary is at its radius.

Modifiers that use lighting derive from a common MeshModifierLightBase class that handles generic point-light lighting, as shown in Listing 14.3.

```
class PointLight
{
    protected:

        float       mColor[4];                  // RGBA color
        float       mXY[2];                     // our position

        float       mScreenRadius;              // 0-1
        float       mScreenRadiusAsDistance;    // world space distance
        float       mFallOffPercent;            // 0-1, multiplies radius

        uint32      mID;                        // our unique ID
};
```

Listing 14.2. Data from a point light.

```cpp
bool MeshModifierLightBase::LightVertex(uint32 inFrameID,
        MeshVertex& ioV, const PointLight *inL)
{
    // Get our distance sq, another closer check.
    float theDistSq = DistanceSqTo(inL->GetX(), inL->GetY(),
        ioV.GetWorldX(), ioV.GetWorldY());
    float theRadiusDistSq = inL->GetRadiusDistSq();
    if (theDistSq > theRadiusDistSq)
    {
        // Too far away to modify this vertex.
        return (false);
    }

    // Now we need the real distance.
    float theDist = FastSqrt(theDistSq);

    // Compute our fall off.
    float fallDist = inL->GetRadiusDist() * inL->GetFallOffPercent();

    // Find our strength (of lighting edge).
    float perStr = 1.0F;
    if (theDist >= fallDist)
    {
        // Inverse left over falloff amount.
        float leftOver = (inL->RadiusAsDist() - fallDist);

        // Compute percent str, clamp it to a 0-1 range.
        perStr = clamp((1.0F - ((theDist - fallDist) / leftOver)),
            0.0F, 1.0F);
    }
    // Blending.
    if (!ioV.TouchedThisColorFrame(inFrameID))
    {
        // First time through, use the base color of our object.
        LerpColors(mMesh->GetRGBA(), inL->GetRGBA(), perStr, ioV.GetRGBA());

        // Mark this vertex as having its color changed this frame.
        ioV.MarkColorFrameID(inFrameID);
    }
    else
    {
```

```
        // Already colored this frame, blend with it.
        LerpColors(ioV.GetRGBA(), inL->GetRGBA(), perStr, ioV.GetRGBA());
    }


    return (true);
}
```

Listing 14.3. Example of using a point light to light a vertex.

During the mesh's UpdateModifiers() phase prior to rendering, each modifier has its ProcessMesh() method called. Our MeshModifierLightGroups::ProcessMesh() method, as shown in Listing 14.4, performs the following steps:

- Examines the dirty flags and updates itself as needed.
- Caches the base light color to be applied.
- Loops over each vertex and does the following:
  - Applies the base color for this modifier or mesh. If the mesh and modifier have different default colors, we mark this vertex as color changed.
  - Determines whether the vertex is in our bounding rectangle (as an optimization).
  - Loops over each point light and calls LightVertex().

```
bool MeshModifierLightGroup::ProcessMesh(uint32 inFrame)
{
    bool res = false;

    // Update dirty.
    UpdateDirtyIfNeeded();

    // Used as assert checking now, but could make thread safe.
    ListLocker theLock(mProtectLightList);

    size_t theLightMax = mLights.size();

    // Virtual call, gets us the modifier (or mesh) color for each vertex.
    const float *theBaseColor = GetBaseColor();

    // If this is a unique base color, mark this as having
    // a different color.
```

```
    bool theDiffBaseColor = !ColorMatches(theBaseColor, mMesh->GetRGBA());

    // Loop over each and process it.
    uint32 theRowSize = mMesh->GetRowSize();
    for (uint32 theX = 0; theX < theRowSize; ++theX)
    {
        for (uint32 theY = 0; theY < theRowSize; ++theY)
        {
            // Modify this vertex by the modifier object.
            MeshVertex& vert = mMesh->GetVertex(theX, theY);

            // Set the default color.
            vert.SetRGBA(theBaseColor);

            // Is the modifier's default color different than the mesh color?
            if (theDiffBaseColor)
                vert.MarkColorFrameID(inFrameID);

            // Is it in the bounds?
            if (!Contains(vert.GetWorldX(), vert.GetWorldY()))
                continue;

            // Yes, this modifier altered the mesh.
            res = true;

            // For each light, light up the vertex.
            for (size_t lightC = 0; lightLoop < lightMax; ++lightC)
                res |= LightVertex(inFrame, vert, mLights[lightC]);
        }
    }

    return (res);
}
```

Listing 14.4. Example of how a modifier with many lights interacts with a mesh.

Ultimately, our mesh makes many effects possible with minimal coding. It's designed around building modifier objects, similar to functors in the standard template library. Certainly, instead of a mesh, unique systems can be generated for many of the effects listed below. Particles work for some, while independently colored quads would suffice for others. In the end, the mesh is about power and convenience, both of which are difference makers, especially for small teams.

Sunrise and Double Rainbows

Many of us wonder what seeing a double rainbow means. We can give our players the chance to ponder that question with a simple lighting modifier effect. This effect requires no texture and only renders colored quads. It consists of adding multiple point lights in an arc to a lighting group modifier, then changing per-light properties over time, altering the color, radius, fall-off, and position of each point light, as shown in Figure 14.6.

This type of effect is incredibly useful in adding a dynamic feel to background images. It can apply to a nighttime sky, a space scene, or the flicker of a sunrise. Incorporating lighting with standard textures is particularly effective. For example, the aforementioned flickering sunrise is implemented as a texture containing a sun in the foreground and a sunbeam-style point light behind it. Both rise together, while the point light modifies the background screen's mesh in seemingly perfect concert with the rising-sun texture. To the player, the sun and the point light behind it are one and the same.

There is almost no limit to what visual pairings work, given a mesh and normal foreground texture. Imagine a rocket whose exhaust trail is a series of point

Figure 14.6. Lighting effects on a mesh, from the mobile phone application *atPeace* [6].

lights spat out from its engines. These lights ejected from the back of the rocket would spin, gently fall, and fade away as the rocket ship sped off into space.

Taking this one step further, you can light multiple objects in a world by having many objects with meshes and adding lighting modifiers on each one. Creating and modifying the point lights outside of the mesh means that they can be applied to multiple mesh objects, in a sense providing a global light source. That light would correctly impact the meshes of objects around it, regardless of whether they each have their own mesh. Sound confusing? Consider that any 2D object can have a mesh and that each mesh has its own modifier. The key is that *any* point light can be passed into each mesh's modifiers. In other words, we can control point lights from the application and change the lighting modifiers on each mesh prior to rendering. That provides global lighting for the scene, which is superb for world lighting consistency.

Flashlight

The mesh has seemingly no limit to what it offers. Instead of moving lights across a mesh and altering colors, it provides us the basis for negative effects, such as the flashlight. Instead of coloring vertices as a light moves through a mesh, we have the ability to hide the texture and use the point light as a way to uncover an image. Figure 14.7 shows several lights that, when applied to an image, uncover it. One could imagine how lights could be used in different strengths and colors such as a hard-beamed flashlight or a dim torch. Indeed, a

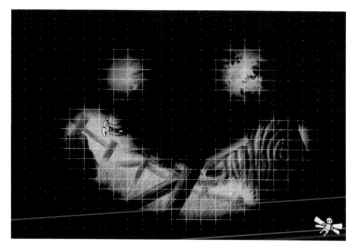

Figure 14.7. Using lights to uncover an image.

casual game could be based largely on this concept of uncovering an image by using different types of lights that have varied shapes, colors, and sizes.

The implementation of this effect takes only a few lines of code since it derives from our lighting-modifier group class. The only change is that every vertex is colored black and our point light color value is pure white. Normal fall-off distances and screen-radius parameters used in our point light class still apply without alteration.

Pinch

Colors are just the beginning. Where the rubber meets the road is in the modification of vertex positions. It's here where we can make an image seem as if it's underwater by applying a water morphing effect, or where we can make our mesh respond as if it's a trampoline. Our pinch example is about as simple as it gets. Given a point, radius, and fall-off function (similar to a point light), we can bend vertices in toward the center of our pinch point as demonstrated in Figure 14.8. Note how the debugging lines rendered on the image show how each affected vertex is bent in toward the pinch point and how we color our vertices in the pinch to indicate depth. The opposite effect would be equally as simple, and we could create bulges in our texture instead.

Figure 14.8. The pinch effect pulls vertices towards the pinch point.

Optimization

As mentioned earlier, when implementing the mesh that accompanies this article, optimization was a low priority. That's not to say it shouldn't be a high priority prior to shipping, since it greatly affects the frame rate. This mesh is fast enough to support many mobile platforms, especially when used selectively as a single-textured sky background. However, given multiple meshes and many modifier objects, the number of vertices that require updating (and rendering) grows substantially. If optimization is a priority, here are some tips to remember:

- As with every optimization, it pays to make sure you're fixing what is broken. Use a profiler to gather real data on bottlenecks.
- A major issue is the number of vertices affected per update or frame. Consider the following:
  - Reduce the mesh points per row or column.
  - Use bounding areas to eliminate large groups of vertices. The best situation is one where we never need to even look at a vertex to know it's unaffected by our modifiers and we can skip updating (or rendering) the vertex. We achieve this by combining the bounding box of all modifiers together. With these combined bounds, we then determine which vertices are unaffected by our modifiers and can optimize accordingly.
  - Since each modifier has the ability to affect every vertex on the mesh, limit the number of modifier objects or have ways to short circuit an update. This is extremely important since every modifier added has the potential to update every vertex on the mesh. Something has to give in order to reduce the world load.

- Consider building the mesh using state-based vertex objects where modifiers exist as "fire-and-forget" objects that modify the vertices of the mesh once, and let each vertex find its way back to their original state.
- Use texture sheets. Rendering objects on the same texture at the same time is an optimization over texture swapping. Avoid loading in many texture sheets that are only required for a small piece of texture. Instead, try to include as many assets on a texture sheet for a given scene as possible.
- Graphics bottlenecks need to be considered. A detailed mesh has many quads. Consider things such as grouping vertices into large quads for unaffected areas and rendering those areas as chunks. Think of it this way: an unmodified mesh is easily rendered with only the typical four corners.

14.8 Conclusion

Life has a way of taking small building blocks and generating colossal structures. We can do the same in 2D if we closely examine our own building blocks in graphics. Just because it's 2D doesn't mean we're limited to simple textures or "sprites." We need to exploit every power they provide. Small development shops building 2D games need to be particularly aggressive in the pursuit of quality and an edge as they often face competition that is bigger and better funded. Using a colored UI or 2D mesh is a perfect place to put some wind in the sails and get a jump on the competition. Big results with minimal effort, an almost magical situation.

Acknowledgements

Special thanks to Rick Bushie for the art in the examples from the game *atPeace*.

References

[Balkan et al. 2003] Aral Balkan, Josh Dura, Anthony Eden, Brian Monnone, James Dean Palmer, Jared Tarbell, and Todd Yard. "Flash 3D Cheats Most Wanted." New York: friends of ED, 2003.

[Dietrich 2000] Sim Dietrich. "Texture Addressing." Nvidia, 2000. Available at http://developer.nvidia.com/object/Texture_Addressing_paper.html.

[Ivanov 2006] Ivan-Assen Ivanov. "Practical Texture Atlases." *Gamasutra*, 2006. Available at http://www.gamasutra.com/features/20060126/ivanov_01.shtml.

[Pipenbrinck 1998] Nils Pipenbrinck. "Hermite Curve Interpolation." 1998. Available at http://www.cubic.org/docs/hermite.htm.

[Porter and Duff 1984] Thomas Porter and Tom Duff. "Compositing Digital Images." *Computer Graphics (Proceedings of SIGGRAPH 84)* 18:3, ACM, pp. 253–259.

Part II

Game Engine Design

Part II

15

High-Performance Programming with Data-Oriented Design

Noel Llopis

Snappy Touch

Common programming wisdom used to encourage delaying optimizations until later in the project, and then optimizing only those parts that were obvious bottlenecks in the profiler. That approach worked well with glaring inefficiencies, like particularly slow algorithms or code that is called many times per frame. In a time when CPU clock cycles were a good indication of performance, that was a good approach to follow. Things have changed a lot in today's hardware, and we have all experienced the situation where, after fixing the obvious culprits, no single function stands out in the profiler but performance remains subpar. Data-oriented design helps address this problem by architecting the game with memory accesses and parallelization from the beginning.

15.1 Modern Hardware

Modern hardware can be characterized by having multiple execution cores and deep memory hierarchies. The reason for the complex memory hierarchies is due to the gap between CPU power and memory access times. Gone are the days when CPU instructions took about the same time as a main memory access. Instead, this gap continues to increase and shows no signs of stopping (see Figure 15.1).

Different parts of the memory hierarchy have different access times. The smaller ones closer to the CPU are the fastest ones, whereas main memory can be really large, but also very slow. Table 15.1 lists some common access times for different levels of the hierarchy on modern platforms.

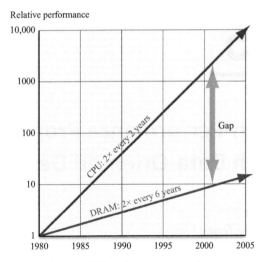

Figure 15.1. Relative CPU and memory performance over time.

With these kinds of access times, it's very likely that the CPU is going to stall waiting to read data from memory. All of a sudden, performance is not determined so much by how efficient the program executing on the CPU is, but how efficiently it uses memory.

Barring a radical technology change, this is not a situation that's about to change anytime soon. We'll continue getting more powerful, wider CPUs and larger memories that are going to make memory access even more problematic in the future.

Looking at code from a memory access point of view, the worst-case situation would be a program accessing heterogeneous trees of data scattered all over memory, executing different code at each node. There we get not just the constant data cache misses but also bad instruction cache utilization because it's calling different functions. Does that sound like a familiar situation? That's how most modern games are architected: large trees of different kinds of objects with polymorphic behavior.

What's even worse is that bad memory access patterns will bring a program down to its metaphorical knees, but that's not a problem that's likely to appear anywhere in the profiler. Instead, it will result in the common situation of everything being slower than we expected, but us not being able to point to a particular spot. That's because there isn't a single place that we can fix. Instead, we need to change the whole architecture, preferably from the beginning, and use a data-oriented approach.

Memory Level	CPU Cycles per Access
Register	1
L1 cache	5–8
L2 cache	30–50
Main memory	500+

Table 15.1. Access times for different levels of the memory hierarchy for modern platforms.

15.2 Principles of Data-Oriented Design

Before we can talk about data-oriented design, we need to step back and think about what a computer program is. One of the most common definitions of a computer program is "a sequence of instructions to perform a task." That's a reasonably good definition, but it concentrates more on the hows rather than on the whys. What are those instructions for? Why are we writing that program?

A more general definition of a computer program is "something that transforms input data into output data." At first glance, some people might disagree with this definition. It might be true for the calculations in a spreadsheet, but is it really a good description for a game? Definitely. In a game, we have a set of input data: the clock value, the game controller state, network packets, and the state of the game during the previous frame. The outputs we're calculating are a new game state, a set of commands for the graphics processor, sound, network packets, etc. It's not very different from a spreadsheet, except that it runs many times per second, at interactive rates.

The emphasis in Computer Science and Engineering is to concentrate on algorithms and code architecture. In particular, procedural programming focuses on procedure and function calls as its main element, while object-oriented programming deals mostly with objects (which are sets of data and the code that works on that data).

Data-oriented design turns that around and considers data first: how it is laid out and how it is read and processed in the program. Then, the code is something written to transform the input data into the output data, but it itself is not the focus.

As a consequence of looking at the input data carefully, we can apply another principle of data-oriented design: where there's one, there are more. How often have you had just one player in the game? Or one enemy? One vehicle? One bullet? Never! Yet somehow, we insist on treating each object separately, in isolation, as if it were the only one in the world. Data-oriented design encourages optimizing for the common case of having multiple objects of the same type.

15.3 Data-Oriented Design Benefits

Data-oriented design has three major performance benefits:

1. *Cache utilization.* This is the big one that motivated us to look at data in the first place. Because we can concentrate on data and memory access instead of the algorithms themselves, we can make sure our programs have as close to an ideal memory access pattern as possible. That means avoiding heterogeneous trees, organizing our data into large sequential blocks of homogeneous memory, and processing it by running the same code on all of its elements. This alone can bring a huge speed-up to our code.

2. *Parallelization.* When we work from the data point of view, it becomes a lot easier to divide work up into parts that different cores can process simultaneously with minimal synchronization. This is true for almost any kind of parallel architecture, whether each core has access to main memory or not.

3. *Less code.* People are often surprised at this one. As a consequence of looking at the data and only writing code to transform input data into output data, there is a lot of code that disappears. Code that before was doing boring bookkeeping, or getter/setters on objects, or even unnecessary abstractions, all go away. And simplifying code is very much like simplifying an algebraic equation: once you make a simplification, you often see other ways to simplify it further and end up with a much smaller equation than you started with.

When a technique promises higher performance, it often comes at a cost in some other department, usually in terms of readability or ease of maintenance. Data-oriented design is pretty unique in that it also has major benefits from a development perspective:

1. *Easier to test.* When your code is something that simply transforms input data into output data, testing it becomes extremely simple. Feed in some test input data, run the code, and verify the output data is what you expected. There are no pesky global variables to deal with, calls to other systems, interaction with other objects, or mocks to write. It really becomes that simple.

2. *Easier to understand.* Having less code means not just higher performance but also less code to maintain, understand, and keep straight in our heads. Also, each function in itself is much simpler to understand. We're never in the situation of having to chase function call after function call to understand all the consequences of one function. Everything you want to know about it is there, without any lower-level systems involved.

To be fair and present all the sides, there are two disadvantages to data-oriented design:

1. *It's different.* So far, data-oriented design isn't taught in Computer Science curricula, and most developers aren't actively using it, so it is foreign to most team members. It also makes it more difficult to integrate with third-party libraries and APIs that are not data-oriented.
2. *Harder to see the big picture.* Because of the emphasis on data and on small functions that transform data, it might be harder to see and express the big picture of the program: When is an operation happening? Why is this data being transformed? This is something that might be addressed with tools, language extensions, or even a new programming language in the future. For now, we'll have to rely on examining the code and the data carefully.

15.4 How to Apply Data-Oriented Design

Let's get more specific and start applying data-oriented design. Eventually, we'd like the entire game to be architected this way, but we need to start somewhere. So pick a subsystem that needs to be optimized: animation, artificial intelligence, physics, etc.

Next, think about all the data involved in that system. Don't worry too much about how it is laid out in memory, just about what's involved. Apart from the explicit inputs and outputs, don't forget about data that the system accesses explicitly, such as a world navigation graph, or global handle managers.

Once you have identified all the data the system needs, carefully think about how each type of data is used and sort them into read-only, read-write, or write-only. Those will become your explicit inputs and outputs. It will also allow you to make better decisions about how to lay out the data.

Also, at this point, it's important to think about the amount of data. Does this system ever process more than one of each type of data? If so, start thinking of it in terms of arrays of data, preferably as contiguous blocks of the same data type that can be processed at once.

Finally, the most important step is to look at the data you've identified as input and figure out how it can be transformed into the output data in an efficient way. How does the input data need to be arranged? Normally, you'll want a large block of the same data type, but perhaps, if there are two data types that need to be processed at the same time, interleaving them might make more sense. Or maybe, the transformation needs two separate passes over the same data type, but the second pass uses some fields that are unused in the first pass. In that case, it might be a good candidate for splitting it up into two types and keeping each of them sequentially in memory.

Once you have decided on the transformation, the only thing left is gathering the inputs from the rest of the system and filling the outputs. When you're transitioning from a more traditional architecture, you might have to perform an explicit gathering step—query some functions or objects and collect the input data in the format you want. You'll have to perform a similar operation with the output, feeding it into the rest of the system. Even though those extra steps represent a performance hit, the benefits gained usually offset any performance costs. As more systems start using the data-oriented approach, you'll be able to feed the output data from one system directly into the input of another, and you'll really be able to reap the benefits.

15.5 Real-World Situations

Homogeneous, Sequential Data

You are probably already applying some of the principles of data-oriented design in your games right now: the particle system. It's intended to handle thousands and thousands of particles. The input data is very carefully designed to be small, aligned, and fit in cache lines, and the output data is also very well defined because it probably feeds directly into the GPU.

Unfortunately for us, we don't have that many situations in game development where we can apply the same principle. It may be possible for some sound or image processing, but most other tasks seem to require much more varied data and lots of different code paths.

Heterogeneous Data

Game entities are the perfect example of why the straightforward particle approach doesn't work in other game subsystems. You probably have dozens of different game entity types. Or maybe you have one game entity, but have dozens, or even hundreds, of components that, when grouped together, give entities their own behavior.

One simple step we can take when dealing with large groups of heterogeneous data like that is to group similar data types together. For example, we would lay out all the health components for all entities in the game one right after the other in the same memory block. Same thing with armor components, and every other type of component.

If you just rearranged them and still updated them one game entity at a time, you wouldn't see any performance improvements. To gain a significant performance boost, you need to change the update from being entity-centric to being component-centric. You need to update all health components first, then all armor components, and proceed with all component types. At that point, your memory access patterns will have improved significantly, and you should be able to see much better performance.

Break and Batch

It turns out that sometimes even updating a single game entity component seems to need a lot of input data, and sometimes it's even unpredictable what it's going to need. That's a sign that we need to break the update into multiple steps, each of them with smaller, more predictable input data.

For example, the navigation component casts several rays into the world. Since the ray casts happen as part of the update, all of the data they touch is considered input data. In this case, it means that potentially the full world and other entities' collision data are part of the input data! Instead of collecting that data ahead of time and feeding it as an input into each component update, we can break the component update into two parts. The initial update figures out what ray casts are required and generates ray-cast queries as part of its output. The second update takes the results from the ray casts requested by the first update and finishes updating the component state.

The crucial step, once again, is the order of the updates. What we want to do is perform the first update on all navigation components and gather all the ray-cast queries. Then we take all those queries, cast all those rays, and save the results. Finally, we do another pass over the navigation components, calling the second update for each of them and feeding them the results of the ray queries.

Notice how once again, we managed to take some code with that tree-like memory access structure and turn it into something that is more linear and works over similar sets of data. The ray-casting step isn't the ideal linear traversal, but at least it's restricted to a single step, and maybe the world collision data might fit in some cache so we won't be getting too many misses to main memory.

Once you have this implemented, if the ray-casting part is still too slow, you could analyze the data and try to speed things up. For example, if you often have lots of grouped ray casts, it might be beneficial to first sort the ray casts spatially, and when they're being resolved, you're more likely to hit data that is already cached.

Conditional Execution

Another common situation is that not all data of the same type needs to be updated the same way. For example, the navigation component doesn't always need to cast the same number of rays. Maybe it normally casts a few rays every half a second, or more rays if other entities are closer by.

In that case, we can let the component decide whether it needs a second-pass update by whether it creates a ray-cast query. Now we're not going to have a fixed number of ray queries per entity, so we'll also need a way to make sure we associate the ray cast with the entity it came from.

After all ray casts are performed, we iterate over the navigation components and only update the ones that requested a ray query. That might save us a bit of CPU time, but chances are that it won't improve performance very much because we're going to be randomly skipping components and missing out on the benefits of accessing memory linearly.

If the amount of data needed by the second update is fairly small, we could copy that data as an output for the first update. That way, whenever we're ready to perform the second update, we only need to access the data generated this way, which is sequentially laid out in memory.

If copying the data isn't practical (there's either too much data or that data needs to be written back to the component itself), we could exploit temporal coherence, if there is any. If components either cast rays or don't, and do so for several frames at a time, we could reorder the components in memory so all navigation components that cast rays are found at the beginning of the memory block. Then, the second update can proceed linearly through the block until the last component that requested a ray cast is updated. To be able to achieve this, we need to make sure that our data is easily relocatable.

Polymorphism

Whenever we're applying data-oriented design, we explicitly traverse sets of data of a known type. Unlike an object-oriented approach, we would never traverse a set of heterogeneous data by calling polymorphic functions in each of them.

Even so, while we're transforming some well-known data, we might need to treat another set of data polymorphically. For example, even though we're updating the bullet data (well-known type), we might want to deliver damage to any entity it hits, independent of the type of that entity. Since using classes and inheritance is not usually a very data-friendly approach, we need to find a better alternative.

There are many different ways to go about this, depending on the kind of game architecture you have. One possibility is to split the common functionality of a game entity into a separate data type. This would probably be a very small set of data: a handle, a type, and possibly some flags or indices to components. If every entity in the world has one corresponding set of data of this type, we can always count on it while dealing with other entities. In this case, the bullet data update could check whether the entity has a damage-handling component, and if so, access it and deliver the damage.

If that last sentence left you a bit uncomfortable, congratulations, you're starting to really get a feel for good data access patterns. If you analyze it, the access patterns are less than ideal: we're updating all the current bullets in the world. That's fine because they're all laid out sequentially in memory. Then, when one of them hits an entity, we need to access that entity's data, and then potentially the damage-handling component. That's two potentially random accesses into memory that are almost guaranteed to be cache misses.

We could improve on this a little bit by having the bullet update not access the entity directly and, instead, create a message packet with the damage it wants to deliver to that entity. After we're done updating all of the bullets, we can make another pass over those messages and apply the damage. That might result in a marginal improvement (it's doubtful that accessing the entity and its component is going to cause any cache misses on the following bullet data), but most importantly, it prevents us from having any meaningful interaction with the entity. Is the entity bullet proof? Maybe that kind of bullet doesn't even hit the entity, and the bullet should go through unnoticed? In that case, we really want to access the entity data during the bullet update.

In the end, it's important to realize that not every data access is going to be ideal. Like with all optimizations, the most important ones are the ones that happen more frequently. A bullet might travel for hundreds of frames, and it will hit something at most in one frame. It's not going to make much of a difference if, during the frame when it hits, we have a few extra memory accesses.

15.6 Parallelization

Improving memory access patterns is only part of the performance benefits provided by data-oriented design. The other half is being able to take advantage of multiple cores very easily.

Normally, to split up tasks on different cores, we need to create a description of the job to perform and some of the inputs to the job. Unfortunately, where things fall down for procedural or object-oriented approaches is that a lot of tasks have implicit inputs: world data, collision data, etc. Developers try to work around it by providing exclusive access to data through locking systems, but that's very error prone and can be a big hit on performance depending on how frequently it happens.

The good news is that once you've architected your code such that you're thinking about the data first and following the guidelines in earlier sections, you're ready to parallelize it with very little effort. All of your inputs are clearly defined, and so are your outputs. You also know which tasks need to be performed before other tasks based on which data they consume and produce.

The only part missing is a scheduler. Once you have all of that information about your data and the transformations that need to happen to it, the scheduler can hand off tasks to individual cores based on what work is available. Each core gets the input data, the address where the output data should go, and what kind of transformation to apply to it.

Because all inputs are clearly defined, and outputs are usually new memory buffers, there is often no need to provide exclusive access to any data. Whenever the output data writes back into an area of memory that was used as an input (for example, entity states), the scheduler can make sure there are no jobs trying to read from that memory while the job that writes the output is executing. And because each job is very "shallow" (in the sense that it doesn't perform cascading function calls), the data each one touches is very limited, making the scheduling relatively easy.

If the entire game has been architected this way, the scheduler can then create a complete directed acyclic graph of data dependencies between jobs for each frame. It can use that information and the timings from each previous frame (assuming some temporal coherency) and estimate what the critical path of data transformation is going to be, giving priority to those jobs whenever possible.

One of the consequences of running a large system of data transformations this way is that there are often a lot of intermediate memory buffers. For platforms with little memory available, the scheduler can trade some speed for extra memory by giving priority to jobs that consume intermediate data instead of ones in the critical path.

This approach works well for any kind of parallel architecture, even if the individual cores don't have access to main memory. Since all of the input data is explicitly listed, it can be easily transferred to the core local memory before the transformation happens.

Also, unlike the lock-based parallelization approaches, this method scales very well to a large number of cores, which is clearly the direction future hardware is going toward.

15.7 Conclusion

Data-oriented design is a departure from traditional code-first thinking. It addresses head-on the two biggest performance problems in modern hardware: memory access and parallelization. By thinking about programs as instructions to transform data and thinking first about how that data should be laid out and worked on, we can get huge performance boosts over more traditional software development approaches.

16

Game Tuning Infrastructure

Wessam Bahnassi
Electronic Arts, Inc.

16.1 Introduction

Every game has to go through a continuous cycle of tuning and tweaking during its development. To support that, many of today's game engines provide some means for editing and tuning "objects" (or "entities") and other settings in the game world. This article provides food for thought on implementing tuning capabilities in a game engine. We cover the infrastructure options available and discuss their particularities and the scenarios that would benefit from each one of them. Finally, we conclude with case studies from a published game and a commercial engine.

16.2 The Need for Tweak

Implementing a tuning system is a challenge not to be underestimated, and making it really convenient and productive is even more challenging. Games involve a lot of data besides textures and 3D geometry: logical data such as the move speed of a tank, reload rate of a weapon, sun position, unit view radius, etc. Such data is obviously better not kept in hard-coded values in the game's executable. As executable build times continuously increase, tuning such values would become a very time-consuming task. A method for accessing, modifying, storing, and managing such data must be put in place. This is one aspect of why editors are needed in modern game engines.

Editors have the advantage of being able to offer powerful and convenient user interfaces to view and modify values for the various entities and parameters in the game. This article focuses only on the back-end facet of engine editors within the context of tuning, as user-interface discussions are a very big topic on their own and are outside the scope of this book.

For a while, editors had to provide only the means to modify game data and build this data to become ready for use by the game run-time code. However, a situation similar to executable build times has risen. Level build times have also become increasingly lengthy, making the turnaround time of visualizing data modifications too long and less productive. The need for a faster method for visualizing changes has thus become necessary, and a new breed of the so-called what-you-see-is-what-you-get (WYSIWYG) editors has also become available.

However, it is important to not get caught up in industry frenzy. Indeed, not all projects need or have the opportunity to utilize an engine with a full-scale editor. Except in the case of using a licensed engine, the game team might not have the time to build anything but the most primitive tool to do data editing and tuning. For such cases, this article also considers approaches that lack the presence of a separate editor application.

16.3 Design Considerations

Since the task of implementing a tuning system can be highly involved, it is important to take a step back and think about the expected goals of such a system within the context of the situation at hand. Considering the following points should help avoid underestimating or overshooting features for the system:

- Acceptable turnaround time for each class of tunable parameters. This is the time wasted between the user making the modification and seeing its effect.
- Convenience and ease of use with regard to the target system users and frequency of usage. A system that is going to be used daily should receive more focus than a system used rarely.
- Amount of development work involved and how intrusive code changes are allowed to be. The engine's code base is a primary factor in this area.
- Potential for system reuse in other projects, or in other words, generality. For example, is it needed to serve one game only? Or games of similar genre? Or any game project in general?
- Type of data to be tuned and its level of sophistication, complexity, and multiplicity (e.g., global settings, per-entity, per-level).
- Cross-platform tuning support—certain parameters are platform-specific and require tuning on the target platform directly.

One additional point that was taken out from the list above is stability and data safety. This should not be a "consideration," but rather a strict requirement. An editor that crashes in the middle of work is not considered a valid piece of software.

Next, we go into detail about the available options for implementing a tuning infrastructure in a game engine. The options are categorized into four main sections: The Tuning Tool, Data Exchange, Schema and Exposure, and Data Storage. A choice from each section can be made and then mixed and matched with choices from the other sections to finally form the definition of the tuning system that would best serve the team's needs. The list of considerations above should be kept in mind when reading through the following sections in order to help make the right decisions.

16.4 The Tuning Tool

The choice of the tuning tool is not limited to building an editor application from scratch. In fact, there is good potential for reuse in this area. The four options we discuss here range from the lowest-tech approach of using the debugger, all the way to implementing a dedicated editor application.

Using the Debugger

We start with the most basic case. A game has a number of values controlling a few aspects, like player walk speed, total health, or camera settings. As programmers who hopefully follow good coding habits, it is expected that such values are at least hard-coded in named preprocessor defines or in constants that reflect the nature of the value, as opposed to inlining the value directly where it's used.

The lowest-level method of tuning such values is to first convert them to global variables with initial values, as shown in Listing 16.1, and then run the game. To tune a value, one can add its global variable to the debugger's Watch window and change it directly in there (some debuggers allow doing this without even halting program execution).

The results are immediate, and the amount of code change is minimal. However, this is limited to people with access to a debugger. For artists and game designers, this would be an issue of convenience. Furthermore, the final values have to be noted and hard-coded again after the tuning session ends, which can be annoying if there are many modified values or if their values are lengthy.

```
#define MAX_PLAYER_HEALTH 100          // This is obviously not tunable.

const float kPlayerWalkSpeed = 2.5F;   // Not tunable either.
float g_CameraDistance = 15.0F;        // This can be tuned at run time.

void Player::Update(bool isShot)
{
    ...

    if (isShot)
        this->health -= 5;    // This is bad! Move to a named value.

    ...
}
```

Listing 16.1. Pseudo C++ code showing the proper setup for interactive global variable tuning.

Using a Console Window

A system similar to that presented by Jensen [2001] can be used to allow team members to tune values on the fly by typing name-value pairs in a console window attached to the game (e.g., see Figure 16.1). Compared to the previous method, a certain additional amount of coding is now necessary to implement the console window.

If the console window is implemented inside the game, then this approach (and the one before) can be conducted on any platform the game targets. This can be quite valuable when tuning values particular to a certain platform (e.g., tuning color correction or mipmap level selection). Console windows are commonly used for debugging and cheating purposes (toggling rendering modes, hiding characters, killing all enemies, etc.), but they could be considered a tuning tool as well.

Integrating with an Existing DCC[1] Tool

Bahnassi [2004] and Woo [2007] provide good examples of DCC tool integration. This can be quite convenient for artists and level designers who spend the majority of their time on the project modeling, texturing, and laying out levels for the game. Instead of forcing those team members to generate assets and then wait

---

[1] *Digital content creation tool* is a term used to refer to programs such as Softimage, 3DS Max, Maya, etc.

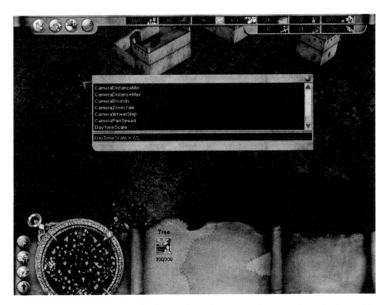

Figure 16.1. Tuning exposed variables using the game's console window. (*Image from the game Quraish, courtesy of Dar Al-Fikr Publishing.*)

for the pipeline to build them in order to view the results in the game's renderer, this approach offers them instant visualization, which can boost productivity.

DCC tools come with a large palette of utilities that can be used to layout a level and art exactly as one wants, thus providing an excellent tuning experience. However, the programming effort required by this technique is not to be underestimated.

There are two methods of approaching this option. One method is to implement the game renderer *inside* the DCC tool (possible in Autodesk Softimage), and the other is to have the tool communicate with the game executable in order to send it updates about changes happening in the DCC tool to directly display it to the user. In this venue, an engine can go as far as implementing asset hot loading so changes to models and textures could be directly visualized in the game, too. The storage of tuned data (models, levels, and custom properties) then becomes the role of the DCC tool.

Dedicated Editor Application

Another sophisticated method is to build a dedicated editor application that works with the game. Such a method puts total control and flexibility in the hands of the team to make the tool tune virtually any value deemed tunable. The widgets and controls for tuning can all be customized to be as suitable to the data as possible (e.g., color pickers, range-limited value sliders, or curve editors). A full-blown implementation can go as far as providing instantaneous live tuning through the actual game's executable, such as in Unreal and CryEngine.

Depending on how sophisticated a team wants to get with this, implementing the user interface and controls for a successful editor can be a difficult task that requires background and experience in developing user interfaces, which is a specialization of its own. If not considered well, the result most probably will be inconvenient (e.g., widgets not behaving in a standard way, bad interface layout, missing shortcuts, or missing undo or redo).

It can now be understood why this method, while being most sophisticated, is most difficult to get right. It can involve a huge amount of programmer work and the dedication of a number of members of the team. Going with third-party solutions can be a wise decision here.

On the implementation side, the editor application, being separate from the game, is free to use its own programming language. .NET languages have been found to be excellent for rapidly developing desktop control-rich applications. However, using a different language than the game can complicate data exchange and communication, but solutions exist [Bahnassi 08].

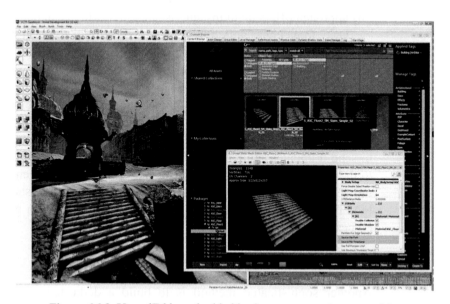

Figure 16.2. UnrealEd is embedded in the same game's executable.

Some implementations, such as the Unreal editor shown in Figure 16.2, involve the editor directly within the actual game's executable. This simplifies the task of data exchange and provides access to the game's most up-to-date rendering features and data structures. But on the other hand, such approaches can complicate the game's code by involving the editor's code in many of the game's internal aspects and forcing it to handle situations that are not faced when launching the game in standalone mode.

Another important issue is stability. By binding the editor to the game code itself, the editor inherits the same level of instability as the game (which can be high during game production), causing grief and lost work to the entire team. Thus, it is advisable to implement the editor as a separate application running in its own address space and using its own codebase. It can be made so that even if the game or visualization tool crashes, then it would still be possible to save data so that no work is lost.

16.5 Data Exchange

As has been shown in the previous section, the tuning tool might have to communicate with a separate visualization tool (be it the game, a DCC tool, or some other application). In this case, a method for exchanging tuning data between the two ends must be established. We discuss the possible approaches in this section.

Direct Value Access

Direct value access is the most straightforward possibility. When the tuning tool shares the same address space as the visualization tool, direct access to the tuned values becomes possible. Sharing the address space means the tuning tool is either implemented in the game executable (such as in Unreal) or is loaded as a dynamic-link library (DLL).

Interprocess Communication

When the tuning tool is implemented as a separate application, it can communicate with the game process through interprocess communication methods. This first requires writing a game-side module to handle the communication and execute the tuning commands. Second, a protocol must be devised to identify tunable values and serialize them between the game and the tuning tool. As is shown in Section 16.6, this can be a complicated task when the tuned data structure is complex or the game does not offer proper identification of the objects living in its world.

Cross-Platform Communication

If an interprocess communication system has been put in place, then it is relatively easy to extend it to handle cross-platform communication. With this feature, the tuning tool can be run on the development platform (e.g., PC) and tuning can occur live inside the game running on the target platform (e.g., PlayStation 3). All major console SDKs provide some means to communicate programmatically with the console from a PC. Implementing this method is the only way for dedicated editors to support on-console tuning.

16.6 Schema and Exposure

Historically, a lot of games have relied on configuration files (usually of type .INI) to expose tunable values to users. As discussed earlier, such a method is not convenient for games that have a large launch overhead. This is even more problematic if the result of tuning can only be seen after going to a certain location in the game or performing a lengthy series of steps to get to it.

Exposing data for tuning involves identifying variables to expose, their types, their value limits, and other possible tuning-related information and user interface options. In this section, we discuss some possibilities available in this area.

Marking Tunable Variables

A very simple method is to give tunable variables a special declaration in the C++ code. For example, variables can be wrapped in a preprocessor define that would expand to register the variable with a tuning service, similar to the method given by Jensen [2001].

This works fine with global variables but needs more thought to be extended to handle instance variables (e.g., the object registers its tunables explicitly upon instantiation). The registration can include the additional tuning information mentioned previously.

Reflection

If C++ had code reflection capabilities, then the task of exposing tunables would have been much simpler. This lack of reflection has strongly influenced some engines to extend their systems with an accompanying reflected language in their engines [Sweeny 1998]. The presence of reflection information simplifies matters a lot and makes for uniform tuning code.

The .NET languages are a good role model in this area. The great news is that C++/CLI has become a well-developed solution that can remedy such a situation elegantly. For example, any class can become reflected if declared as a CLR type; and the rest of the class can be left untouched in frequent cases. This reflection can be easily turned off for the final game executable if the dependency on .NET is to be avoided in the final product. The tuning tool can then read the structure of the tuned object and deal directly with its members.

Explicit Definition

Another alternative is to define the properties of tunable game objects in a separate place outside the game code and rely on a tool to generate data definition and access information for both the editor and the game to use. For example, a data schema can be written once and then passed to a code generation utility to generate a C++ class for in-game use, along with an accompanying reflection information class usable by the tuning tool. The code generation utility can go all the way to even implementing the necessary serialization code between the native object and its tuning service proxy.

16.7 Data Storage

Once the tunable values are exposed and under convenient control, the question becomes how to store value changes in such a way that subsequent game sessions remember those changes and how to properly share such changes with the rest of the team. This section gives advice about several aspects of data storage.

Text or Binary

Text file storage is highly recommended. First, the tunable values are usually not overwhelmingly dense, which drops the necessity of storing them in a binary format. Second, such a human readable format allows for easy differencing to see a history of changes that went through a file. This is a valuable feature for debugging.

If loading efficiency is to be maximized, a text-to-binary conversion process can optionally be supported for the release version of the game. Interestingly, some games actually prefer to ship those files in a human readable format for players to tweak and modify.

Divide and Conquer

Another production-related piece of advice is to avoid storing tuned data in single monolithic configuration files. Such a setup prevents parallel work when multiple team members are involved in tuning different areas of the game because they are forced to wait for each other to release the file and continue working. Although it is usually possible to merge changes from different versions of the file and thus restore the capability of parallel work, it is not a good idea to involve yet an additional step in the workflow of game tuning. Thus, it is advisable that tunable values be stored in different files in accordance with logical sectioning, thus reducing the possibility of conflicts and the need to resolve such conflicts.

Supporting Structure Changes

When writing tuned data to a file, it is possible to fall into the trap of writing them in a serial manner and assuming an implicit ordering that matches the internal structure holding this data in memory. It is a trap because this could compromise all tuned data if any changes occurred to the internal data structure, and the team then has to write version-aware code to deal with files not written using the new data structure layout.

An alternative is to name tuned data as it is written to file as illustrated by Listing 16.2. This way, regardless of where that field falls in the internal data structure layout, it can still be loaded correctly, without having to consider any changes that might occur later in time.

File Format

There are well-known formats that can be used for storing tuned data. Those include the .INI file format (sectioned name-value pairs) and the more expressive XML format. Adopting a well-known format can save development time by using libraries that are already available for reading and writing such file formats.

```
[weapon]
name = "Tachyon Gun"
sustainTime = 10
chargeRate = 6

[upgrade]
name = "Heat Dispenser"
effect = sustainTime
value = +5

[upgrade]
```

```
name = "Fission Battery"
value = +2
effect = chargeRate
```

Listing 16.2. Sample tuned data stored with explicit naming. The order of data appearance is irrelevant in this format.

Writing Data

One final aspect of data storage is considering what it is that actually writes the data. Depending on all the possibilities mentioned above, a game might have a dedicated PC-side tool communicating with it. In this case, storage can be initiated by the PC tool, requesting values for all tunable parameters and writing them down in the project's directory on the PC. The developer can then submit this updated file to the source control system to be shared with the rest of the team. This actually might be the sole possible option for some console systems.

Alternatively, for systems lacking interprocess communication but still running on console platforms, the game itself can write the values to its local folder, and then the developer can manually copy the generated files by hand from the console's file system to the project's directory.

16.8 Case Studies

Here, we present two case studies. The first is a simple one suitable as a guide for smaller teams, and the second is more sophisticated and suitable for larger teams.

Quraish

Quraish is a 3D real-time strategy game, shown in Figure 16.3, developed in from 2003 to 2006 for the PC platform by a small team made up of one game designer, one level designer, three core programmers, and about eight artists and animators. The game engine offers a separate world editor application for building maps, as well as for editing all entity properties and defining new races.

Both the game and the world editor share the core rendering and animation engine code base. Thus, the editor provides an exact instantaneous view of the game world when building maps. Since the core code is shared, data exchange is very easy because the same C++ structures had to be written only once and then used in both the game and the editor verbatim. The editor was written in C++ using the MFC framework.

Figure 16.3. Image from the game Quraish. (*Image courtesy of Dar Al-Fikr Publishing.*)

The editor stores all of its data in a single custom database file that can be loaded in sections by the game, and those sections are packed and prepared for game use ahead of time. A pointer fix-up pass is the only post-load operation needed for the data to become usable (maps were also stored similarly).

A major amount of tuning code was written in the editor in a game-specific way. That is, there is no generality towards code structures. Editors had to be written explicitly for each entity type (e.g., units, weapons, buildings, animals). This was feasible because the number of entity types was quite limited and manageable. However, this could have become a problem if the number of entity types grew to something above seven or eight.

In-game tuning was possible through the use of a console system built into the game (see Figure 16.1) and it connected to C++ functions and variables defined to be tunable manually in code through a registration process.

Tunable values could be written to a C-script file, like the example shown in Listing 16.3. The game can execute both compiled and interpreted C-script files using a text-parsing system like that described by Boer [2001]. This was to allow players to tweak and modify the game in a flexible manner. The files that were not to be exposed to the players were compiled, and the rest of them were left open in text format.

```
#include "DataBase\\QurHighAIDif.c"


ApplyWaterTax(SecondsToAITime(25), 1);
DayTimeScale = 0.1;


// Calling frequency of each need.
// Decreasing this makes the CPU more aware, not smarter!
BuildFrequency = SecondsToAITime(10000);
GenerateFrequency = SecondsToAITime(10000);
WorkFrequency = SecondsToAITime(50);
KillFrequency = SecondsToAITime(100000);


PolicyWith(2, AI_POLICY_ENEMIES);    // Friend to the player.
PolicyWith(0, AI_POLICY_ALLIES);


CompleteUpgrade(AI_UPGRADEID_MONEYPRINT);
CompleteUpgrade(AI_UPGRADEID_ANAGLYPH);


SetMilitaryDistribution(0, 0, 50, 30, 40);
SetResourcesImportance(20, 60, 0, 20);
SetCaravansCount(0);
```

Listing 16.3. Code listing of a C-script file containing tuned values.

Unreal 3/Unreal Development Kit

At the heart of the Unreal 3 engine is UnrealScript, an object-oriented C++-like language that is used to express all engine classes in need of reflection and garbage collection services. The engine's editor, UnrealEd, relies heavily on UnrealScript's reflection information and annotations to provide convenient editing capabilities.

Interestingly, UnrealScript is a compiled language, and the compiler is the game executable itself. This can create situations with a circular dependency if not handled with care since the game executable also relies on C++ header files generated during script compilation.

Both the game and the editor are compiled in one executable that can launch either of them through command line options. This allows the editor to inherit all game features easily and instantly (e.g., rendering features) and to exchange tuning data seamlessly. The disadvantage is that the editor can become very vulnerable to bugs and crashes due to continuous development work occurring in the game's code.

The engine uses binary package files to store its information, which is necessary for geometry and texture data, but it suffers from the binary data issues mentioned earlier in Section 16.7. Still, although the data is stored in binary format, the engine is capable of handling data structure changes conveniently. For example, variable order changes do not corrupt the data, and removing existing variables or adding new ones works just fine, without any additional effort needed to patch existing package files.

The engine also allows in-game tuning outside of the editor through a console window that can access virtually any UnrealScript property or function. Values tuned this way are not meant to be persistent, though; they exist only for temporary experimentation. Additionally, the engine uses .INI files to read settings across all areas of the game and the editor. The intention seems to be that .INI files are used for global settings, and UnrealScript and UnrealEd are used for per-object data.

16.9 Final Words

Provided with all of the possibilities and combinations in this chapter, a game engine architect has to wisely choose the approach most suitable to the case at hand, away from external influences that have no relevance to the project's benefit. At such a level, decisions must be based on sound reasoning. It is easy nowadays to get carried away by industry trends, but such trends will not work for all studios and game projects. Deciding what level of tuning is needed for the game is one requirement for shaping up a game's engine architecture. Be wise!

Acknowledgements

I would like to thank Homam Bahnassi, Abdul Rahman Lahham, and Eric Lengyel for proofreading this article and helping me out with its ideas.

References

[Bahnassi 2005] Homam Bahnassi and Wessam Bahnassi, "Shader Visualization Systems for the Art Pipeline." *ShaderX3*, edited by Wolfgang Engel. Boston: Charles River Media, 2005.

[Bahnassi 2008] Wessam Bahnassi. "3D Engine Tools with C++/CLI." *ShaderX6*, edited by Wolfgang Engel. Boston: Cengage Learning, 2008.

[Boer 2001] James Boer, "A Flexible Text Parsing System." *Game Programming Gems 2*, edited by Mark DeLoura. Hingham, MA: Charles River Media, 2001.

[Jensen 2001] Lasse Staff Jensen. "A Generic Tweaker." *Game Programming Gems 2*, edited by Mark DeLoura. Hingham, MA: Charles River Media, 2001.

[Sweeny 1998] Tim Sweeny. UnrealScript Language Reference. Available at http://unreal.epicgames.com/UnrealScript.htm.

[Woo 2007] Kim Hyoun Woo. "Shader System Integration: Nebula2 and 3ds Max." *ShaderX5*, edited by Wolfgang Engel. Boston: Charles River Media, 2007.

17

Placeholders beyond Static Art Replacement

Olivier Vaillancourt
Richard Egli
Centre MOIVRE, Université de Sherbrooke

Placeholder assets are temporary resources used during the development of a game in place of final resources that haven't been created yet. Even though they are an integral part of game development, placeholders are often overlooked or perceived negatively rather than being seen as useful development tools. The objective of this gem is to have a discussion on the use, integration, and construction of efficient placeholder systems in a game engine and provide a concrete example of such a system.

The first part of the chapter begins with a brief discussion underlining the technical advantages a well-thought-out placeholder system can provide beyond simple art replacement. This leads to the second part of the chapter, which consists of a detailed and technical presentation of a procedural method that automatically generates articulated placeholder meshes from animation skeletons. The last part of the chapter discusses how to transparently integrate the technique in the various pipelines of a game engine.

17.1 Placeholder Assets in a Game

Programmer art, prototype assets, mock objects, or placeholders—name them as you wish—all these resources have one thing in common: none of them should remain when the game goes gold. These temporary game assets, which we will refer to as *placeholders*, are simplified game resources that are uniquely used during development. They are created to fill a void when the real asset hasn't yet been created. Since they appear in pretty much every production, placeholders are a common sight for many developers. They usually require very little time to produce, and little effort is made to make them look or behave like the final asset. It can also be said that the majority of placeholders seen during development involve 3D game objects in some ways, since the construction of this particular type of asset requires a large amount of work from multiple trades (modeling, texturing, animation, game logic programming, sound editing, etc.). However, it's important to keep in mind that placeholders aren't limited to 3D objects and also extend to other assets types such as sound, text, or even code.

The motivation behind placeholders arises from the fact that multiple developers of different trades often work on the exact same part of the game at different moments in time, or they iterate over a given task at different paces. In these cases, some of the developers involved in a task usually have to wait until it attains a certain level of completion before being able to start working on it. These small wait times can add up and cause important production bottlenecks when considered as a whole. Fortunately, in the cases where these dependencies involve the creation of game assets, the wait time can be greatly reduced by the creation of placeholder assets. They provide a minimal working resource to the developers needing it, without overly affecting the schedule of the rest of the team. In the end, the developers might still have to revisit their work to ensure that it fits the final asset, but they nonetheless save more time than if they had remained idle while waiting for the completed resource.

Unfortunately, despite having a generally positive effect on development time, placeholders can become a source of confusion when they are incorrectly crafted. For example, a common practice is to reuse an existing asset from the production or a previous production instead of building a dedicated placeholder asset. While this has the advantage of inserting more attractive and functional temporary objects, it greatly increases the risk of forgetting that the asset is, in fact, a placeholder. In some extreme cases, licensed or branded content from a previous game could even make its way through to another client's game. Comparatively, the presence of temporary assets from previous productions in a game sometimes proves to be an important source of distraction during testing. The fact that a resource doesn't match the game's style or quality, but still appears final, can end up being more confusing than anything else for an unaware tester. In these cases, the testers often waste time reporting art problems with the placeholder, thinking they are part of the game's assets. This becomes especially true in the case of focus group tests not involving professional testers. Obviously, the testers can be explicitly told that some of the art isn't final, but this reduces their confidence in reporting legitimate art problems. There are ways to "tweak" these reused assets to remove potential confusion, but at this point, the time involvement in creating and tweaking placeholders on a case-by-case basis becomes too much to be efficient.

To fix these potential problems, many developers choose to go the easy way and use oversimplified placeholders to eliminate sources of confusion. Therefore, large pink boxes and checkerboard textures often come to mind when discussing placeholders. While they sure have the merit of being obvious and quick to create, they do relatively few things to actually help development. The hard truth is that we, as developers, rarely see placeholders as an opportunity to increase productivity or prevent problems. Most of the time, they're seen as an annoying reminder that someone is waiting for something. Ignoring them becomes somewhat natural as they tend to be negatively (and falsely) associated with unfinished work. A better way to approach placeholders is to consider them to be a part of the normal production process and to see them as productivity and communication tools rather than as unfinished work.

Placeholders as Development Productivity Tools

As stated earlier, placeholders mostly serve as a way to diminish coupling among design, art, and programming. However, creating a good, well-crafted, and well-thought-out placeholder system can have other positive effects on a development team, elevating it to a useful production and communication tool. To find out how to do this, the first thing to do is to start looking back at the idea of placeholders: what should be their primary purpose, and how well do they fill this purpose?

First and foremost, placeholders can be seen as a means of communication between a developer who needs something to be done and a developer who needs to do something. Sure, the placeholder is only there to fill a void, but it is the one thing inside the game that connects these two developers. For instance, when a modeler starts the art pass on a game level, the temporary geometry placed by the designer becomes a way to communicate to the artist where to place walls, floors, or obstacles in the level. What if the designer could encode supplementary information on the collision geometry to better communicate his intentions to the modeler? This could reduce the iteration count between both parties, which would remove most of the back and forth due to misunderstandings from the artist regarding the designer's intentions. Even more interestingly, such a system would allow the placement of important information in the level, information that could have been forgotten if the art pass was done several months after the level design. In this mindset, well-crafted placeholders would not only increase communication but also ensure that information is kept and remembered over long periods throughout the development.

Some occurrences of such well-thought-out and battle-tested placeholder systems can be seen in the videogame industry. A classic example is the popular "orange map" [Speyrer and Jacobson 2006] system used by Valve Software to facilitate communication between designers and artists. Level designers usually create the basic level geometry before handing it to the artists. Where the orange map comes into play is that some predefined textures are available to the designer to place on the map's surfaces. The textures indicate some pretty useful information, such as the basic desired material properties; window, stairway, railings or door dimensions; or default player height. More interestingly, the dimensions aren't only encoded in the texture motif but are also literally written on them. This helps the designer to remain homogeneous in his level design dimensions and helps the communication with the artist.

Continuing with our previous reasoning, placeholders have another important purpose and that is to emulate missing resources. However, to what extent does a given placeholder truly represent that missing resource? More often than not, it is built following a view that is almost exclusively game-oriented. For example, a large collision box might temporarily replace a game object, giving a rough approximation of the object's dimensions, location, and geometry, properties that are important for level design and gameplay. While this might give some clues to an artist or other designers concerning the size and scale of the object, it does few things to help the programmer estimate the memory and performance cost of the object in the scene. Nor does it help to verify that the object's logic is respected or that the physics are working correctly. As far as boxes go, it barely even gives any indication of the object's purpose in the level (unless the object is really a box, in which case, you are in business). To eliminate this shortcoming of classic placeholders, a solution would be to start building more complex representations that can emulate visual appearance, geometric complexity, or even physical properties. While this might seem like basically recreating the desired final asset, it's important to keep in mind that the objective here is to concentrate on emulating properties that are relevant not only to designers but also to artists and programmers. Computing or memory costs, general visual appearance, general material properties, or physical simulation collisions are all aspects of a resource that can be emulated easily with simple automatic generation processes. In the end, having the exact desired polygon count or working with a perfectly animated placeholder might not be within reach, but the importance here is to have something that gives a correct estimate of the true asset.

To give a proper example, a placeholder system that concentrates on emulating assets over a broader range of properties is presented in great detail in the second part of this gem.

Desired Features of a Placeholder System

What can be considered a "perfect" placeholder? Does such a thing even exist? The simple answer is no, mainly because most features of a placeholder system are decided through compromise. Do you select a very high-fidelity placeholder that almost perfectly estimates the final asset but is easily mistaken for it, or do you build a rough estimate that eliminates confusion but hardly helps programmers? Do you go for an automated and fast temporary asset generator that produces lower-quality placeholders or a human-built placeholder that does exactly what you asked for, but requires half a day to create? It all amounts to what is important in your very own case. Usually, trying to exceed in one facet results in failing in another. Thus, creating a system that aims for the lowest common denominator of all your needs is probably the right idea. It will not be perfect everywhere, but it will do the trick in every situation.

Below are some of the desired aspects of what we consider to be the "perfect placeholder." Be wary though—it's pretty much impossible to put everything together in a perfect manner. The points below are mostly guidelines, which if followed, give satisfying results.

- *Reduce coupling as much as possible.* Most placeholders decouple the work among art, design, and programming. Try to push the system further by decoupling even the smaller tasks, such as modeling, animation, and texturing. This reduces intradepartmental bottlenecks.

- *Reduce confusion and distraction while remaining pleasing to the eye.* As we discussed earlier, placeholders that look like final assets have the tendency to confuse other developers or testers, or they might end up in the final product. Naturally, ensuring that they're obvious doesn't mean that you must go to extreme measures to have them noticed. At that point, they only become annoying and negatively perceived by the team.

- *Provide an accurate estimate of the technical cost of the final resource.* Without being pin-point accurate about the correct final resource cost, having a reliable estimate might prevent some headaches at later stages of production. This won't prevent the final megabyte grind in fitting everything on the DVD or the polygonal witch hunt in reaching 30 frames per seconds. It will, however, guarantee that you never find yourself losing half of your frame rate or blasting through your memory budget twofold because these hundreds of innocent-looking pink boxes suddenly became fully animated and detailed 3D henchmen.

■ *Provide an accurate estimate of the look and behavior of the final asset.* This one is a bit tricky and can be considered a bit more loosely. The idea of "accurate estimate" in terms of look and behavior differs a lot from one project to another. Using a cylinder with a sphere sitting on top of it might be enough to picture a character in a real-time strategy game. To express the behavior of the final resource, a simple arrow can be added to indicate in which direction the unit is facing. However, in a platformer or an adventure game, the placeholder character might need to be a bit more detailed. If the animations are already available, that character could even be articulated and animated. The amount of work that needs to be done to correctly represent the look and behavior of the final asset amounts to the quantity of details required to understand the placeholder.

■ *Can be created or generated rapidly with no particular skill required.* By definition, placeholders are temporary and do not involve detailed work at all. If creating them becomes a time burden or requires fine tuning, the whole point of building a placeholder gets somewhat lost. Moreover, if special skills are required to create the placeholder, such as artistic or programming skills, the idea of reducing coupling by creating them also gets lost in the process, since someone depends on the expertise and schedule of someone else to have the placeholder created. Ideally, everyone on a development team should be able to create a placeholder within a matter of minutes.

■ *Facilitate communication between team members.* While a placeholder will never replace a design document or a good old explanation, it does not necessarily mean it cannot help on the communication side of things. Placeholders have a shape or a surface through which information can be conveyed. What information to convey and how to do it should be one of the primary questions when building a placeholder system.

■ *Integrate easily in the already existing resource pipeline.* One of the important aspects of a placeholder system is for it to have a small footprint on production time. This applies to the creation of the placeholder itself but also to the creation of the system. If integrating placeholders into your 3D pipeline requires three months of refactoring, you might want to rethink your original placeholder idea (or rethink the flexibility of your resource pipeline, but that's another discussion).

■ *Scalable and flexible enough to work on multiple platforms.* Placeholders are a bit like code. If they're done correctly, they should be able to scale and adapt to any kind of platform. This is especially true now, since game studios have become more and more likely to build games on multiple platforms at the same time. Some studios have even started working on game engines that run on PCs, consoles, handhelds, and cell phones alike. Avoid using platform-specific features when creating placeholders, and ensure that the techniques used to generate them can be scaled and adapted to multiple platforms.

As we conclude this part of the chapter, what remains important to remember is that taking the time to create a good placeholder system for your game certainly helps development time in the long run. The effects are a bit subtle and hard to gauge unless you go back to an environment that doesn't use placeholders efficiently, but it remains there and smooths development time, removing much of the stress and delays associated with schedule coupling.

17.2 Preaching by Example: The Articulated Placeholder Model

The previous discussion wouldn't be very valuable without an example to back it up. To further develop the topic of placeholders, the remainder of this gem focuses on a placeholder system that generates articulated (and therefore animated) placeholder meshes from animation skeletons. The reasoning behind this particular choice of placeholder is that animation skeletons are often reused throughout multiple models that have roughly the same shape or are developed using prototype geometry, while the final model will be completed many months later during development [Lally 2003]. Various approaches can be used to fill in the missing animated meshes that have yet to be produced. The first approach is to reuse the animation rig or prototype model geometry as a placeholder. This is a correct approach since the placeholder nature of the rig geometry is easily recognizable and the rig remains an acceptable visual representation of the final mesh. However, the rig geometry is often very simple and provides a poor approximation of the required rendering, physics, and animation costs of the final resource. Therefore, the idea remains good for prototype development but has its limitations for production purposes. Another approach is to reuse a previously modeled mesh that has the same skeleton. Unfortunately, as we've discussed earlier, this can create confusion and should be avoided.

To bypass the previously mentioned shortcomings, we present a method that generates a placeholder mesh using implicit surfaces based exclusively on an animation skeleton, as shown in Figure 17.1. Once the mesh is generated, the skeleton's bone weights are automatically assigned to the mesh vertices, which can then be animated using standard skeletal animation skinning methods. Every step

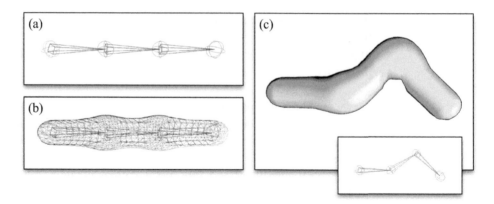

Figure 17.1. Depiction of the articulated placeholder generation technique. (a) The animation skeleton is used as the basis for the (b) mesh generation using implicit surface. (c) The mesh is skinned to the skeleton, which can then be animated, animating the articulated placeholder mesh itself.

employs popular computer graphics techniques that are very easy to implement and well documented. Special care was also put into ensuring that the placeholder construction uses popular rendering engine infrastructures when possible to further reduce the time required to integrate the system into a production pipeline. The generation process is sufficiently detailed to ensure that a user with no prior experience with skeletal animation and implicit surface generation can still build the system. In this regard, advanced users can feel comfortable skipping some of the entry-level explanations.

Skeleton Construction

The first step is to create the animation skeleton itself. The animation skeleton is the root of a widely used animation technique called "skeletal animation" [Kavan and Žára 2003], which is arguably the most popular animation technique currently used in the videogame industry.

Skeletal animation is based on a two-facet representation of an animated 3D mesh: the skin and the skeleton. The skin is the visual representation of the object to be animated, which consists of a surface representation of the object. This surface is often, but not always, made of tightly knit polygons called a mesh. The other part of skeletal animation is the skeleton, which is a representation of the underlying articulated structure of the model to animate. By animating the skeleton and then binding the surface to it, it is possible to animate the surface itself.

Before delving into the intricacies of the latter part of the process, we first begin by looking at the construction of the skeleton.

The skeleton consists of a hierarchical set of primitives called *bones*. A bone is made up of a 3D transformation (position, scale, and orientation) and a reference to a parent bone. (The parent bone is optional, and some bones, such as the topmost bone of the hierarchy, have no parent.) The parent-child relationship between bones creates the skeletal hierarchy that determines how a bone is transformed. For example, if an upper arm bone is rotated upward, the forearm bone rotates upward accordingly since it is attached (through the hierarchy) to the upper arm bone.

Even if the final bone animation only has one transformation, the bones, during the animation creation, must have two separate transforms: The *bone-space transform* B^{-1} and the *pose transform P*. Since these transforms apply to a particular bone of the skeleton, we identify the bones as the j-th bone of the skeleton, and denote the transforms by B_j^{-1} and P_j. These transforms can be represented as a 4×4 homogeneous matrix having the general form

$$T_j = \begin{bmatrix} T_j^{\text{rot}} & \mathbf{T}_j^{\text{trans}} \\ \mathbf{0} & 1 \end{bmatrix}, \tag{17.1}$$

where T_j^{rot} is a 3×3 rotation matrix, $\mathbf{0}$ is a 1×3 zero vector, and $\mathbf{T}_j^{\text{trans}}$ is a 3×1 position vector. Another approach, which is more compact, is to represent T_j as a quaternion-translation pair. This, however, prevents the inclusion of a nonuniform scale in the transformation, which might be a problem if it is needed.

The bone-space transform B_j^{-1} is the inverse of the bone's transformation in world coordinate space when it is in its initial pose, also known as the *bind pose* (the pose in which the skeleton was created before being animated). The pose transform P_j is the combination (product) of a given bone's parents' transforms and its own local transformation. The matrices P_j and B_j^{-1}, and their effects on a given bone are represented in Figure 17.2. The final and unique transformation M_j for a given bone is found by multiplying the bone-space and pose transformations:

$$M_j = P_j B_j^{-1}. \tag{17.2}$$

The matrix M_j is used later to transform the mesh vertices according to the animation during the skinning process.

At this point, you should have the required components and mathematics to build your own animation skeleton. However, it remains a hierarchy of 3D trans-

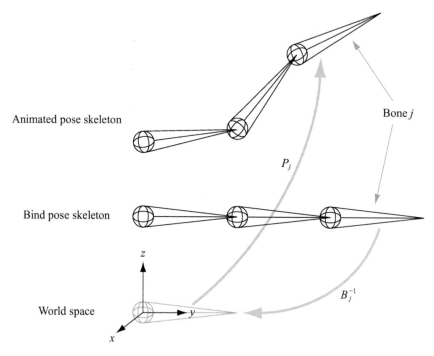

Figure 17.2. Bone j from the skeleton is brought to the origin by the transformation B_j^{-1}. The transformation P_j then transforms the bone j to its animated pose.

formations, and this can be rather hard to visualize. To help in viewing your skeleton and ensuring that your transformations are applied correctly, it is suggested that you visually represent a skeleton as a set of bones and joints, where the difference between the translations for two transformation matrices (the bones) are modeled as line segments, and the connections between bones (the joints) are visualized as spheres. If you already work in a development environment, this representation is probably already built into your 3D modeling software or into your engine's debugging view.

Placeholder Mesh Generation with Implicit Surfaces

With the skeleton constructed, the placeholder mesh can now be created. The objective here is to find a technique that generates the mesh in such a way that it visually resembles the desired final asset. Moreover, we'd like the technique to generate a surface that can be easily animated to emulate the final asset's behavior. On top of this, the technique must provide control over the amount of geometry (triangle count) that constructs the mesh, to emulate animation and rendering costs. Finally, it must be able to generate the mesh rapidly and automatically. The technique we suggest to satisfy all of these requirements is to use *implicit surfaces*, also known as *level sets*, to generate the mesh geometry.

A level set, generally speaking, is a set of points $(x_1,...,x_n)$ for which a real-valued function f of n variables satisfies the constraint $f(x_1,...,x_n) = c$, where c is a constant. The entire set is then

$$\{(x_1,...,x_n) \mid f(x_1,...,x_n) = c\}. \tag{17.3}$$

When $n = 3$, the set of points defined by the constraint creates what is called a *level surface*, also commonly known as an *isosurface* or an *implicitly defined surface*. A classic example of a 3D implicit surface is the sphere, which can be defined as a set of points (x,y,z) through the general equation

$$\{(x,y,z) \mid f(x,y,z) = x^2 + y^2 + z^2 = r^2\}. \tag{17.4}$$

In this case, our constant c would be r^2 and our implicit surface would be defined as all the 3D points (x,y,z) located at a distance r from the origin, effectively creating a sphere.

Another way to picture an implicit surface is to imagine that the function $f(x_1,...,x_n)$ defines a density field in the space where it is located and that all the points having a certain density value c within the field are part of the set (which is a surface in the 3D case). Finding the density value at a certain point is done simply by evaluating f with the coordinates of that point.

In the case of our placeholder system, we define a mesh by generating an implicit surface from the skeleton itself. The idea is to define a density field based on an inverse squared distance function defined from the bones of the skeleton. As we saw earlier, a bone only consists of a transformation matrix and a reference to its parent. However, to create a distance function for a given bone, the bone must be mathematically represented as a line segment rather than a single transformation. To obtain a line segment from the bone, we use the bone-space transform. The translation of the bone $\mathbf{B}_j^{\text{trans}}$ is used to create one end of the line segment, while the parent's translation $\mathbf{B}_{\text{parent}(j)}^{\text{trans}}$ is used as the other end of the segment, where $\text{parent}(j)$ maps the index j to the index of the parent bone. (Note that we're using B and not B^{-1} here.) With this formulation, the squared distance function can be defined as

$$\text{dist}(\mathbf{p}, j) = \begin{cases} \left\| \mathbf{p} - \mathbf{B}_{\text{parent}(j)}^{\text{trans}} \right\|^2, & \text{if } \left(\mathbf{p} - \mathbf{B}_{\text{parent}(j)}^{\text{trans}} \right) \cdot \left(\mathbf{B}_{\text{parent}(j)}^{\text{trans}} - \mathbf{B}_j^{\text{trans}} \right) > 0; \\ \left\| \mathbf{p} - \mathbf{B}_j^{\text{trans}} \right\|^2, & \text{if } \left(\mathbf{p} - \mathbf{B}_j^{\text{trans}} \right) \cdot \left(\mathbf{B}_j^{\text{trans}} - \mathbf{B}_{\text{parent}(j)}^{\text{trans}} \right) > 0; \qquad (17.5) \\ d^2, & \text{otherwise,} \end{cases}$$

where $\mathbf{p} = (x, y, z)$ and

$$d = \frac{\left\| \left(\mathbf{B}_{\text{parent}(j)}^{\text{trans}} - \mathbf{B}_j^{\text{trans}} \right) \times \left(\mathbf{B}_j^{\text{trans}} - \mathbf{p} \right) \right\|}{\left\| \mathbf{B}_{\text{parent}(j)}^{\text{trans}} - \mathbf{B}_j^{\text{trans}} \right\|}. \qquad (17.6)$$

Once the distance can be calculated for a single bone, evaluating the density field at a certain point amounts to summing the distance function for all bones and determining whether that point is on the surface by setting a certain distance threshold d. The implicit surface generated from our skeleton can therefore be formulated using the distance function and the set

$$\left\{ \mathbf{p} \mid f(\mathbf{p}) = \sum_j \text{dist}(\mathbf{p}, j) = d \right\}. \qquad (17.7)$$

One way to define the whole 3D surface from the field generated by the skeleton would be to sample every point in our 3D space and build a polygonal surface from those that happen to fall directly on the threshold d. This, however, would be extremely inefficient and would most likely produce an irregular surface unless the sampling is very precise. A smarter way to generate an implicit surface from a density field is to use the marching cubes algorithm[1] [Lorensen and Cline 1987] or its triangular equivalent, marching tetrahedrons [Doi and Koide 1991, Müller and Wehle 1997], which we favor here for its simplicity and robustness.

The marching tetrahedrons algorithm, depicted in Figure 17.3, starts by sampling the density field at regular intervals, creating a 3D grid. Every discrete point is then evaluated using the density function and flagged as inside or outside depending on the computed value of the density field in regard to the threshold (inside if the value is higher than the threshold and outside if the value is lower). The whole 3D grid can be seen as an array of cells defined from groups of eight points ($2 \times 2 \times 2$ points) forming a cube. Each of these cells can be further divided in six tetrahedrons that are the base primitives defining our polygonal implicit

---

[1] The marching cubes algorithm is explained in detail in the chapter "Volumetric Representation of Virtual Environments" in *Game Engine Gems 1* [Williams 2010].

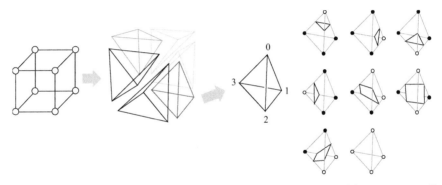

Figure 17.3. Representation of the marching tetrahedrons algorithm. Every cell of $2 \times 2 \times 2$ points is split into six tetrahedrons. Each of these tetrahedrons is matched to one of the eight cases depicted on the right, and the geometry is generated accordingly.

surface. Every tetrahedron is built with four different points, and each of these points can have two different states (inside or outside). Therefore, there exist 16 possible different configurations of inside and outside values for each tetrahedron. (These 16 configurations can be reduced to two mirrored sets of eight configurations.) At this point, all of the tetrahedrons containing a transition (a transition occurs when some points of the tetrahedron are higher than the threshold and other points are lower) create one or two triangles based on their configuration. When all of the tetrahedrons have been evaluated, have been classified,

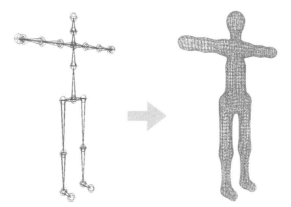

Figure 17.4. From the skeleton's bone, the geometry is generated using the marching tetrahedrons algorithm. For the above example, the bone's density was reduced depending on the skeletal distance from the root bone, giving the limb extremities a smaller size.

and have generated their triangles, the resulting geometry is a fully defined implicit surface based on the density field. In our case, with a density field generated from the squared distance to our skeleton, the implicit surface should look like a party balloon having roughly the shape of our skeleton, as shown in Figure 17.4.

One of the advantages of using this technique is that the amount of generated geometry can be easily adjusted by increasing or decreasing the sampling precision. A larger sampling grid generates shapes with less geometry, whereas a finer sampling grid increases the polygon count. This comes in handy and generates a placeholder mesh that provides similar performance to the final desired asset. It also becomes particularly useful when working under polygon constraints for game objects, as the generated mesh provides a good performance estimate.

Automatic Vertex Weight Assignment

At this point, we have an animated skeleton and a mesh generated from the skeleton's bind pose. This is a good start, but nothing actually binds the mesh to the skeleton. This means that even if the skeleton is animated, the mesh will not follow with the movements. The process of binding the mesh to the skeleton is called *vertex weight assignment* or simply *vertex assignment* and consists of defining which vertices are affected by a bone or multiple bones and what influence, or weight, each of these bones has in affecting the vertices' positions. The weight is usually normalized between zero and one, and it represents the percentage of that particular bone's transformation that is used for the animation.

For example, imagine that you have a mesh that needs to be bound to a skeleton having only one bone j. In this particular case, all the vertices of the mesh would be assigned to the bone j with a weight equal to one. In a more complex case, say for an arm, all the vertices that are directly in the middle of the elbow joint would be equally affected by the forearm bone and the upper arm bone. The weight would therefore be 0.5 for the forearm and 0.5 for the upper arm. When animated, these vertices would blend half of the forearm's and half of the upper arm's transforms to obtain their final positions. (More to come on this topic in the next section.)

In 3D modeling and animation, this whole weight assignment process is usually done by selecting a bone in the animation skeleton and painting the importance of that bone directly on the vertices of the mesh. This practice is often referred to as *weight painting*, and it remains relatively time consuming. In this regard, this approach is hardly practicable for our placeholder system since we need something fast and completely automated. Luckily for us, many 3D CAD applications already offer a feature to automatically assign weights. While artists usually don't use the feature due to the poor weight maps it produces, it remains good enough for a placeholder. If you decide to generate your placeholders directly in the 3D CAD software, your best bet might be to directly use the software's automatic weight system (remember that the faster your placeholder system is up, the better). Otherwise, we'll draw inspiration from these systems and devise a small and simple algorithm that remains fast and produces acceptable results.

Most of the existing automatic weight assignment algorithms use a distance calculation of some sort to determine which bones affect a certain vertex. If the animation system supports n bones per vertex, then the algorithm finds the closest n bones or fewer within a certain threshold and assigns a weight to each bone based on the distance. The distance itself is what usually differentiates one algorithm from another. Some algorithms use geodesic distance on the surface, and others use a heat diffusion simulation [Rosen 2009] to generate a volumetric distance map by filling the volume defined by the mesh with 3D voxels. The most advanced algorithms even use mesh segmentations or other mesh analysis methods to produce the weight maps [Baran and Popović 2007, Aguiar et al. 2008]. To keep things simple, we don't delve into any of this and simply use a skeleton-aware Euclidean distance computation to generate the map.

What we mean by "skeleton-aware" distance is that it considers the skeleton's structure when calculating distances. A pure Euclidean distance that doesn't consider bone connectivity would most likely lead to strange artifacts. For example, the two feet of a character are often close to each other in space, but far apart in the skeleton's structure. By using only a Euclidean distance, the left foot's vertices would be partly bound to the left foot bone and partly bound to the right foot bone since they're close to each other. This would undoubtedly create strange artifacts when the feet are animated.

To prevent this problem, our weight assignment algorithm functions in two parts and is performed on every vertex of the mesh. The first part finds the closest bone to a particular vertex and keeps the distance from that vertex to the bone in memory, reusing the vertex-to-segment distance computation given by Equation (17.5). During the second step, the nearest bone's parent and children are recursively explored. If their distances are lower than a certain threshold t, then the distance and bone ID is saved in a distance-ordered list, and the parent and children of that new bone that haven't been visited yet are recursively submitted to the same process. If the distance is higher than the threshold, then the exploration of this branch of the skeleton stops there. Once all of the bones within the threshold are found, the n closest bones are assigned to the vertex, and their weights are given relative to their distance. (It is advisable to use the cubed or squared distance because they give better visual results than the direct Euclidean distance.) In some cases, the number of bones within the threshold might be lower than n, which is normal if fewer bones are located near the vertex. This whole two-step process ensures that spatially close bones that are far in the skeletal structure, and thus are physiologically unrelated, do not share weights all over the mesh. To recapitulate, the automatic weight assignment algorithm requires the following steps:

1. Loop through all the vertices.
2. Find the closest bone for each vertex.
3. Recursively find all the bone's parents and children that are below the distance threshold for that vertex. Only explore those that haven't been visited and add them to the assignment list.
4. Trim the list to only keep the n closest bones to the vertex. Don't forget that the list might have fewer than n entries.
5. Normalize the bone's distance between zero and one and use it as the bone's weight for that vertex.

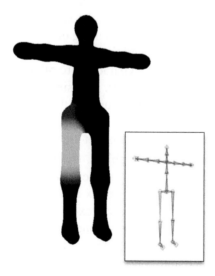

Figure 17.5. Weight distribution for a nontrivial case in an animation skeleton. The presented weight distribution is for the right femoral bone of the skeleton (highlighted in the inset). Brighter colors on the mesh surface indicate a greater influence of the middle bone, whereas completely dark parts of the surface represent no influence.

The automatic weight assignment algorithm is illustrated in Listing 17.1. After the algorithm has been run on the mesh, the vertex weight assignment is complete. The obtained result is a weight map for our articulated placeholder. The weight map is far from perfect but entirely sufficient for a placeholder asset. As a reference, Figure 17.5 shows the weight distribution for the vertices affected by a given bone.

```cpp
map<float, BoneAssignement> VertexAssignement(vec3 vertex,
        const vector<Bone *>& bones, float threshold,
        int maxSupportedBones)
{
    map<int, float>      assignationMap;
    float nearestDist = 0.0F;

    // We start by finding the nearest bone. The NearestDist argument
    // is returned by the FindNearestBone function and returns the
    // nearest squared distance.
    Bone *nearestBone = FindNearestBone(bones, vertex, nearestDist);

    AssignationMap[NearestDist].bone = nearestBone;
    AssignationMap[NearestDist].dist = nearestDist;

    nearestBone->SetVisited(true);

    // We recursively search through the nearest bone's parents
    // and children.
    AssignRecur(vertex, nearestBone, threshold, assignationMap);

    // We trim the obtained list to maxSupportedBones and normalize
    // the squared distances.
    assignationMap.trim(maxSupportedBones);

    float distSum = 0.0F;
    for (map<int, float>::iterator it = assignationMap.begin();
        it != assignationMap.end(); ++it)
    {
        distSum += it->Dist;
    }
```

```
    for (map<int, float>::iterator it = assignationMap.begin();
        it != assignationMap.end(); ++it)
    {
        it->dist /= distSum;
    }


    return (assignationMap);
}


AssignRecur(vec3 vertex, const Bone *bone, float threshold,
        map<float, BoneAssignement>& assignationMap)
{
    // Go through all the children of the bone and get those that
    // haven't been visited and are lower than the threshold.
    for (int i = 0; i < bone->ChildrenCount(); ++i)
    {
        float dist = distance(bone->Child[i], vertex);
        if (!bone->Child[i]->Visited() && dist < threshold)
        {
            assignationMap[dist].bone = bone->Child[i];
            assignationMap[dist].dist = dist;
            bone->Child[i]->SetVisited(true);

            AssignRecur(vertex, bone->Child[i], threshold,
                assignationMap);
        }
    }

    float dist = distance(bone->Parent(), vertex);
    if (!bone->Parent()->Visited() && dist < threshold)
    {
        assignationMap[dist].bone = bone->Parent();
        assignationMap[dist].dist = dist;
        bone->Parent()->SetVisited(true);

        AssignRecur(vertex, bone->Parent(), threshold, assignationMap);
    }
}
```

Listing 17.1. Implementation of the weight assignment algorithm.

Skinning

At this point in the process, we now know which bone affects a particular vertex and to what extent it affects it. The only thing left to do is to actually grab the bones' matrices and apply them to our mesh's vertices in order to transform the mesh and animate it. The act of deforming a mesh to fit on a given skeleton's animation is called *skinning*. Multiple skinning techniques exist in the industry, the most popular being linear-blend skinning [Kavan and Žára 2003] and spherical-blend skinning [Kavan and Žára 2005]. Both of the techniques have been implemented with a programmable vertex shader in the demo code on the website, but only the linear-blend skinning technique is explained in this section. Spherical-blend skinning requires some more advanced mathematics and could be the subject of a whole gem by itself. However, keep in mind that if you can afford it, spherical-blend skinning often provides better visual quality than does linear-blend skinning. Again, also note that if you are already operating in a 3D development environment, skinning techniques are almost certainly already available and reusing them is preferable, as stated in our list of desired placeholder features.

Linear-blend skinning is a very simple and widely popular technique that has been in use since the Jurassic era of computer animation. While it has some visible rendering artifacts, it has proven to be a very efficient and robust technique that adapts very well to a wide array of graphics hardware. The details given here apply to programmable hardware but can be easily adapted for nonprogrammable GPUs where the same work can be entirely performed on the CPU.

The idea behind linear-blend skinning is to linearly blend the transformation matrices. This amounts to multiplying every bone's matrix by its weight for a given vertex and then summing the multiplied matrices together. The whole process can be expressed with the equation

$$\mathbf{v}_i' = \sum_j w_{ij} M_j \mathbf{v}_j, \tag{17.8}$$

where $\mathbf{v}_i$ is the i-th untransformed vertex of the mesh in its bind pose, $\mathbf{v}_i'$ is the transformed vertex after it has been skinned to the skeleton, M_j is the transformation matrix of the j-th bone, and w_{ij} is the weight of bone j when applied to vertex i. (In the case where only the n closest bones are kept, you can view all the w_{ij} as being set to zero except for the n closest ones.)

Implementing linear-blend skinning on programmable rendering hardware remains equally straightforward and can be completely done in the vertex shader stage of the pipeline. Before looking at the code, and to ensure that the imple-

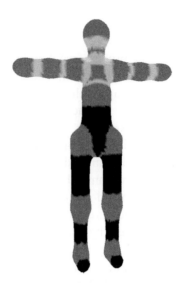

Figure 17.6. Bone count per vertex rarely goes above four. The bone count is depicted here with a color map on a humanoid mesh. The brighter spots near the neck and shoulders have four bone influences, the dark spots on the legs and the abdomen have only a single bone influence. Colors in between have two or three bone influences.

mentation remains simple, the number of bones that can affect a single vertex has to be limited to four. (The bone count affecting a vertex rarely goes higher than four in practice, as shown in Figure 17.6.)

The standard input values for the linear-blend skinning vertex shader are the vertex position (attribute), the model-view-projection matrix (uniform), and the bone transformation matrices (uniforms). (The normal vector and texture coordinates aren't required if you only want to perform linear-blend skinning.) Three more inputs dedicated to skinning have to be added to the shader. The first one is a uniform array of 4×4 matrices that contains the transformation matrix for each of the animation skeleton's bones (the M_j matrices) at the animated pose to be rendered. The size of the array determines how many bones the animation skeleton can have (40 is usually a good number). The second input is a four-dimensional integer vector attribute that encodes the IDs of the four closest bones for each vertex. If the influencing bone count is lower than four, the supplementary matrix identifiers and weights can be set to zero, which ensures they have no influence on the final vertex position. The last input is a four-dimensional floating-point vector attribute storing the weight values for the four closest bones. The weights stored in this vector must be in the same order as the corresponding bone

```glsl
// Input attributes
attribute vec4    vertex;
attribute vec3    normal;
attribute vec4    weight;
attribute vec4    boneId;

// Skeleton bones transformation matrix
uniform mat4      bonesMatrices[40];

void main(void)
{
    vec4 Position = vertex;

    vec4 DefV = vec4(0,0,0,0);
    vec4 DefN = vec4(0,0,0,0);

    Position.w = 1.0;

    for (int i = 0; i < 4; ++i)
    {
        mat4 BoneMat = bonesMatrices[int(boneId[i])];

        DefV += BoneMat * Position * weight[i];
        DefN += BoneMat * vec4(normal,0) * weight[i];

        gl_Position = gl_ModelViewProjectionMatrix * DefV;
    }
}
```

Listing 17.2. GLSL implementation of the linear blend skinning vertex shader.

IDs in the previous vector. With this information, the linear-blend skinning given by Equation (17.8) can be directly implemented, giving us the skinning shader shown in Listing 17.2 (with an added calculation for the normal vector).

Skinning now integrated, the mesh can be loaded in your game, the animations can be run, and you should now be able to see what looks like an articulated balloon running in your scene. The completed technique can be seen in Figure 17.7. From this point on, it should be pretty obvious that the placeholder is not a final asset, other programmers are able to test their game logic by using

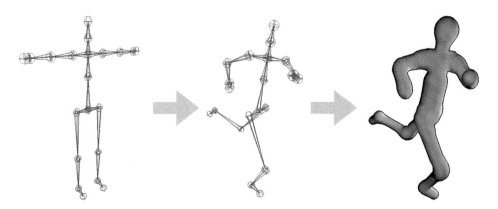

Figure 17.7. Articulated placeholder skinned to an animation pose. The placeholder mesh is articulated and follows the animation.

existing animations, and they will most likely be able to accomplish relatively accurate performance tests by tweaking the amount of geometry and using different skinning methods on the placeholder. Most importantly, the addition of this new placeholder in your production environment should decouple the work of modelers, animators, and programmers since only a very basic skeleton is now required to obtain convenient and useful placeholder geometry.

Limitations

The method presented above should give results of sufficient quality for a placeholder type of asset. However, it has some shortcomings that should be underscored. The weakest part of the process, the automatic weight assignment, provides good results for a placeholder asset but in no way replaces the precise and detailed work a good animator or modeler would do. The problem becomes mostly apparent in regions with multiple bone influences (three or four), where subtle tearing can be seen in the mesh. The automatic weight assignment could be improved to reduce this artifact by using heat diffusion methods or geodesic distance methods on the mesh.

Another artifact that should be mentioned is the "ballooning" effect of the skeleton joints on the generated mesh. As can be noticed, round artifacts appear in the geometry around the skeleton's articulation. This is because each skeleton bone is considered individually, and their density fields add up near the articulations. This artifact can be attenuated by reducing the intensity of the density field near the extremities of a bone. Keep in mind, though, that some people tend to prefer this look since it clearly defines the presence of an articulation.

Finally, the generated mesh might appear a bit flat, with undefined features, if every bone has the same effect on the density field. A possible improvement to this is to add a simple "influence" property to every bone and manually increase the influence of some of them (such as the head of a humanoid character to make it appear rounder). This is, however, a purely aesthetic improvement and otherwise affects the system in no way.

17.3 Integration in a Production Environment

Placeholder systems are usually far down the priority list, or nonexistent, for a production or technology team unless very heavy prototyping is required for a game. Therefore, the gain of creating such a system often does not seem to be worth the risk of disrupting an already stable pipeline. In this regard, it is important to reduce the footprint of this new system by eliminating as many potential risks as possible, and this means reusing as much of the existing infrastructure as possible. This begs multiple questions: In which part of the resource pipeline should the placeholder be created? Who should be the person in charge of creating it? How can it be integrated into existing pipelines? These questions, unfortunately, do not have a single clear answer and mostly depend on your own production environment. We do, however, attempt to give some answers by taking our articulated placeholder model as an example and discussing how and where it should be integrated in an existing engine. Hopefully, this should give you an idea of the things to look for and the things to consider when integrating a placeholder system in a production environment.

Who Makes It, and Where Do You Make It?

One of the most natural questions to ask is, "Who makes the placeholder?" The answer is straightforward: "As many people as possible." What truly results depends on where you decide to integrate the placeholder into the assets pipeline and what the production rules are for your team or studio.

Taking the articulated placeholder as an example, if it is added too far into the asset pipeline (for example, if it is generated exclusively in code at run time based on an animation skeleton), then you limit the placeholder creation to the programmers. This obviously doesn't help in decoupling the work since the responsibility of creating placeholders falls on the shoulders of a small part of your team. The opposite case, generating the placeholder at the beginning (e.g., as a plug-in in the 3D modeling software), might be a good solution if everyone on your team has basic knowledge of these tools and knows how to generate the placeholder from them. However, some studios do not install these applications on game programmer or designer stations to ensure that they don't start producing artistic assets or simply to reduce operational costs. In this case, a plug-in for the 3D software might not be the best solution.

The common ground that seems to be the best compromise is to perform the generation directly in the game asset auditing tools (the game editor) or through a custom tool that falls between the game editor and the artistic production software. These tools are often available to everyone and created in-house by tool programmers, so training remains easy, and this ensures that you always have all the required support and control over the placeholder generation system. On the downside, you might not be able to reuse features that were available in other production applications. On the bright side though, this should give you access to the shared features of your game engine that are available both in-game and in the game tools.

Ideally, you should try to maximize the number of different people and departments that can generate the placeholder. If you have studio rules limiting access to some of your software, try to target a spot in the pipeline that opens it to the most people. In practice, the beginning of the asset pipeline or the game editor is usually the best place to ensure that. Finally, refrain from creating a purely code-oriented system. Programmers should not be the only team members to know how to use the system or benefit from it.

Game Pipeline Integration

We have seen the different options as to *where* a placeholder system can be integrated in the asset pipeline. However, we haven't seen *how* it can be integrated. What we suggest here is to treat the placeholder as a real asset as much as possible. This allows you to reuse all the available features of your existing asset pipeline, and it preserves homogeneity throughout your resource base. Creating special cases to manage placeholder assets differently only makes asset management more complicated and creates confusion during development.

In the case of our articulated placeholder, the best idea is to generate an articulated mesh out of the skeleton and output it with the animated mesh format you normally use for animated assets. Using this approach, the placeholder behaves and appears exactly as any other asset from an asset pipeline point of view, which greatly reduces the amount of work required to integrate it in the game.

Physics, rendering, and sound pipeline integration depends in great part on your asset pipeline integration. If your placeholder is generated with the same file formats as your standard assets, integrating it in game pipelines should require absolutely no cost. For the articulated placeholder, the system could also be extended to add support for your physics and sound pipelines. Collision primitive geometry could be generated from the skeleton and then exported to be used in the physics engine. As for the sound pipeline, if your system uses the same constructs and infrastructures as the usual mesh or collision geometry, sound events and collision sounds should be easy to add with the tools your audio designer already uses.

All in all, the idea remains the same: integrate your placeholder in the most transparent fashion in your game engine and reuse as much of it as possible to reduce the time required to integrate the system in your production environment.

17.4 In the End, Is It Really Needed?

The question probably seems obvious, and you might wonder why it is asked so late in this chapter. The reason is simply because it's hard to answer it without having in mind the possibilities and advantages provided by a placeholder system. The true answer is that unless you have a perfect team that never has a single delay and organizes things so well that no one ever waits for anyone's work, building placeholder systems where they are needed is definitely a plus and will positively impact your development speed for many productions to come. But yet, is it really always needed? Yes, but the complexity varies greatly and is often so low that you might not even consider it a placeholder "system."

The articulated placeholder system we present above is what we could consider complex, which is correct since animated models are relatively complex resources. For the case of simpler assets, such as text strings, you probably don't need to have something as complex. Generating a temporary string that displays standard information, such as the current language and string ID, is fine enough and truly does the trick. There is no need to start overthinking the whole system and begin building a huge *lorem ipsum* generator for such simple assets. As long as the placeholder you devise follows the guidelines mentioned earlier, you can consider it efficient. It remains important to keep in mind that adding a new placeholder system takes some time and that this overhead should be considered when deciding if it is really needed or not. Moreover, keep in mind that a complex system isn't always the way to go, and simpler placeholders often do the trick.

Placeholders will not revolutionize the way you build games, and many people on your team might never realize someone had to actually build the placeholder systems they're using. The positive effect of these systems is more subtle and only appears in the long run when you look at the big picture. The challenges of today's game industry are all about efficiency, creating more in less time, with fewer people, and at lower costs. Thinking about how development can be rendered more efficient plays a big part in a team's success, and every developer has to do his share of work on this front. Study your everyday routine, identify your production bottlenecks, and verify whether a simple placeholder system could alleviate the pressure. They will not make their way into the final build, but these placeholders will definitely show their true value before that.

17.5 Implementation

The code provided on the website allows you to compile and use the articulated placeholder system described in this chapter. A Visual Studio solution with the required library and header files is located on the website, and it contains everything necessary to compile the project.

References

[Aguiar et al. 2008] Edilson de Aguiar, Christian Theobalt, Sebastian Thrun, and Hans-Peter Seidel. "Automatic Conversion of Mesh Animations into Skeleton-Based Animations." *Computer Graphics Forum* 27:2 (April 2008), pp. 389–397.

[Baran and Popović 2007] Ilya Baran and Jovan Popović. "Automatic Rigging and Animation of 3D Characters." *ACM Transactions on Graphics* 26:3 (July 2007).

[Doi and Koide 1991] Akio Doi and Akio Koide. "An Efficient Method of Triangulating Equi-Valued Surface by Using Tetrahedral Cells." *IEICE Transactions* E74:1 (January 1991), pp. 214–224.

[Kavan and Žára 2003] Ladislav Kavan and Jiří Žára. "Real Time Skin Deformation with Bones Blending." *WSCG Short Papers Proceedings*, 2003.

[Kavan and Žára 2005] Ladislav Kavan and Jiří Žára. "Spherical Blend Skinning: A Real-time Deformation of Articulated Models." *Proceedings of the 2005 Symposium on Interactive 3D Graphics and Games* 1 (2005), pp. 9–17.

[Lally 2003] John Lally. "Giving Life to Ratchet & Clank." *Gamasutra*. February 11, 2003. Available at http://www.gamasutra.com/view/feature/2899/giving_life_to_ratchet__clank_.php.

[Lorensen and Cline 1987] William E. Lorensen and Harvey E. Cline. "Marching Cubes: A High Resolution 3D Surface Construction Algorithm." *Computer Graphics (Proceedings of SIGGRAPH 87)* 21:4, ACM, pp. 163–169.

[Müller and Wehle 1997] Heinrich Müller and Michael Wehle. "Visualization of Implicit Surfaces Using Adaptive Tetrahedrizations." *Dagstuhl '97 Proceedings of the Conference on Scientific Visualization*, pp. 243–250.

[Rosen 2009] David Rosen. "Volumetric Heat Diffusion Skinning." *Gamasutra* Blogs, November 24, 2009. Available at http://www.gamasutra.com/blogs/DavidRosen/20091124/3642/Volumetric_Heat_Diffusion_Skinning.php.

[Speyrer and Jacobson 2006] David Speyrer and Brian Jacobson. "Valve's Design Process for Creating Half-Life 2." *Game Developers Conference*, 2006.

[Williams 2010] David Williams. "Volumetric Representation of Virtual Environments." *Game Engine Gems 1*, edited by Eric Lengyel. Sudbury, MA: Jones and Bartlett, 2010.

18

Believable Dead Reckoning for Networked Games

Curtiss Murphy

Alion Science and Technology

18.1 Introduction

Your team's producer decides that it's time to release a networked game, saying "We can publish across a network, right?" Bob kicks off a few internet searches and replies, "Doesn't look that hard." He dives into the code, and before long, Bob is ready to begin testing. Then, he stares in bewilderment as the characters jerk and warp across the screen and the vehicles hop, bounce, and sink into the ground. Thus begins the nightmare that will be the next few months of Bob's life, as he attempts to implement dead reckoning "just one more tweak" at a time.

This gem describes everything needed to add believable, stable, and efficient dead reckoning to a networked game. It covers the fundamental theory, compares algorithms, and makes a case for a new technique. It explains what's tricky about dead reckoning, addresses common myths, and provides a clear implementation path. The topics are demonstrated with a working networked game that includes source code. This gem will help you avoid countless struggles and dead ends so that you don't end up like Bob.

18.2 Fundamentals

Bob isn't a bad developer; he just made some reasonable, but misguided, assumptions. After all, the basic concept is pretty straight forward. *Dead reckoning* is the process of predicting where an actor is right now by using its last known position, velocity, and acceleration. It applies to almost any type of moving actor, including cars, missiles, monsters, helicopters, and characters on foot. For each

remote actor being controlled somewhere else on the network, we receive updates about its kinematic state that include its position, velocity, acceleration,
orientation, and angular velocity. In the simplest implementation, we take the last
position we received on the network and project it forward in time. Then, on the
next update, we do some sort of blending and start the process all over again.
Bob is right that the fundamentals aren't that complex, but making it believable is
a different story.

Myth Busting—Ground Truth

Let's start with the following fact: there is no such thing as ground truth in a
networked environment. "Ground truth" implies that you have perfect
knowledge of the state of all actors at all times. Surely, you can't know the
exact state of all remote actors without sending updates every frame in a zero
packet loss, zero latency environment. What you have instead is your own
perceived truth. Thus, the goal becomes believable estimation, as opposed to
perfect re-creation.

Basic Math

To derive the math, we start with the simplest case: a new actor comes across the
network. In this case, one of our opponents is driving a tank, and we received our
first kinematic state update as it came into view. From here, dead reckoning is a
straightforward linear physics problem, as described by Aronson [1997]. Using
the values from the message, we put the vehicle at position $\mathbf{P}'_0$ and begin moving
it at velocity $\mathbf{V}'_0$ with acceleration $\mathbf{A}'_0$, as shown in Figure 18.1. The dead-
reckoned position $\mathbf{Q}_t$ at a specific time T is calculated with the equation

$$\mathbf{Q}_t = \mathbf{P}'_0 + \mathbf{V}'_0 T + \frac{1}{2}\mathbf{A}'_0 T^2.$$

Figure 18.1. The first update is simple.

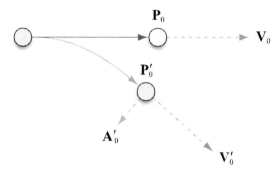

Figure 18.2. The next update creates two realities. The red line is the estimated path, and the green curve is the actual path.

Continuing our scenario, the opponent saw us, slowed his tank, and took a hard right. Soon, we receive a message updating his kinematic state. At this point, we have conflicting realities. The first reality is the position Q_t where we guessed he would be using the previous formula. The second reality is where he actually went, our new $\mathbf{P}'_0$, which we refer to as the *last known* state because it's the last thing we know to be correct. This dual state is the beginning of Bob's nightmares. Since there are two versions of each value, we use the prime notation (e.g., $\mathbf{P}'_0$) to indicate the last known, as shown in Figure 18.2.

To resolve the two realities, we need to create a believable curve between where we thought the tank would be, and where we estimate it will be in the future. Don't bother to path the remote tank through its last known position, $\mathbf{P}'_0$. Instead, just move it from where it is now, $\mathbf{P}_0$, to where we think it is supposed to be in the future, $\mathbf{P}'_1$.

Myth Busting—Discontinuities Are Not Minor

The human brain is amazing at recognizing patterns [Koster 2005] and, more importantly, changes in patterns, such as when the tiniest piece of fuzz moves past our peripheral vision. What this means is that players will notice subtle discontinuities in a vehicle path long before they realize the vehicle is in the wrong location. Therefore, discontinuities such as hops, warps, wobbles, and shimmies are the enemy.

18.3 Pick an Algorithm, Any Algorithm

If you crack open any good 3D math textbook, you'll find a variety of algorithms for defining a curve. Fortunately, we can discard most of them right away because they are too CPU intensive or are not appropriate (e.g., B-splines do not pass through the control points). For dead reckoning, we have the additional requirement that the algorithm must work well for a single segment of a curve passing through two points: our current location $\mathbf{P}_0$ and the estimated future location $\mathbf{P}'_1$. Given all these requirements, we can narrow the selection down to a few types of curves: cubic Bézier splines, Catmull-Rom splines, and Hermite curves [Lengyel 2004, Van Verth and Bishop 2008].

These curves perform pretty well and follow smooth, continuous paths. However, they also tend to create minor repetitive oscillations. The oscillations are relatively small, but noticeable, especially when the actor is making a lot of changes (e.g., moving in a circle). In addition, the oscillations tend to become worse when running at inconsistent frame rates or when network updates don't come at regular intervals. In short, they are too wiggly.

Projective Velocity Blending

Let's try a different approach. Our basic problem is that we need to resolve two realities (the current $\mathbf{P}_0$ and the last known $\mathbf{P}'_0$). Instead of creating a spline segment, let's try a straightforward blend. We create two projections, one with the current and one with the last known kinematic state. Then, we simply blend the two together using a standard linear interpolation (lerp). The first attempt looks like this:

$$\mathbf{P}_t = \mathbf{P}_0 + \mathbf{V}_0 T_t + \frac{1}{2}\mathbf{A}'_0 T_t^2 \qquad \text{(projecting from where we were)},$$

$$\mathbf{P}'_t = \mathbf{P}'_0 + \mathbf{V}'_0 T_t + \frac{1}{2}\mathbf{A}'_0 T_t^2 \qquad \text{(projecting from last known)},$$

$$\mathbf{Q}_t = \mathbf{P}_t + (\mathbf{P}'_t - \mathbf{P}_t)\hat{T} \qquad \text{(combined)}.$$

This gives $\mathbf{Q}_t$, the dead-reckoned location at a specified time. (Time values such as T_t and $\hat{T}$ are explained in Section 18.4.) Note that both projection equations above use the last known value of acceleration $\mathbf{A}'_0$. In theory, the current projection $\mathbf{P}_t$ should use the previous acceleration $\mathbf{A}_0$ to maintain C^2 continuity. However, in practice, $\mathbf{A}'_0$ converges to the true path much quicker and reduces oscillation.

This technique actually works pretty well. It is simple and gives a nice curve between our points. Unfortunately, it has oscillations that are as bad as or worse than the spline techniques. Upon inspection, it turns out that with all of these techniques, the oscillations are caused by the changes in velocity ($\mathbf{V}_0$ and $\mathbf{V}_0'$). Maybe if we do something with the velocity, we can reduce the oscillations. So, let's try it again, with a tweak. This time, we compute a linear interpolation between the old velocity $\mathbf{V}_0$ and the last known velocity $\mathbf{V}_0'$ to create a new blended velocity $\mathbf{V}_b$. Then, we use this to project forward from where we were.

The technique, *projective velocity blending*, works like this:

$$\mathbf{V}_b = \mathbf{V}_0 + (\mathbf{V}_0' - \mathbf{V}_0)\hat{T} \qquad \text{(velocity blending)},$$

$$\mathbf{P}_t = \mathbf{P}_0 + \mathbf{V}_b T_t + \frac{1}{2}\mathbf{A}_0' T_t^2 \qquad \text{(projecting from where we were)},$$

$$\mathbf{P}_t' = \mathbf{P}_0' + \mathbf{V}_0' T_t + \frac{1}{2}\mathbf{A}_0' T_t^2 \qquad \text{(projecting from last known)},$$

$$\mathbf{Q}_t = \mathbf{P}_t + (\mathbf{P}_t' - \mathbf{P}_t)\hat{T} \qquad \text{(combined)}.$$

And the red lines in Figure 18.3 show what it looks like in action.

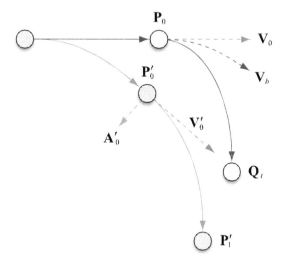

Figure 18.3. Dead reckoning with projective velocity blending shown in red.

In practice, this works out magnificently! The blended velocity and change of acceleration significantly reduce the oscillations. In addition, this technique is the most forgiving of both inconsistent network update rates and changes in frame rates.

Prove It!

So it sounds good in theory, but let's get some proof. We can perform a basic test by driving a vehicle in a repeatable pattern (e.g., a circle). By subtracting the real location from the dead-reckoned location, we can determine the error. The images in Figure 18.4 and statistics in Table 18.1 show the clear result. The projective velocity blending is roughly five to seven percent more accurate than cubic Bézier splines. That ratio improves a bit more when you can't publish acceleration. If you want to test it yourself, the demo application on the website has implementations of both projective velocity blending and cubic Bézier splines.

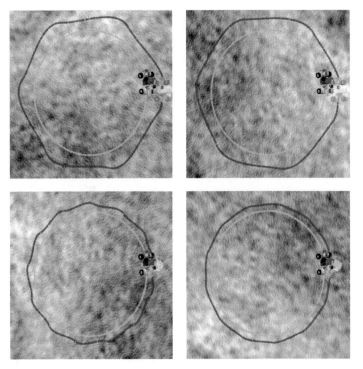

Figure 18.4. Cubic Bézier splines (left) versus projective velocity blending (right), with acceleration (top) and without acceleration (bottom).

Update Rate	Cubic Bézier (DR Error)	Projective Velocity (DR Error)	Improvement
1 update/sec	1.5723 m	1.4584 m	7.24% closer
3 updates/sec	0.1041 m	0.1112 m	6.38% closer
5 updates/sec	0.0574 m	0.0542 m	5.57% closer

Table 18.1. Improvement using projective velocity blending. Deck-reckoning (DR) error is measured in meters.

As a final note, if you decide to implement a spline behavior instead of projective velocity blending, you might consider the cubic Bézier splines [Van Verth and Bishop 2008]. They are slightly easier to implement because the control points can simply be derived from the velocities V_0 and V_0'. The source code on the website includes a full implementation.

18.4 Time for *T*

So far, we've glossed over time. That's okay for an introduction, but, once you begin coding, the concept of time gets twisted up in knots. So, let's talk about T.

What Time Is It?

The goal is to construct a smooth path that an actor can follow between two moments in time T_0 and T_1. These two times mark the exact beginning and end of the curve and are defined by locations P_0 and P_1', respectively. The third time T_t is how much time has elapsed since T_0. The final time $\hat{T}$ represents how far the actor has traveled along the path as a normalized value, with $0.0 \leq \hat{T} \leq 1.0$.

T_0 is easy. It's the time stamp when the last known values were updated. Basically, it's "now" at the time of the update. If you've seen the movie *Spaceballs*, then T_0 is "now, now." When we process a new network update, we mark T_0 as now and set T_t back to zero. The slate is wiped clean, and we start a whole new curve, regardless of where we were.

If T_0 is now, then T_1 must be in the future. But how far into the future, T_Δ, should the projection go? Well, if we knew that the actor updates were coming at regular intervals, then we could just use the inverse update rate. So, for three updates per second, $T_\Delta = 0.333$ s. Even though network updates won't always be perfectly spaced out, it still gives a stable and consistent behavior. Naturally, the update rate varies significantly depending on the type of game, the network conditions, and the expected actor behavior. As a general rule of thumb, an update rate of three per second looks decent and five or more per second looks great.

Time to Put It Together

From an implementation perspective, normalized time values from zero to one aren't terribly useful. In many engines, you typically get a time T_f since the last frame. We can easily add this up each frame to give the total time since the last update T_t. Once we know T_t, we can compute our normalized time $\hat{T}$ as follows:

$$T_t \leftarrow T_t + T_f$$

$$\hat{T} = \frac{T_t}{T_\Delta}.$$

Now we have all the times we need to compute the projective velocity blending equations. That leaves just one final wrinkle in time. It happens when we go past T_Δ (i.e., $T_t > T_\Delta$). This is a very common case that can happen if we miss an update, have any bit of latency, or even have minor changes in frame rate. From earlier,

$$\mathbf{Q}_t = \mathbf{P}_t + (\mathbf{P}'_t - \mathbf{P}_t)\hat{T}.$$

Because $\hat{T}$ is clamped at one, the $\mathbf{P}_t$ drops out, leaving the original equation

$$\mathbf{Q}_t = \mathbf{P}'_0 + \mathbf{V}'_0 T_t + \frac{1}{2}\mathbf{A}'_0 T_t^2.$$

The math simplifies quite nicely and continues to work for any value of $\hat{T} \geq 1.0$.

Just in Time Notes

Here are a few tips to consider:

- Due to the nature of networking, you can receive updates at any time, early or late. In order to maintain C^1 continuity, you need to calculate the instantaneous velocity between this frame's and the last frame's dead-reckoned position, $(\mathbf{P}_t - \mathbf{P}_{t-1})/T_f$. When you get the next update and start the new curve, use this instantaneous velocity for $\mathbf{V}_0$. Without this, you will see noticeable changes in velocity at each update.

- Actors send updates at different times based on many factors, including creation time, behavior, server throttling, latency, and whether they are moving. Therefore, track the various times separately for each actor (local and remote).

- If deciding your publish rate in advance is problematic, you could calculate a run-time average of how often you have been receiving network updates and use that for T_Δ. This works okay but is less stable than a predetermined rate.

- In general, the location and orientation get updated at the same time. However, if they are published separately, you'll need separate time variables for each.

- It is possible to receive multiple updates in a single frame. In practice, let the last update win. For performance reasons, perform the dead reckoning calculations later in the game loop, after the network messages are processed. Ideally, you will run all the dead reckoning in a single component that can split the work across multiple worker threads.

- For most games, it is not necessary to use time stamps to sync the clocks between clients/servers in order to achieve believable dead reckoning.

18.5 Publish or Perish

So far, the focus has been on handling network updates for remote actors. However, as with most things, garbage in means garbage out. Therefore, we need to take a look at the publishing side of things. In this section, forget about the actors coming in over the network and instead focus on the locally controlled actors.

When to Publish?

Let's go back and consider the original tank scenario from the opponent's perspective. The tank is now a local actor and is responsible for publishing updates on the network. Since network bandwidth is a precious resource, we should reduce traffic if possible. So the first optimization is to decide *when* we need to publish. Naturally, there are times when players are making frequent changes in direction and speed and five or more updates per second are necessary. However, there are many more times when the player's path is stable and easy to predict. For instance, the tank might be lazily patrolling, might be heading back from a respawn, or even sitting still (e.g., the player is chatting).

The first optimization is to only publish when necessary. Earlier, we learned that it is better to have a constant publish rate (e.g., three per second) because it keeps the remote dead reckoning smooth. However, before blindly publishing every time it's allowed (e.g., every 0.333 s), we first check to see if it's neces-

```
bool ShouldForceUpdate(const Vec3& pos, const Vec3& rot)
{
    bool forceUpdateResult = false;
    if (enoughTimeHasPassed)
    {
        Vec3 posMoved = pos - mCurDeadReckoned_Pos;
        Vec3 rotTurned = rot - mCurDeadReckoned_Rot;

        if ((posMoved.length2() > mPosThreshold2) ||
            (rotTurned.length2() > mRotThreshold2))
        {
            // Rot.length2 is a fast approx (i.e., not a quaternion).
            forceUpdateResult = true;
        }

        // ... Can use other checks such as velocity and accel.
    }

    return (forceUpdateResult);
}
```

Listing 18.1. Publish—is an update necessary?

sary. To figure that out, we perform the dead reckoning as if the vehicle was re-mote. Then, we compare the real and the dead-reckoned states. If they differ by a set threshold, then we go ahead and publish. If the real position is still really close to the dead-reckoned position, then we hold off. Since the dead reckoning algorithm on the remote side already handles $T_t > T_\Delta$, it'll be fine if we don't up-date right away. This simple check, shown in Listing 18.1, can significantly re-duce network traffic.

What to Publish

Clearly, we need to publish each actor's kinematic state, which includes the posi-tion, velocity, acceleration, orientation, and angular velocity. But there are a few things to consider. The first, and least obvious, is the need to separate the actor's real location and orientation from its last known location and orientation. Hope-fully, your engine has an actor property system [Campbell 2006] that enables you to control which properties get published. If so, you need to be absolutely sure you never publish (or receive) the actual properties used to render location and orientation. If you do, the remote actors will get an update and render the last known values instead of the results of dead reckoning. It's an easy thing to overlook and results in a massive one-frame discontinuity (a.k.a. blip). Instead, create publishable properties for the last known values (i.e., location, velocity, acceleration, orientation, and angular velocity) that are distinct from the real values.

The second consideration is partial actor updates, messages that only contain a few actor properties. To obtain believable dead reckoning, the values in the kinematic state need to be published frequently. However, the rest of the actor's properties usually don't change that much, so the publishing code needs a way to swap between a partial and full update. Most of the time, we just send the kinematic properties. Then, as needed, we send other properties that have changed and periodically (e.g., every ten seconds) send out a heartbeat that contains everything. The heartbeat can help keep servers and clients in sync.

> ## Myth Busting—Acceleration Is Not Always Your Friend
>
> In the quest to create believable dead reckoning, acceleration can be a huge advantage, but be warned that some physics engines give inconsistent (a.k.a. spiky) readings for linear acceleration, especially when looked at in a single frame as an instantaneous value. Because acceleration is difficult to predict and is based on the square of time, it can sometimes make things worse by introducing noticeable under- and overcompensations. For example, this can be a problem with highly jointed vehicles for which the forces are competing on a frame-by-frame basis or with actors that intentionally bounce or vibrate.

With this in mind, the third consideration is determining what the last known values should be. The last known location and orientation come directly from the actor's current render values. However, if the velocity and acceleration values from the physics engine are giving bad results, try calculating an instantaneous velocity and acceleration instead. In extreme cases, try blending the velocity over two or three frames to average out some of the sharp instantaneous changes.

Publishing Tips

Below are some final tips for publishing:

- Published values can be quantized or compressed to reduce bandwidth [Sayood 2006].

- If an actor isn't stable at speeds near zero due to physics, consider publishing a zero velocity and/or acceleration instead. The projective velocity blend will resolve the small translation change anyway.

- If publishing regular heartbeats, be sure to sync them with the partial updates to keep the updates regular. Also, try staggering the heartbeat time by a random amount to prevent clumps of full updates caused by map loading.

- Some types of actors don't really move (e.g., a building or static light). Improve performance by using a static mode that simply teleports actors.

- In some games, the orientation might matter more than the location, or vice versa. Consider publishing them separately and at different rates.

- To reduce the bandwidth using `ShouldForceUpdate()`, you need to dead reckon the local actors in order to check against the threshold values.

- Evaluate the order of operations in the game loop to ensure published values are computed correctly. An example order might include: handle user input, tick local (process incoming messages and actor behaviors), tick remote (perform dead reckoning), publish dead reckoning, start physics (background for next frame), update cameras, render, finish physics. A bad order will cause all sorts of hard-to-debug dead reckoning anomalies.

- There is an optional damping technique that can help reduce oscillations when the acceleration is changing rapidly (e.g., zigzagging). Take the current and previous acceleration vectors and normalize them. Then, compute the dot product between them and treat it as a scalar to reduce the acceleration before publishing (shown in the `ComputeCurrentVelocity()` function in Listing 18.2).

- Acceleration in the up/down direction can sometimes cause floating or sinking. Consider publishing a zero instead.

The Whole Story

When all the pieces are put together, the code looks roughly like Listing 18.2.

```
void OnTickRemote(const TickMessage& tickMessage)
{
    // This is for local actors, but happens during Tick Remote.
    double elapsedTime = tickMessage.GetDeltaSimTime();
    bool forceUpdate = false, fullUpdate = false;

    Vec3 rot = GetRotation();
    Vec3 pos = GetTranslation();
```

```
        mSecsSinceLastUpdateSent += elapsedTime;
        mTimeUntilHeartBeat -= elapsedTime;

        // Have to update instant velocity even if we don't publish.
        ComputeCurrentVelocity(elapsedTime, pos, rot);

        if ((mTimeUntilHeartBeat <= 0.0F) || (IsFullUpdateNeeded()))
        {
            fullUpdate = true;
            forceUpdate = true;
        }
        else
        {
            forceUpdate = ShouldForceUpdate(pos, rot);
            fullUpdate = (mTimeUntilHeartBeat < HEARTBEAT_TIME * 0.1F);
        }

        if (forceUpdate)
        {
            SetLastKnownValuesBeforePublish(pos, rot);
            if (fullUpdate)
            {
                mTimeUntilHeartBeat = HEARTBEAT_TIME;  // +/- random offset
                NotifyFullActorUpdate();
            }
            else
            {
                NotifyPartialActorUpdate();
            }

            mSecsSinceLastUpdateSent = 0.0F;
        }
    }

    void SetLastKnownValuesBeforePublish(const Vec3& pos, const Vec3& rot)
    {
        SetLastKnownTranslation(pos);
        SetLastKnownRotation(rot);
        SetLastKnownVelocity(ClampTinyValues(GetCurrentVel()));
        SetLastKnownAngularVel(ClampTinyValues(GetCurrentAngularVel()));
```

```
    // (OPTIONAL!) ACCELERATION dampen to prevent wild swings.
    // Normalize current accel. Dot with accel from last update. Use
    // the product to scale our current Acceleration.
    Vec3 curAccel = GetCurrentAccel();
    curAccel.normalize();

    float accelScale = curAccel * mAccelOfLastPublish;
    mAccelOfLastPublish = curAccel;      // (pre-normalized)
    SetLastKnownAccel(GetCurrentAccel() * Max(0.0F, accelScale));
}

void ComputeCurrentVelocity(float deltaTime, const Vec3& pos,
                                   const Vec3& rot)
{
    if ((mPrevFrameTime > 0.0F) && (mLastPos.length2() > 0.0F))
    {
        Vec3 prevComputedLinearVel = mComputedLinearVel;
        Vec3 distanceMoved = pos - mLastPos;
        mComputedLinearVel = distanceMoved / mPrevFrameTime;
        ClampTinyValues(mComputedLinearVel);

        // accel = the instantaneous differential of the velocity.
        Vec3 deltaVel = mComputedLinearVel - prevComputedLinearVel;
        Vec3 computedAccel = deltaVel / mPrevDeltaFrameTime;
        computedAccel.z() = 0.0F;    // up/down accel isn't always helpful.

        SetCurrentAcceleration(computedAccel);
        SetCurrentVelocity(mComputedLinearVel);
    }
    mLastPos = pos;
    mPrevFrameTime = deltaTime;
}
```

Listing 18.2. Publish—the whole story.

18.6 Ground Clamping

No matter how awesome your dead reckoning algorithm becomes, at some point, the problem of ground clamping is going to come up. The easiest way to visualize the problem is to drop a vehicle off of a ledge. When it impacts the ground, the velocity is going to project the dead-reckoned position under the ground. Few things are as disconcerting as watching a tank disappear halfway into the dirt. As an example, the demo on the website allows mines to fall under ground.

Can We Fix It?

As with many dead reckoning problems, there isn't one perfect solution. However, some simple ground clamping can make a big difference, especially for far away actors. Ground clamping is adjusting an actor's vertical position and orientation to make it follow the ground. The most important thing to remember about ground clamping is that it happens *after* the rest of the dead reckoning. Do everything else first.

The following is one example of a ground clamping technique. Using the final dead reckoned position and orientation, pick three points on the bounding surface of the actor. Perform a ray cast starting above those points and directed downward. Then, for each point, check for hits and clamp the final point if appropriate. Compute the average height H of the final points $\mathbf{Q}_0$, $\mathbf{Q}_1$, and $\mathbf{Q}_2$, and compute the normal $\mathbf{N}$ of the triangle through those points as follows:

$$H = \frac{(\mathbf{Q}_0)_z + (\mathbf{Q}_1)_z + (\mathbf{Q}_2)_z}{3}$$

$$\mathbf{N} = (\mathbf{Q}_1 - \mathbf{Q}_0) \times (\mathbf{Q}_2 - \mathbf{Q}_0).$$

Use H as the final clamped ground height for the actor and use the normal to determine the final orientation. While not appropriate for all cases, this technique is fast and easy to implement, making it ideal for distant objects.

Other Considerations

- Another possible solution for this problem is to use the physics engine to prevent interpenetration. This has the benefit of avoiding surface penetration in all directions, but it can impact performance. It can also create new problems, such as warping the position, the need for additional blends, and sharp discontinuities.
- Another way to minimize ground penetration is to have local actors project their velocities and accelerations into the future before publishing. Then, damp the values as needed so that penetration will not occur on remote actors (a method known as predictive prevention). This simple trick can improve behavior in all directions and may eliminate the need to check for interpenetration.

- When working with lots of actors, consider adjusting the ground clamping based on distance to improve performance. You can replace the three-point ray multicast with a single point and adjust the height directly using the intersection normal for orientation. Further, you can clamp intermittently and use the offset from prior ground clamps.
- For character models, it is probably sufficient to use single-point ground clamping. Single-point clamping is faster, and you don't need to adjust the orientation.
- Consider supporting several ground clamp modes. For flying or underwater actors, there should be a "no clamping" mode. For vehicles that can jump, consider an "only clamp up" mode. The last mode, "always clamp to ground," would force the clamp both up and down.

18.7 Orientation

Orientation is a critical part of dead reckoning. Fortunately, the basics of orientation are similar to what was discussed for position. We still have two realities to resolve: the current drawn orientation and the last known orientation we just received. And, instead of velocity, there is angular velocity. But that's where the similarities end.

Hypothetically, orientation should have the same problems that location had. In reality, actors generally turn in simpler patterns than they move. Some actors turn slowly (e.g., cars) and others turn extremely quickly (e.g., characters). Either way, the turns are fairly simplistic, and oscillations are rarely a problem. This means C^1 and C^2 continuity is less important and explains why many engines don't bother with angular acceleration.

Myth Busting—Quaternions

Your engine might use HPR (heading, pitch, roll), *XYZ* vectors, or full rotation matrices to define an orientation. However, when it comes to dead reckoning, you'll be rotating and blending angles in three dimensions, and there is simply no getting around quaternions [Hanson 2006]. Fortunately, quaternions are easier to implement than they are to understand [Van Verth and Bishop 2008]. So, if your engine doesn't support them, do yourself a favor and code up a quaternion class. Make sure it has the ability to create a quaternion from an axis/angle pair and can perform spherical linear interpolations (slerp). A basic implementation of quaternions is provided with the demo code on the website.

```
Vec3 angVelAxis(mLastKnownAngularVelocityVector);

// normalize() returns length.
float angVelMagnitude = angVelAxis.normalize();

// Rotation around the axis is magnitude of ang vel * time.
float rotationAngle = angVelMagnitude * actualRotationTime;
Quat rotationFromAngVel(rotationAngle, angVelAxis);
```

Listing 18.3. Computing rotational change.

With this in mind, dead reckoning the orientation becomes pretty simple: project both realities and then blend between them. To project the orientation, we need to calculate the rotational change from the angular velocity. Angular velocity is just like linear velocity; it is the amount of change per unit time and is usually represented as an axis of rotation whose magnitude corresponds to the rate of rotation about that axis. It typically comes from the physics engine, but it can be calculated by dividing the change in orientation by time. In either case, once you have the angular velocity vector, the rotational change $R'_{\Delta t}$ is computed as shown in Listing 18.3.

If you also have angular acceleration, just add it to `rotationAngle`. Next, compute the two projections and blend using a spherical linear interpolation. Use the last known angular velocity in both projections, just as the last known acceleration was used for both equations in the projective velocity blending technique:

$$R'_{\Delta t} = \text{quat}\left(R'_{\text{mag}}T_t, R'_{\text{dir}}\right) \qquad \text{(impact of angular velocity)},$$

$$R_t = R'_{\Delta t}R_0 \qquad \text{(rotated from where we were)},$$

$$R'_t = R'_{\Delta t}R'_0 \qquad \text{(rotated from last known)},$$

$$S_t = \text{slerp}\left(\hat{T}, R_t, R'_t\right) \qquad \text{(combined)}.$$

This holds true for $\hat{T} < 1.0$. Once again, $\hat{T}$ is clamped at one, so the math simplifies when $\hat{T} \geq 1.0$:

$$S_t = R'_{\Delta t}R'_0 \qquad \text{(rotated from last known)}.$$

Two Wrong Turns Don't Make a Right

This technique may not be sufficient for some types of actors. For example, the orientation of a car and its direction of movement are directly linked. Unfortunately, the dead-reckoned version is just an approximation with two sources of error. The first is that the orientation is obviously a blended approximation that will be behind and slightly off. But, even if you had a perfect orientation, the remote vehicle is following a dead-reckoned path that is already an approximation. Hopefully, you can publish fast enough that neither of these becomes a problem. If not, you may need some custom actor logic that can reverse engineer the orientation from the dead-reckoned values; that is, estimate an orientation that would make sense given the dead-reckoned velocity. Another possible trick is to publish multiple points along your vehicle (e.g., one at front and one in back). Then, dead reckon the points and use them to orient the vehicle (e.g., bind the points to a joint).

18.8 Advanced Topics

This last section introduces a variety of advanced topics that impact dead reckoning. The details of these topics are generally outside the scope of this gem, but, in each case, there are specific considerations that are relevant to dead reckoning.

Integrating Physics with Dead Reckoning

Some engines use physics for both the local and the remote objects. The idea is to improve believability by re-creating the physics for remote actors, either as a replacement for or in addition to the dead reckoning. There are even a few techniques that take this a step further by allowing clients to take ownership of actors so that the remote actors become local actors, and vice versa [Feidler 2009]. In either of these cases, combining physics with dead reckoning gets pretty complex. However, the take away is that even with great physics, you'll end up with cases where the two kinematic states don't perfectly match. At that point, use the techniques in this gem to resolve the two realities.

Server Validation

Dead reckoning can be very useful for server validation of client behavior. The server should always maintain a dead-reckoned state for each player or actor. With each update from the clients, the server can use the previous last known state, the current last known state, and the ongoing results of dead reckoning as input for its validation check. Compare those values against the actor's expected behavior to help identify cheaters.

Who Hit Who?

Imagine player A (local) shoots a pistol at player B (remote, slow update). If implemented poorly, player A has to "lead" the shot ahead or behind player B based on the ping time to the server. A good dead reckoning algorithm can really help here. As an example, client A can use the current dead-reckoned location to determine that player B was hit and then send a hit request over to the server. In turn, the server can use the dead-reckoned information for both players, along with ping times, to validate that client A's hit request is valid from client A's perspective. This technique can be combined with server validation to prevent abuse. For player A, the game feels responsive, seems dependent on skill, and plays well regardless of server lag.

Articulations

Complex actors often have *articulations*, which are attached objects that have their own independent range of motion and rotation. Articulations can generally be lumped into one of two groups: real or fake. Real articulations are objects whose state has significant meaning, such as the turret that's pointing directly at you! For real articulations, use the same techniques as if it were a full actor. Fortunately, many articulations, such as turrets, can only rotate, which removes the overhead of positional blending and ground clamping. Fake articulations are things like tires and steering wheels, where the state is either less precise or changes to match the dead-reckoned state. For those, you may need to implement custom behaviors, such as for turning the front tires to approximate the velocity calculated by the dead reckoning.

Path-Based Dead Reckoning

Some actors just need to follow a specified path, such as a road, a predefined route, or the results of an artificial intelligence plan. In essence, this is not much different from the techniques described above. Except, instead of curving between two points, the actor is moving between the beginning and end of a specified path. If the client knows how to recreate the path, then the actor just needs to publish how far along the path it is, $\hat{T}$, as well as how fast time is changing, T_v. When applicable, this technique can significantly reduce bandwidth. Moyer and Speicher [2005] have a detailed exploration of this topic.

Delayed Dead Reckoning

The first myth this gem addresses is that there is no ground truth. However, one technique, delayed dead reckoning, can nearly re-create it, albeit by working in the past. With delayed dead reckoning, the client buffers network updates until it has enough future data to re-create a path. This eliminates the need to project into the future because the future has already arrived. It simplifies to a basic curve problem. The upside is that actors can almost perfectly re-create the original path. The obvious downside is that everything is late, making it a poor choice for most real-time actors. This technique can be useful when interactive response time is not the critical factor, such as with distant objects (e.g., missiles), slow-moving system actors (e.g., merchant NPCs), or when playing back a recording. Note that delayed dead reckoning can also be useful for articulations.

Subscription Zones

Online games that support thousands of actors sometimes use a subscription-zoning technique to reduce rendering time, network traffic, and CPU load [Cado 2007]. Zoning is quite complex but has several impacts on dead reckoning. One significant difference is the addition of dead reckoning modes that swap between simpler or more complex dead reckoning algorithms. Actors that are far away or unimportant can use a low-priority mode with infrequent updates, minimized ground clamping, quantized data, or simpler math and may take advantage of delayed dead reckoning. The high-priority actors are the only ones doing frequent updates, articulations, and projective velocity blending. Clients are still responsible for publishing normally, but the server needs to be aware of which clients are receiving what information for which modes and publish data accordingly.

18.9 Conclusion

Dead reckoning becomes a major consideration the moment your game becomes networked. Unfortunately, there is no one-size-fits-all technique. The games industry is incredibly diverse and the needs of a first-person MMO, a top-down RPG, and a high-speed racing game are all different. Even within a single game, different types of actors might require different techniques.

The underlying concepts described in this gem should provide a solid foundation for adding dead reckoning to your own game regardless of the genre. Even so, dead reckoning is full of traps and can be difficult to debug. Errors can occur anywhere, including the basic math, the publishing process, the data sent over the network, or plain old latency, lag, and packet issues. Many times, there are multiple problems going on at once and they can come from unexpected places, such as bad values coming from the physics engine or uninitialized variables. When you get stuck, refer back to the tips in each section and avoid making assumptions about what is and is not working. Believable dead reckoning is tricky to achieve, but the techniques in this gem will help make the process as easy as it can be.

Acknowledgements

Special thanks to David Guthrie for all of his contributions.

References

[Aronson 1997] Jesse Aronson. "Dead Reckoning: Latency Hiding for Networked Games." *Gamasutra*, September 19, 1997. Available at http://www.gamasutra.com/view/feature/3230/dead_reckoning_latency_hiding_for_.php.

[Cado 2007] Olivier Cado. "Propagation of Visual Entity Properties Under Bandwidth Constraints." *Gamasutra*, May 24, 2007. Available at http://www.gamasutra.com/view/feature/1421/propagation_of_visual_entity_.php.

[Campbell 2006] Matt Campbell and Curtiss Murphy. "Exposing Actor Properties Using Nonintrusive Proxies." *Game Programming Gems 6*, edited by Michael Dickheiser. Boston: Charles River Media, 2006.

[Feidler 2009] Glenn Fiedler. "Drop in COOP for Open World Games." *Game Developer's Conference*, 2009.

[Hanson 2006] Andrew Hanson. *Visualizing Quaternions*. San Francisco: Morgan Kaufmann, 2006.

[Koster 2005] Raph Koster. *A Theory of Fun for Game Design*. Paraglyph Press, 2005.

[Lengyel 2004] Eric Lengyel. *Mathematics for 3D Game Programming & Computer Graphics*, Second Edition. Hingham, MA: Charles River Media, 2004.

[Moyer and Speicher 2005] Dale Moyer and Dan Speicher. "A Road-Based Algorithm for Dead Reckoning." *Interservice/Industry Training, Simulation, and Education Conference*, 2005.

[Sayood 2006] Khalid Sayood. *Introduction to Data Compression*, Third Edition. San Francisco: Morgan Kaufmann, 2006.

[Van Verth and Bishop 2008] James Van Verth and Lars Bishop. *Essential Mathematics in Games and Interactive Applications: A Programmer's Guide*, Second Edition. San Francisco: Morgan Kaufmann, 2008.

19

An Egocentric Motion Management System

Michael Ramsey

Ramsey Research, LLC

Between the motion
And the act
Falls the shadow.
—T. S. Eliot

The egocentric motion management system (ECMMS) is both a model for agent movement and an application of a behavioral theory. Any game that features agents (e.g., animals, soldiers, or tanks) that move around in a 3D scene has a need for an agent movement solution. A typical movement solution provides mechanisms that allow for an agent to move through a scene, avoiding geometry, all the while executing some sort of behavior.

This article discusses not only how focusing on the agent drives the immediate interactions with the environment but also, more importantly, that by gathering some information about the environment during locomotion, we gain the ability to generate spatial semantics for use by the agent's behavior system. Portions of the ECMMS were used in a cross-platform game entitled *World of Zoo* (WOZ), shown in Figure 19.1. WOZ is an animal simulator that requires various zoo animals to move through their environments in an incredibly compelling manner while the players constantly alter the environment. So the proving ground for this system was in an environment that could be changed around the agents at any particular moment.

Figure 19.1. Screenshots from *World of Zoo*.

Convincing behavior is not a one-way street—what the agent does is just as important as what is perceived by the user. The immediate question that comes to mind is how we control the perception of an agent in a scene in such a way that facilitates *perceived intent*. Burmedez [2007] delineates between simple mindreading and perceptual mindreading. Simple mindreading is fundamentally behavioral coordination—this gets us nowhere, as we are obviously not going to expect the user to mimic the agent's behavior in order to understand it. However, perceptual mindreading is slightly different in that the focus is on the perceptual states of others, and we accomplish this by providing mechanisms to inform the user of not only the agent's intended state but its eventual state (i.e., goal-directed behavior, also known as propositional attitudes). This is critical because humans have a criterion for understanding behavior. If they witness the attribution of desire to belief, then it is likely that the behavior is justifiable. It's simply not enough to represent the states of a behavior—humans need to understand how they fit together.

One of the higher-order goals when designing a motion management system is that we ideally would like to observe an agent in a scene responding to stimuli in an appropriate manner. What is appropriate is open to interpretation, but what is not open for interpretation is the desire that *we* perceive the agent acting in a purposeful manner. To help facilitate this, we need to understand how physics, animation, and artificial intelligence are interwoven into a shadowy substance to imbue our agents with these characteristics. As such, this chapter focuses on the components that form the system and its resultant end product, but the takeaway should be about the process through which these results were obtained.

19.1 Fundamental Components of the ECMMS

The ECMMS provides a mechanism for agents to plan and interact with the environment in a manner that is focused not only on the typical utilitarian tasks of locomotion but also allows for the intent of the agent's behavior to be more believable by providing specific environmental cues to the behavioral system. What we ideally would like to have is a system that allows for different types of agents to interact with the environment based upon how *they* interpret the scene, and a major component of understanding the environment comes from the acknowledgment and the use of relative spatial orientations. To help accomplish this, the ECMMS is composed of several elements that allow an agent to receive spatial information about its environment: they are collision sensors and the query space. Every model has, at its core, a fundamental, atomic component that is essentially the enabler on which the higher-order systems are built. The ECMMS is no different—the collision sensor is the element that allows for information to be received from the environment.

19.2 Collision Sensors

As shown in Figure 19.2, the collision sensor is a simple primitive that acts as a callback mechanism that is triggered when specific collisions are registered with it. As we mention below, the individual pieces of collision geometry are assigned to a collision layer. The significance of the collision layer is that it allows for an agent to detect surfaces that may be of interest to it—behavioral interest. Gibson [1986] defines an affordance as "what it offers an animal, what it provides," where "it" is an object. This important quote can be attributed to our system because it means that just because our collision sensor receives a callback about an overlap with some object, it doesn't mean we need to interact with it, just that it's available for possible interaction. What helps us determine whether we want to

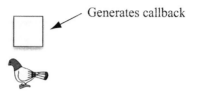

Generates callback

Figure 19.2. A top view of a bird with one collision sensor. When an object overlaps with the collision sensor, a callback is invoked to register the object with the agent.

interact with the object is the following: the current behavior, the available space for an action, and the *ability* of the agent to interact with the object. We discuss this later on, but note that this system is predicated on a very fundamental dictum, and agents need to influence behavior in an egocentric manner. Behaviors need to be understood from the perspective of the agent—what may be usable by an elephant may have an entirely different use or purpose for a mouse.

19.3 Query Space

The query space is a construct that is used to group collision sensors into sets that are not only organizationally consistent for your game's behavioral system but are also used to generate the semantics for the situation. The collision sensors allow for the generation of a syntax for the query space, and then the query space allows for the attribution of semantics to a situation.

The collision sensors can be authored and grouped in a query space in almost any manner. In an early version of WOZ, the collision sensors were procedurally constructed around the animals, but we settled upon authoring the collision sensor cubes inside 3DS Max. Figure 19.3 shows a typical query space configuration around an agent (a bird) that moves on the ground and occasionally jumps. When designing a query space for your entity, you should be aware of not only the desired behavior of your agent but, more importantly, the behavioral constraints that the agent potentially exhibits. These constraints are important because if collision sensors are authored only near an agent, then that agent is unable to pick up information that is spatially distal.

Figure 19.4 shows a general design for a query space that is usable for a variety of agents. Typically, a query space is composed of an immediate ring of collision sensors surrounded by an outer ring of collisions sensors. This allows for the

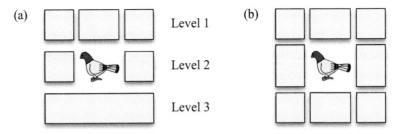

Figure 19.3. (a) A side view of a query space with three levels of collision sensors. (b) The same query space viewed from directly above.

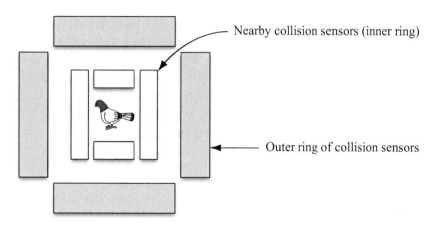

Figure 19.4. A query space template.

agent to not only receive callbacks due to spatially proximal bodies but also allows for the agent to receive more distal callbacks that could influence its behavior. As mentioned above, the query space is used to generate the semantics for the agent relative to the geometry in the environment (which we discuss in more detail in Section 19.7), but what the query space fundamentally allows for is the ability to ask behavioral questions such as, "Is there anything on my right?" or "Do I need to modify my gait to keep pace with the pack?" These are questions that the agent should ask itself based upon its orientation with objects in the environment.

Query Space Allows for an Understanding of Space

Modeling of an environment doesn't just stop with the modeling of the geometric entities [Ramsey 2009a]. It must also be extended to the modeling of space within that environment (or have it procedurally generated). What the query space enables is for the behavior engine to ask at any moment certain fundamental questions such as, "Do I have enough space to execute this particular action?" It's important to separate the term *action* (which in our case means animation) from that of a *behavior*. When combined (as they routinely are in game development), its eventual consequence is that of a muddied concept. So action means animation, while behavior means something that occurs over time. That is a very important distinction that ECMMS makes use of, since at any particular moment in time, an agent may execute an action, but it may be executing that action as part of a more encompassing behavior. What we mean by a more encompassing behavior is a series of events that may be externally visible to the player (or perhaps not), but nonetheless influences the agent's intention. Now it makes obvious sense to make as much as possible visible to the player, so agents should provide visual cues to what is transpiring in their minds (e.g., facial movements to more exaggerated actions like head looks or body shifts).

19.4 Modeling the Environment

How we view an object in the world serves as an important basis for how the objects are acted upon because our perception is only one of many possible perspectives. In the ECMMS, we not only wanted to model an asset, but we also wanted to provide the data in a form such that different agents have the ability to understand an object *relative to themselves*. When we look at a chair, we typically understand what that chair could do for us, but if I'm a dog and I look at that chair, then I have a whole set of other affordances available to me! It's just a matter of perspective, and that perspective guides our eventual behavioral responses.

While authoring assets for the ECMMS, the only strict requirement is that collidable objects have a collision layer associated with them. A collision layer is a device that controls whether overlapping objects generate collision callbacks for the agent. This collision layer is assigned to the object as a whole, or it can be done on a per-surface basis. Assigning multiple collision layers to an object makes sense if an agent can interact with a specific surface on that object differently than it can with the object as a whole. Referring to Figure 19.5, we see a rocky outcropping that has the top surface tagged as jumpable, while the rest of the rock is tagged as standard collidable. What the jumpable tag signifies is that when this surface is overlapping a collision sensor, it affords the possible interaction of jumping to the agent.

By applying tags to specific surfaces, we are essentially assigning affordances for the potential types of interaction that a surface allows for. To handle the

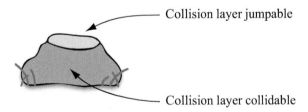

Collision layer jumpable

Collision layer collidable

Figure 19.5. A rock showing the various tags that can be applied to an asset.

different types of interpretations that a surface may have to differing types of agents, we use an affordance mapper. The affordance mapper determines the type of interaction that the object allows to the agent. This allows the modelers to label the surfaces to an *objectified agent*, and then the animators populate the various interactions for the different agent types. For example, a surface may be jumpable for a small animal, but only serves as a stepping surface for a very large animal.

19.5 The ECMMS Architecture

Figure 19.6 shows the various components of the framework that forms the ECMMS. The primary system that facilitates access to the core navigation components (e.g., spatial representation and pathfinding) is the navigation manager. When an agent needs to move through the environment, it calls through the animal planner into the navigation manager. The navigation manager then generates a coarse route [Ramsey 2009a]. The pathfinder uses a modified A* algorithm [Hart et al. 1968]. As noted by Ramsey [2009a], the layout of the navigable spatial representation for the environment may be nonuniform, and the agent's locomotion model may also need to factor in the agent's facing and velocity in order to generate a realistic motion path. The agent utilizes a planner to handle coarse routing through the environment while factoring in the behavioral constraints of nearby agents. The behavioral controller handles the interaction of the agent with any predefined contexts, as well as implements the behavioral response algorithm (see Section 19.9). The behavioral controller also interfaces with the ECMMS manager. The ECMMS manager deals with the creation of the query space, handling of collision callbacks, generation of spatial semantics (see Section 19.7), and animation validation (see Section 19.8).

19.6 Modeling an ECMMS-Enabled Agent

Creating an agent for the ECMMS requires three different representations: a coarse collision representation, a ragdoll representation, and the collision sensor layout. The coarse collision representation for an agent can be anything from a simple primitive to a convex hull that is a rough approximation of the agent's physique. The coarse representation is used during the physics simulation step to generate the contact points for the agent's position. This is just the agent's rough position in the world, as we can still perform inverse kinematics to fix up an agent's foot positions. The ragdoll representation is a more accurate depiction of an agent's physical makeup. Typically, a ragdoll is created by associating rigid

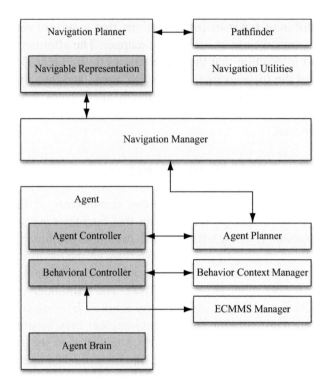

Figure 19.6. The ECMMS layout.

bodies with the bones of an agent's skeleton. Then when the agent animates, the rigid bodies are keyframed to their respective bone positions. This in itself allows for nice granular interactions with dynamic game objects. The collision sensors placed around an agent do and should differ based upon aspects such as the agent's size, turning radius, and speed. The layout of the query space needs to be done in conjunction with the knowledge of corresponding animation information. If an agent is intended to jump long distances, then the query space generally needs to be built such that the collision sensors receive callbacks from overlapping geometry in time to not only determine the validity of the actions but also the intended behavioral result.

19.7 Generating a Behavior Model with the ECMMS

A full behavioral model is beyond the scope of this chapter, but in this section, we cover the underlying components and processes that the ECMMS provides so that you can build your own behavioral system. An agent's observed behavior provides significant insight into its emotional state, attitudes, and attention, and as a result, a considerable amount of perceived behavior originates from how an agent moves through the world relative to not only the objects but also to the available space within that environment. How an agent makes use of space has been covered [Ramsey 2009a, Ramsey 2009b]—here we focus on how the ECMMS provides the underpinnings for a behavioral model that embraces egocentric spatial awareness.

As we've mentioned before, the ECMMS is a system that allows for an agent to gather information about its environment as it moves through it. What an agent needs is the ability to classify this information and generate what we call *spatial semantics*; spatial semantics allow the higher-order systems to make both short-term as well as long-term decisions based upon the spatial orientation of an agent relative to the geometry in the scene. Spatial semantics signifies an important distinction from the typical approach of agent classification in games, where they rely upon methods of perhaps too fine a granularity to drive the immediate action of an agent. To that end, we want to build the basis for informing behavioral decisions from one aspect of the situation at hand, that being the relative spatial orientation of an agent with the elements in its environment.

Figure 19.7 shows an example of an agent that is next to a wall, as well as what its query space looks like. In general, we came up with a series of fundamental categories that allowed us to generate a meaning from the raw collision sensor information. The syntax we allowed consisted of SOLeft, SORight, SOBehind, SOInFront, SOAbove, SOBelow, SONear, and SOFar. If something was im-

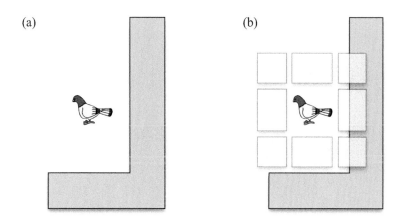

Figure 19.7. (a) A bird next to a wall. (b) The bird's query space.

peding movement in a particular direction, we said that direction was blocked. We were also able to quantify the coverage of the collision sensors relative to how much of a given object was within the query space of an agent. Quantification fell into three categories: QSmall, QMedium, or QLarge. A couple of ancillary benefits are that the quantification process would allow a more advanced artificial intelligence system to not only generate quantities but also generate proportions relative to past results, as well as adjectives for the quantifications such as QNarrow or QLong.

Since the ECMMS is focusing on the agent, the quantification process needs to be relative to the agent's size. So to handle the quantification mapping, a simple function scales the quantified object relative to the agent's size. This allows us to objectively quantify a result and then allow the agent to requantify the object based upon its perception, which is mainly due to its size.

Now that we have the ability to quantify objects that overlap with the collision sensors and the ability to generate a syntax for the specific situation at hand, we need the ability to generate a resultant semantic. Figure 19.8 shows an example of a bird between two walls. The interpretation that would be made by the agent is that there is a blocking object behind it as well as to the left and right of it. The resultant spatial semantic for the bird would be one of being cornered, so the avenue for available actions would comprise either some forward animation or flight. One of the uses of a spatial semantic is for processing what action, specifically what animation, to play on the agent. For example, by looking at the spatial semantic generated from a situation, we can favor animations that exhibit tight turns in places that are geometrically restrictive for the agent.

(a)
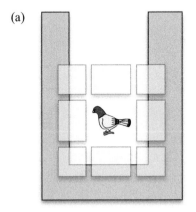

(b) Forward = Unblocked   (c) Move forward
     Backward = Blocked
     Left = Blocked
     Right = Blocked

Figure 19.8. (a) A bird in a corner. (b) The bird's generated syntax. (c) The semantic of the spatial situation that the bird is in.

19.8 Animation Validation

It's important to know whether a selected animation will succeed. By succeed, we mean whether the desired animator end result is achieved. Take for example a jump animation with three specific variations: a short, medium, or long jump (we ignore the motion in the up direction for the time being). Depending on the current situation in the environment, it may be important that the agent jumps immediately, as opposed to taking a few more strides and then jumping. What we need to determine is, if the agent jumped immediately, will it be able to land on the target surface? To accomplish this, we simulate the desired animation using the agent's ragdoll rigid bodies to determine whether the animation results with the primary support bones (e.g., feet) on the surface or whether the animation results in any large amount of geometric penetration. If these bones penetrate any geometry that is tagged as collidable, then the animation could be disallowed since the motion of the animation will have taken the agent into the geometry. The ECMMS does allow for percentages of penetration to occur, but in general, we disallow any animation to be executed if the result is substantial clipping.

One of the interesting properties of how animation validation is performed is that since the motions of the rigid bodies are simulated through time, we not only know if a body penetrated another body but, more importantly, we know by how much. This is ideal because we can use a simple threshold test to determine whether to fail an animation completely or to suggest a subsequent animation for the agent, such as a stumble. A stumble suggestion is just that, a suggestion to the behavior graph that a natural event might have occurred. What this adds is an extremely dynamic look and feel to the agent in the scene. It conveys a real sense of "I almost made it!" to the user, and that means we conveyed the intentionality of the agent to the user. The intentionality of the agent is important because it imparts a sense of purposefulness of the agent relative to its environment. Why does the stumble succeed, and why isn't it labeled a failure? It's because we expected the agent to stumble if the jump barely failed—the agent acted out of purposefulness, and the result did not violate our expectations of its intent.

19.9 A Single Agent Behavioral Response Algorithm and Example

While a fully implemented behavior system is beyond the scope of this chapter, we can provide an algorithm and an example of how the ECMMS can be used to inform a behavior system. This allows for a single system to prioritize both the

Puppy Responses
Get food
Get water
Go outside
Play with user
Solicit for toy/interaction

Table 19.1. Response pool for the puppy's actions.

agent's internal attributes along with any spatial semantics generated from the ECMMS. The single agent behavioral response algorithm selects a response to a perceived situation in the scene; the responses are already associated with specific macro situations that may occur in a scene. For our example, we have a simple puppy game where the puppy has three attributes: hunger, water, and fun. Table 19.1 shows the puppy's response pool, and Table 19.2 lists the perceived situations that a puppy can find itself in.

The following is a straightforward algorithm for classifying responses generated from not only typical behavioral considerations but also from the spatial orientations that the ECMMS has provided.

Perceived Situation
Next to sofa
Next to wall
Cornered
Playing with user
Hungry
Thirsty
Play
Explore

Table 19.2. Situation pool that contains behavioral characteristics, as well as possible spatial orientations that a puppy might find itself in.

1. An agent inserts its internal needs into a perceived situations list.
2. The ECMMS generates the spatial semantics, which get inserted into the perceived situations list.
3. Using Table 19.3, the agent generates responses to all perceived situations.
4. The behavior system scores each situation and response.
5. The behavior system selects the possible assignment with the highest score.
6. If the response is appropriate, the behavior system executes the action; otherwise, it selects the next possible assignment.

The single-agent response algorithm allows for the prioritization of the puppy's spatial situations and its needs at the same time. This allows for the environment to have an immediate influence on the puppy's behavior. The response algorithm's perceived situations list is initially populated by information from the agent itself (this would include how hungry the puppy is, how thirsty, etc.). The ECMMS then inserts the situational semantics into the list (this may include information similar to: wall on the right, there is a wall behind me, I'm in a corner, etc.). The puppy then scores each of the situation and response entries (in this case, there is some game code that evaluates the entries and generates a priority), and the list is sorted. The behavior system decides whether the highest-priority entry is appropriate, and if so, executes the response. It is expected that not every situation will have a response, and this is definitely okay because there are (and should be) several default behaviors that the puppy goes into.

Situation	Generated Response
Next to sofa	Explore
Next to wall	Explore
Cornered	Solicit for toy/interaction
Playing with user	Play
Hungry	Get food
Thirsty	Get water
Want to play	Solicit for toy/interaction

Table 19.3. Response to situation mapper. This mapping has the responses to the situation without any contextual prioritization factored in.

The responses in Table 19.3 contain both a priority and an objective. For the example in our puppy game, food and water would receive a higher priority over activities such as play, but then again, that choice is context dependent since the game may have a specific area just for extensive play periods where we don't want our puppy to get hungry or thirsty. So it makes sense to have response priorities contextually modifiable. The generated response also has an objective that is used to fulfill that specific response; this is another area in which the ECMMS can aid the behavioral model by providing a list of suitable objectives that satisfy the response, in essence creating some variability as opposed to always executing the same response. If no suitable objects are within the query space of the puppy, then the ECMMS can suggest to the behavioral model to seek out the desired objective. What this behavioral example provides is an agent that is exhibiting a nice ebb and flow between itself and its environment, as well as providing the players with an interesting window into the agent's perceived and intended behavior.

References

[Bermudez 2007] Jose Luis Bermudez. *Thinking Without Words*. Oxford University Press, 2007.

[Gibson 1986] James J. Gibson. *The Ecological Approach to Visual Perception*. Hillsdale, NJ: Lawrence Erlbaum Associates, 1986.

[Hart et al. 1968] Peter E. Hart, Nils J. Nilsson, and Bertram Raphael. "A Formal Basis for the Heuristic Determination of Minimum Cost Paths." *IEEE Transactions on Systems Science and Cybernetics SSC4* 4:2 (July 1968), pp. 100–107.

[Ramsey 2009a] Michael Ramsey. "A Unified Spatial Representation for Navigation Systems." *Proceedings of The Fifth AAAI Artificial Intelligence and Interactive Digital Entertainment Conference*, 2009, pp. 119–122.

[Ramsey 2009b] Michael Ramsey. "A Practical Spatial Architecture for Animal and Agent Navigation." *Game Programming Gems 8*, edited by Adam Lake. Boston: Charles River Media, 2010.

20

Pointer Patching Assets

Jason Hughes
Steel Penny Games, Inc.

20.1 Introduction

Console development has never been harder. The clock speeds of processors keep getting higher, the number of processors is increasing, the number of megabytes of memory available is staggering, even the storage size of optical media has ballooned to over 30 GB. Doesn't that make development easier, you ask? Well, there's a catch. Transfer rates from optical media have not improved one bit and are stuck in the dark ages at 8 to 9 MB/s. That means that in the best possible case of a single contiguous read request, it still takes almost a full minute to fill 512 MB of memory. Even with an optimistic 60% compression, that's around 20 seconds.

As long as 20 seconds sounds, it is hard to achieve without careful planning. Most engines, particularly PC engines ported to consoles, tend to have the following issues that hurt loading performance even further:

- Inter- or intra-file disk seeks can take as much as 1/20th of a second.
- Time is spent on the CPU processing assets synchronously after loading each chunk of data.

Bad Solutions

There are many ways to address these problems. One popular old-school way to improve the disk seeks between files is to log out all the file requests and rearrange the file layout on the final media so that seeks are always *forward* on the disk. CD-ROM and DVD drives typically perform seeks forward more quickly than backward, so this is a solution that only partially addresses the heart of the problem and does nothing to handle the time wasted processing the data after each load occurs. In fact, loading individual files encourages a single-threaded mentality that not only hurts performance but does not scale well with modern multithreaded development.

The next iteration is to combine all the files into a giant metafile for a level, retaining a familiar file access interface, like the FILE type, fopen() function, and so on, but adding a large read-ahead buffer. This helps cut down further on the bandwidth stalls, but again, suffers from a single-threaded mentality when processing data, particularly when certain files contain other filenames that need to be queued up for reading. This spider web of dependencies exacerbates the optimization of file I/O.

The next iteration in a system like this is to make it multithreaded. This basically requires some accounting mechanism using threads and callbacks. In this system, the order of operations cannot be assured because threads may be executed in any order, and some data processing occurs faster for some items than others. While this does indeed allow for continuous streaming in parallel with the loaded data initialization, it also requires a far more complicated scheme of accounting for objects that have been created but are not yet "live" in the game because they depend on other objects that are not yet live. In the end, there is a single object called a *level* that has explicit dependencies on all the subelements, and they on their subelements, recursively, which is allowed to become live only after everything is loaded and initialized. This undertaking requires clever management of reference counts, completion callbacks, initialization threads, and a lot of implicit dependencies that have to be turned into explicit dependencies.

Analysis

We've written all of the above solutions, and shipped multiple games with each, but cannot in good faith recommend any of them. In our opinion, they are bandages on top of a deeper-rooted architectural problem, one that is rooted in a failure to practice a clean separation between what is run-time code and what is tools code.

How do we get into these situations? Usually, the first thing that happens on a project, especially when an engine is developed on the PC with a fast hard disk drive holding files, is that data needs to be loaded into memory. The fastest and easiest way to do that is to open a file and read it. Before long, all levels of the engine are doing so, directly accessing files as they see fit. Porting the engine to a console then requires writing wrappers for the file system and redirecting the calls to the provided file I/O system. Performance is poor, but it's working. Later, some sad optimization engineer is tasked with getting the load times down from six minutes to the industry standard 20 seconds. He's faced with two choices:

1. Track down and rewrite all of the places in the entire engine and game where file access is taking place, and implement something custom and appropriate for each asset type. This involves making changes to the offline tools pipeline, outputting data in a completely different way, and sometimes grafting existing run-time code out of the engine and into tools. Then, deal with the inevitable ream of bugs introduced at apparently random places in code that has been working for months or years.
2. Make the existing file access system run faster.

Oh, and one more thing—his manager says there are six weeks left before the game ships. There's no question it's too late to pick choice #1, so the intrepid engineer begins the journey down the road that leads to the Bad Solution. The game ships, but the engine's asset-loading pipeline is forever in a state of maintenance as new data files make their way into the system.

An Alternative, High-Performance, Solution

The technique that top studios use to get the most out of their streaming performance is to use pointer patching of assets. The core concept is simple. Many blocks of data that are read from many separate files end up in adjacent memory at run time, frequently in exactly the same order, and often with pointers to each other. Simply move this to an offline process where these chunks of memory are loaded into memory and baked out as a single large file. This has multiple benefits, particularly that the removal of all disk seeks that are normally paid for at run time are now moved to a tools process, and that as much data as possible is preprocessed so that there is extremely minimal operations required to use the data once it's in memory. This is as much a philosophical adjustment for studios as it is a mechanical one.

Considerations

Unfortunately, pointer patching of assets is relatively hard to retrofit into existing engines. The tools pipeline must be written to support outputting data in this format. The run time code must be changed to expect data in this format, generally implying the removal of a lot of initialization code, but more often than not, it requires breaking explicit dependencies on loading other files directly during construction and converting that to a tools-side procedure. This sort of dynamic loading scheme tends to map cleanly onto run-time lookup of symbolic pointers, for example. In essence, it requires detangling assets from their disk access entirely and relegating that chore to a handful of higher-level systems.

If you're considering retrofitting an existing engine, follow the 80/20 rule. If 20 percent of the assets take up 80 percent of the load time, concentrate on those first. Generally, these should be textures, meshes, and certain other large assets that engines tend to manipulate directly. However, some highly-processed data sets may prove fruitful to convert as well. State machines, graphs, and trees tend to have a lot of pointers and a lot of small allocations, all of which can be done offline to dramatically improve initialization performance when moved to a pointer patching pipeline.

20.2 Overview of the Technique

The basic offline tools process is the following:

1. Load run-time data into memory as it is represented during run time.
2. Use a special serialization interface to dump each structure to the pointer patching system, which carefully notes the locations of all pointers.
3. When a coherent set of data relating to an asset has been dumped, finalize the asset. Coherent means there are no unresolved external dependencies. Finalization is the only complicated part of the system, and it comprises the following steps:

   (a) Concatenate all the structures into a single contiguous block of memory, remembering the location to which each structure was relocated.
   (b) Iterate through all the pointers in each relocated structure and convert the raw addresses stored into offsets into the concatenated block. See Figure 20.1 for an example of how relative offsets are used as relocatable pointers. Any pointers that cannot be resolved within the block are indications that the serialized data structure is not coherent.
   (c) Append a table of pointer locations within the block that will need to be fixed up after the load has completed.
   (d) Append a table of name/value pairs that allows the run-time code to locate the important starting points of structures.

4. Write out the block to disk for run-time consumption.

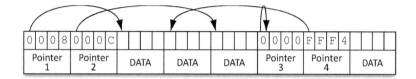

Figure 20.1. This shows all the flavors of pointers stored in a block of data on disk. Forward pointers are relative to the pointer's address and have positive values. Backward pointers are relative to the pointer's address and have negative values. Null pointers store a zero offset, which is a special case when patched and is left zero.

Desirable Properties

There are many ways to get from the above description to actual working code. We've implemented this three different ways over the years and each approach has had different drawbacks and features. The implementation details of the following are left as an exercise to the reader, but here are some high-level properties that should be considered as part of your design when creating your own pointer patching asset system:

- *Relocatable assets.* If you can patch pointers, you can *unpatch* them as well. This affords you the ability to defragment memory, among other things—the holy grail of memory stability.
- *General compression.* This reduces load times dramatically and should be supported generically.
- *Custom compression for certain structures.* A couple of obvious candidates are triangle list compression and swizzled JPEG compression for textures. There are two key points to consider here. First, perform custom compression before general compression for maximum benefit and simplest implementation. This is also necessary because custom compression (like JPEG/DCT for images) is often lossy and general compression (such as LZ77 with Huffman or arithmetic encoding) is not. Second, perform in-place decompression, which requires the custom compressed region to take the same total amount of space, only backfilled with zeros that compress very well in the general pass. You won't want to be moving memory around during the decompression phase, especially if you want to kick off the decompressor functions into separate jobs/threads.
- *Offline block linking.* It is very useful to be able to handle assets generically and even combine them with other assets to form level-specific packages, without having to reprocess the source data to produce the pointer patched

asset. This can lead to a powerful, optimized tools pipeline with minimal load times and great flexibility.

■ *Symbolic pointers with delayed bindings.* Rather than all pointers having physical memory addresses, some studios use named addresses that can be patched at run time after loading is complete. This way you can have pointers that point to the player's object or some other level-specific data without needing to write custom support for each case.

■ *Generic run-time asset caches.* Once loading is handled largely through pointer-patched blocks with named assets, it is fairly simple to build a generic asset caching system on top of this, even allowing dynamic reloading of assets at run time with minimal effort.

■ *Simple tools interface that handles byte swapping and recursion.* Writing out data should be painless and natural, allowing for recursive traversals of live data structures and minimal intrusion.

■ *Special pointer patching consideration for virtual tables.* A method for virtual table patching may be necessary to refresh class instances that have virtual functions. Or choose ways to represent your data without using virtual functions.

■ *Offline introspection tools.* Not only is it very useful for debugging the asset pipeline, but a generic set of introspection tools can help perform vital analyses about the game's memory consumption based on asset type, globally, and without even loading the game on the final platform!

■ *Propagate memory alignment requirements.* Careful attention to data alignment allows hardware to receive data in the fastest possible way. Design your writing interface and linking tools to preserve and propagate alignments so that all the thinking is done in tools, even if it means inserting some wasted space to keep structures starting on the right address. All the run-time code should need to know is the alignment of the block as a whole.

While all the above properties have their merits, a complete and full-featured system is an investment for studios to undertake on their own. A very basic system is provided on the website. It supports byte swapping of atomic types, pointer patching, and a clean and simple interface for recursion.

20.3 A Brief Example

Here is a small example of a simple tree structure. Trees are traditionally very slow because they require many memory allocations and are somewhat challenging to serialize due to the amount of pointer work required to reconstruct them at run time. While this example only contains a few nodes, it is nonetheless a nontrivial example given the comparative complexity of any reasonable and flexible alternative.

Tools Side

The first order of business is to declare the data structure we want in the run-time code. Note that there are no requirements placed on the declarations—minimal intrusion makes for easier adoption. You should be able to use the actual run-time structure declaration in those cases where there are limited external dependencies.

Listing 20.1 shows a simple example of how to dump out a live tree data structure into a pointer-patched block in just a few lines of code. As you can see, the `WriteTree()` function just iterates over the entire structure—any order is actually fine—and submits the contents of each node to the writer object. Each call to `ttw.Write*()` is copying some data from the `Node` into the writer's memory layout, which matches exactly what the run-time code will use. As written, `WriteTree()` simply starts writing a tree recursively down both the left and right branches until it exhausts the data. The writer interface is designed to handle data in random order and has an explicit finalization stage where addresses of structures are hooked to pointers that were written out. This dramatically improves the flexibility of the tools code.

```
struct Node
{
    Node(float v) : mLeft(NULL), mRight(NULL), mValue(v) {}

    Node    *mLeft;
    Node    *mRight;
    float   mValue;
};

void WriteTree(const Node *n, ToolsTimeWriter &ttw)
{
    ttw.StartStruct(n);
    ttw.WritePtr();

    if (n->mLeft)
        WriteTree(n->mLeft, ttw);
    ttw.WritePtr();
```

```
    if (n->mRight)
        WriteTree(n->mRight, ttw);
    ttw.Write4();
    ttw.EndStruct();
}

// First, we construct a handful of nodes into a tree.
Node *root = new Node(3.14F);
root->mLeft = new Node(5.0F);
root->mRight = new Node(777.0F);
root->mLeft->mRight = new Node(1.0F);
root->mLeft->mRight->mLeft = new Node(0.01F);

ToolsTimeWriter ttw(false);
ttw.StartAsset("TreeOfNodes", root);
WriteTree(root, ttw);
std::vector<unsigned char> packedData = ttw.Finalize();
```

Listing 20.1. This is the structure declaration for the sample tree.

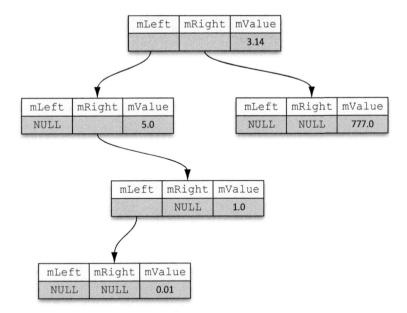

Figure 20.2. This is the in-memory layout of our sample tree. Notice it requires five memory allocations and various pointer traversals to configure.

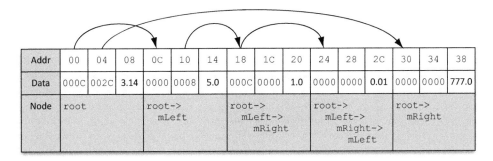

Addr	00	04	08	0C	10	14	18	1C	20	24	28	2C	30	34	38
Data	000C	002C	3.14	0000	0008	5.0	000C	0000	1.0	0000	0000	0.01	0000	0000	777.0
Node	root			root-> mLeft			root-> mLeft-> mRight			root-> mLeft-> mRight-> mLeft			root-> mRight		

Figure 20.3. This is the finalized data from the writer, minus the header and format details. Notice that all offsets are relative to the pointer's address, rather than an absolute index. This vastly simplifies bookkeeping when merging multiple blocks together.

Next, there is a block of code that creates a live tree structure using typical allocation methods you would use in tools code, graphically depicted in Figure 20.2. Finally, we write out the live tree structure and finalize it down to a single block of opaque data. This is done by declaring a writer object, marking the start of the specific structure that the run-time code will later find by name, calling our data writing function above, and then retrieving the baked-down block of bytes that should be written to disk. Amazingly, Figure 20.3 shows the entire structure in just a handful of bytes.

The `ToolsTimeWriter` class shown in Listing 20.2 is only a toy implementation. Some basic limitations in this implementation are that it only supports 32-bit pointers, doesn't handle alignment requirements per structure, etc. Still, it is educational to see the approach taken by this one of many possible interfaces.

```
class ToolsTimeWriter
{
    public:

        ToolsTimeWriter(bool byteSwap);

        // Add an entry to the asset table in the header.
        void StartAsset(char const *name, void *s);

        // Starting a struct allows for a recursion context stack.
        void StartStruct(void *s);
```

```
        // Once a struct has been started, you call these to pump out
        // data. These are needed to handle byte swapping and to
        // measure struct sizes.
        void Write1(void);
        void Write2(void);
        void Write4(void);
        void Write8(void);
        void WritePtr(void);
        void WriteRaw(int numBytes);

        // This pops the recursion context off the stack.
        void EndStruct(void);

        std::vector<unsigned char> Finalize(void);
};
```

Listing 20.2. This basic interface is close to the minimum requirements for a pointer patching asset interface.

The Run-Time Pointer Patching Process

The primary feature of a pointer patching system is that almost all of the work is done once offline, which means there isn't a great deal to discuss about the run-time code. The basic set of operations is to load a coherent chunk of data into memory, then fix up all of the pointers once.

An exceedingly clever implementation might segregate all the ancillary data, such as the pointer table, the asset table, and the format header, into a metadata block that can be thrown away once the pointers are patched and the assets are registered with their respective caches. However, with the metadata gone, this data can no longer be easily relocated in memory. The metadata is generally quite small, so we recommend keeping it. An example file format can be seen in Figure 20.4.

Listings 20.3 and 20.4 show how incredibly simple it is to set up pointer-patched blocks at run time. By adding more metadata, you can set up the loader so that it dispatches types to asset handlers in the engine rather than making your code search for each asset one by one. Then the job of figuring out what content should be loaded can be completely handled in tools.

Pointer Patch Table Header	
int	Num pointers
int	Offset to pointer patch table
int	Num assets
int	Offset to asset table

Pointer Patch Table	
int	Offset to pointer
. . .	...
int	Offset to pointer

Asset Table	
int	Offset to asset name string
int	Offset to asset pointer
. . .	...
int	Offset to asset name string
int	Offset to asset pointer

Data Block	
. . .	Data bytes

Asset Names Block	
char[]	Null-terminated strings

Figure 20.4. This is a very simple file format and is fully loaded into RAM in a single read before the fix-up phase occurs.

One important characteristic is that the fix-up process only modifies the pointers stored within the data block and not in the metadata. This is partly because patching is generally done only once, and therefore, no value is causing a lot of cache lines to be flushed out to RAM by writing them out. It is also partly because the pointer patch table, among other relative pointers, has to be handled outside the normal patching mechanism (otherwise the addresses of the pointers in the table would need to be in the table, see?). If they already must be dealt with explicitly, there would need to be a flag indicating whether the pointers are patched or not, and there would need to be two code paths, depending on whether patching is necessary. So, we leave them unpatched all the time and enjoy reduced code complexity and the strongest possible run-time performance while patching pointers.

```
void DoFixups(PointerPatchTableHeader *header)
{
    PointerPatchTable *ppt = (PointerPatchTable *)
            ((int) header->mPatchTable + header->mPatchTable);
    for (int i = 0; i < header->mNumPointers; i++)
    {
        int *pointerToFixup = (int *) ((int) &ppt[i] + ppt[i]);

        // Special case: if the offset is zero, it would be a pointer
        // to itself, which we assume really means NULL.
        if (*pointerToFixup)
            *pointerToFixup = *pointerToFixup + (int) pointerToFixup;
    }
}
```

Listing 20.3. The following function simply walks the table of offsets to pointers, then adds each pointer's address to the offset stored at that location, constructing a properly patched pointer.

```
void *dataBlock = loadedFromDiskPtr;
DoFixups((PointerPatchTableHeader *) dataBlock);
Node *root = (Node *) FindAssetByName((PointerPatchTableHeader *)
            dataBlock, "TreeOfNodes");
```

Listing 20.4. Assuming that the variable `loadedFromDiskPtr` points to the address of a coherent block stored on disk, these two lines of code are all that is necessary to reconstruct a full tree data structure.

21

Data-Driven Sound Pack Loading and Organization

Simon Franco
The Creative Assembly

21.1 Introduction

Typically, a game's audio data, much like all other game data, cannot be fully stored within the available memory of a gaming device. Therefore, we need to develop strategies for managing the loading of our audio data. This is so we only store in memory the audio data that is currently needed by the game. Large audio files that are too big to fit into memory, such as a piece of music or a long sound effect that only has one instance playing at a time, can potentially be streamed in. Streaming a file in this way results in only the currently needed portion of the file being loaded into memory. However, this does not solve the problem of how to handle audio files that may require multiple instances of an audio file to be played at the same time, such as gunfire or footsteps.

Sound effects such as these need to be fully loaded into memory. This is so the sound engine can play multiple copies of the same sound effect, often at slightly different times, without needing to stream the same file multiple times and using up the limited bandwidth of the storage media.

To minimize the number of file operations performed, we typically organize our audio files into sound packs. Each sound pack is a collection of audio files that either need to be played together at the same time or within a short time period of each other.

Previously, we would package up our audio files into simplified sound packs. These would typically have been organized into a global sound pack, character sound packs, and level sound packs. The global sound pack would contain all audio files used by global sound events that occur across all levels of a game.

This would typically have been player and user interface sound events. Character sounds would typically be organized so that there would be one sound pack per type of character. Each character sound pack would contain all the audio files used by that character. Level sound packs would contain only the audio files only used by sound events found on that particular level.

However, this method of organization is no longer applicable, as a single level's worth of audio data can easily exceed the amount of sound RAM available. Therefore, we must break up our level sound packs into several smaller packs so that we can fit the audio data needed by the current section of the game into memory. Each of these smaller sound packs contain audio data for a well-defined small portion of the game. We then load in these smaller sound packs and release them from memory as the player progresses through the game. An example of a small sound pack would be a sound pack containing woodland sounds comprising bird calls, trees rustling in the wind, forest animals, etc.

The problem with these smaller sound packs is how we decide which audio files are to be stored in each sound pack. Typically, sound packs, such as the example woodlands sound, pack are hand-organized in a logical manner by the sound designer. However, this can lead to wasting memory as sounds are grouped by their perceived relation to each other, rather than an actual requirement that they be bound together into a sound pack.

The second problem is how to decide when to load in a sound pack. Previously, a designer would place a load request for a sound pack into the game script. This would then be triggered when either the player walks into an area or after a particular event happens. An example of this would be loading a burning noises sound pack and having this ready to be used after the player has finished a cut-scene where they set fire to a building.

This chapter discusses methods to automate both of these processes. It allows the sound designer to place sounds into the world and have the system generate the sound packs and loading triggers. It also alerts the sound designer if they have overpopulated an area and need to reduce either the number of variations of a sound or reduce the sample quality.

21.2 Constructing a Sound Map

We examine the game world's sound emitters and their associated sound events in order to generate a sound map. The sound map contains the information we require to build a sound event table. The sound event table provides us with the information needed to construct our sound packs and their loading triggers.

Sound emitters are typically points within 3D space representing the positions from which a sound will play. As well as having a position, an emitter also contains a reference to a sound event. A sound event is a collection of data dictating which audio file or files to play, along with information on how to control how these audio files are played. For example, a sound event would typically store information on playback volume, pitch shifts, and any fade-in or fade-out durations.

Sound designers typically populate the game world with sound emitters in order to build up a game's soundscape. Sound emitters may be scripted directly by the sound designer or may be automatically generated by other game objects. For example, an animating door may automatically generate `wooden_door_open` and `wooden_door_close` sound emitters.

Once all the sound emitters have been placed within the game world, we can begin our data collection. This process should be done offline as part of the process for building a game's level or world data.

Each sound event has an audible range defined by its sound event data. This audible range is used to calculate both the volume of the sound and whether the sound is within audible range by comparing against the listener's position. The listener is the logical representation of the player's ear in the world—it's typically the same as the game's camera position. We use the audible range property of sound emitters to see which sound emitters are overlapping.

We construct a sound map to store an entry for each sound emitter found within the level data. The sound map can be thought of as a three-dimensional space representing the game world. The sound map contains only the sound emitters we've found when processing the level data. Each sound emitter is stored in the sound map as a sphere, with the sphere's radius being the audible distance of the sound emitter's event data.

Once the sound map is generated, we can construct an event table containing an entry for each type of sound event found in the level. For each entry in the table, we must mark how many instances of that sound event there are within the sound map, which other sound events overlap with them (including other instances of the same sound event), and the number of instances in which those events overlap. For example, if a single `Bird_Chirp` sound emitter overlaps with two other sound emitters playing the `Crickets` sound event, then that would be recorded as a single occurrence of an overlap between `Bird_Chirp` and `Crickets` for the `Bird_Chirp` entry. For the `Crickets` entry, it would be recorded as two instances of an overlap. An example table generated for a sample sound map is shown in Table 21.1. From this data, we can begin constructing our sound packs.

Sound Event	Instance Count	Overlapped Events	Occurrences of Overlap
Bird_Chirp	5	Crickets, Bird_Chirp	4, 2
Waterfall	1	None	0
Water_Mill	1	NPC_Speech1	1
NPC_Speech1	1	Water_Mill	1
Crickets	10	Bird_Chirp, Crickets, Frogs	7, 6, 1
Frogs	5	Crickets, Frogs	5, 5

Table 21.1. Sound events discovered within a level and details of any other overlapped events.

21.3 Constructing Sound Packs by Analyzing the Event Table

The information presented in our newly generated event table allows us to start planning the construction of our sound packs. We use the information gathered from analyzing the sound map to construct the sound packs that are to be loaded into memory and to identify those that should be streamed instead.

There are several factors to take into consideration when choosing whether to integrate a sound event's audio data into a sound pack or to stream the data for that sound event while it is playing. For example, we must consider how our storage medium is used by other game systems. This is vital because we need to determine what streaming resources are available. If there are many other systems streaming data from the storage media, then this reduces the number of streamed samples we can play and places greater emphasis on having sounds become part of loaded sound packs, even if those sound packs only contain a single audio file that is played once.

Determining Which Sound Events to Stream

Our first pass through the event table identifies sound events that should be streamed. If we decide to stream a sound event, then it should be removed from the event table and added to a streamed event table. Each entry in the streamed event table is formatted in a manner similar to the original event table. Each entry in the streamed sound event table contains the name of a streamed sound event and a list of other streamed sound events whose sound emitters overlap with a sound emitter having this type of sound event. To decide whether a sound event should be streamed, we must take the following rules into account:

- Is there only one instance of the sound event in the table? If so, then streaming it would be more efficient. The exception to this rule is if the size of the audio file used by the event is too small. In this eventuality, we should load the audio file instead. This is so our streaming resources are reserved for more suitable files. A file is considered too small if its file size is smaller than the size of one of your streaming audio buffers.
- Does the sound event overlap with other copies of itself? If not, then it can be streamed because the audio data only needs to be processed once at any given time.
- Does the audio file used by the sound event have a file size bigger than the available amount of audio RAM? If so, then it must be streamed.
- Are sufficient streaming resources available, such as read bandwidth to the storage media, to stream data for this sound event? This is done after the other tests have been passed because we need to see which other streamed sound events may be playing at the same time. If too many are playing, then we need to prioritize the sound events. Larger files should be streamed, and smaller-sized audio files should be loaded into memory.

Using the data from Table 21.1, we can extract a streamed sound event table similar to that shown in Table 21.2. In Table 21.2, we have three sound events that we believe to be suitable candidates for streaming. Since `Water_Mill` and `NPC_Speech1` overlap, we should make sure that the audio files for these events are placed close to each other on the storage media. This reduces seek times when reading the audio data for these events.

Sound Event	Instance Count	Overlapped Events	Occurrences of Overlap
Waterfall	1	None	0
Water_Mill	1	NPC_Speech1	1
NPC_Speech1	1	Water_Mill	1

Table 21.2. Sound events found suitable for streaming.

Sound Event	Instance Count	Overlapped Events	Occurrences of Overlap
Bird_Chirp	5	Crickets, Bird_Chirp	4, 2
Crickets	10	Bird_Chirp, Crickets, Frogs	7, 6, 1
Frogs	5	Crickets, Frogs	5, 5

Table 21.3. Remaining sound events that need to be placed into sound packs.

Constructing Sound Packs

Now that we have removed the streamed sound events from our original event table, we can begin analyzing which audio files should be combined into sound packs. Table 21.3 shows the remaining entries that we are going to analyze from our original event table.

We must consider the following simple rules when determining which audio files should be placed into a sound pack:

- Sound events that overlap approximately two-thirds of the time are potential candidates for having their audio files placed into the same sound pack. The overlap count should be a high percentage both ways so that sound event A overlaps with sound event B a high number of times, and vice versa. Otherwise, we may end up with redundant audio data being loaded frequently for one of the sound events.
- All audio files used by a sound event should be placed into the same sound pack.
- The file size of a sound pack should not exceed the amount of available sound RAM.
- The ratio between the size of a sound event's audio data and the file size of the sound pack should closely match the percentage of event overlaps. For instance, if we have a sound event whose audio data occupies 80 percent of a sound pack's file size, but is only used 10 percent of the time, then it should be placed in its own sound pack.

Table 21.3 illustrates an interesting example. In this fictional sound map, we have a cluster of frogs next to a single cricket. Therefore, we have all five frogs next to each other (five occurrences of an overlap) and next to a single cricket (five occurrences of an overlap with `Crickets`) in our table. For the `Crickets` entry, we have only a single cricket that was next to the frogs, so there is only one instance of an overlap.

There were no instances of the `Bird_Chirp` event overlapping with the `Frogs` event, so we should put `Bird_Chirp` and `Crickets` into a single sound pack and put `Frogs` into a separate sound pack.

21.4 Constructing and Using Sound Loading Triggers

Now that the sound packs are created, we can take a further look at the sound emitter data set and generate our sound pack loading triggers from it. For each sound emitter, we create a sound loading trigger. Each sound loader has a reference to the sound pack that it is responsible for loading and a loading area. The loading area is initially set to be the same size as the sound emitter's audible distance. When the listener is within the loading area, the sound pack that it references is loaded into memory, if it has not already been loaded.

We have to increase the size of the loading area depending on the bandwidth of the storage media, the file size of the sound pack, the number of other streamed resources taking place, and the maximum distance the listener can move in a game tick. When generating these numbers for your media, you should take into account the worst possible performance cases.

For example, if you have a minimum bandwidth of 1 MB/s (due to other systems accessing the media at the same time), a sound pack that is 2.5 MB in size, and a listener having a maximum travel speed of 3 m/s, then you need to add a minimum of 7.5 meters to the loading area size because it takes 2.5 seconds for the data to load and the listener could travel 7.5 meters in that time.

We unload a sound pack when we find the listener is not within the area of a sound loader referencing that sound pack.

Optimizing Sound Loaders

Not all sound emitters require a separate sound loader to be constructed for them. Most sound emitters do not move, and so there are optimizations that can be made by merging similar sound loaders that overlap.

If two or more sound loaders reference the same sound pack, don't move, and are overlapping, then they should be merged into a single sound loader. This new sound loader must contain both of the original sound loading areas.

Out of Memory Detection

Now that we have our sound loading triggers created, we can check for sound loaders that overlap. For an overlapping set of sound loaders, we can tally up the memory used by the sound packs. If the memory used by several sound packs exceeds the amount of available sound memory, then a warning can be generated to direct the audio designer to investigate the section of the world that has been flagged. He can then make changes to the sound events, such as removing the number of audio files used by a sound event or reducing the quality of the audio files in order to have that section of the game fit into memory.

Loading Sound Packs While Running the Game

When the game is running, it needs to check the listener position each frame and determine whether the listener has entered a sound loading area. Listing 21.1 shows pseudocode that handles the loading of sound packs.

```
mark_all_sound_packs_for_removal(m_loaded_sound_pack_list);

get_list_of_packs_needed(required_pack_list, m_listener_position);

/*
    Iterate through all the loaded sound packs and retain
    those which are in the required list.
*/
for (each required_pack in required_pack_list)
{
    for (each loaded_pack in m_loaded_sound_pack_list)
    {
        if (loaded_pack == required_pack)
        {
            retain_sound_pack(loaded_pack);
            remove_pack_from_required_list(required_pack);
            break;
        }
    }
}

unload_sound_packs_not_retained(m_loaded_sound_pack_list);
```

```
/*
    Now all the sound packs remaining in the required_sound_packs
    list are those which are not yet loaded.
 */
for (each required_pack in required_sound_pack_list)
{
    load_sound_pack_and_add_to_loaded_list(required_pack);
}
```

Listing 21.1. Pseudocode for loading sound packs.

21.5 Conclusion

While not all sound events can have their data packaged up by the process described in this chapter, it still helps simplify the task of constructing and managing sound packs used by a game's environment. Sound events that typically can't
take advantage of this process are global sounds, such as the player being hit or a
projectile being fired, because they can occur at any time during a level.

22

GPGPU Cloth Simulation Using GLSL, OpenCL, and CUDA

Marco Fratarcangeli

Taitus Software Italia

22.1 Introduction

This chapter provides a comparison study between three popular platforms for generic programming on the GPU, namely, GLSL, CUDA, and OpenCL. These technologies are used for implementing an interactive physically-based method that simulates a piece of cloth colliding with simple primitives like spheres, cylinders, and planes (see Figure 22.1). We assess the advantages and the drawbacks of each different technology in terms of usability and performance.

Figure 22.1. A piece of cloth falls under the influence of gravity while colliding with a sphere at interactive rates. The cloth is composed of 780,000 springs connecting 65,000 particles.

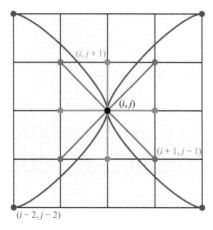

Figure 22.2. A 4×4 grid of particle vertices and the springs for one of the particles.

22.2 Numerical Algorithm

This section provides a brief overview of the theory behind the algorithm used in computing the cloth simulation. A straightforward way to implement elastic networks of particles is by using a mass-spring system. Given a set of evenly spaced particles on a grid, each particle is connected to its neighbors through simulated springs, as depicted in Figure 22.2. Each spring applies to the connected particles a force $\mathbf{F}_{\text{spring}}$:

$$\mathbf{F}_{\text{spring}} = -k \left(\mathbf{l} - \mathbf{l}_0 \right) - b \dot{\mathbf{x}},$$

where $\mathbf{l}$ represents the current length of the spring (i.e., its magnitude is the distance between the connected particles), $\mathbf{l}_0$ represents the rest length of the spring at the beginning of the simulation, k is the stiffness constant, $\dot{\mathbf{x}}$ is the velocity of the particle, and b is the damping constant. This equation means that a spring always applies a force that brings the distance between the connected particles back to its initial rest length. The more the current distance diverges from the rest length, then the larger is the applied force. This force is damped proportionally to the current velocity of the particles by the last term in the equation. The blue springs in Figure 22.2 simulate the stretch stress of the cloth, while the longer red ones simulate the shear and bend stresses.

For each particle, the numerical algorithm that computes its dynamics is schematically illustrated in Figure 22.3. For each step of the dynamic simulation,

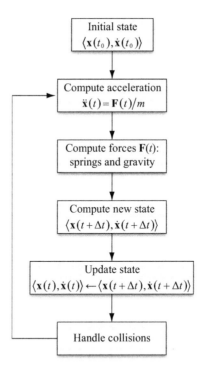

Figure 22.3. Numerical algorithm for computing the cloth simulation.

the spring forces and other external forces (e.g., gravity) are applied to the particles, and then their dynamics are computed according to the Verlet method [Müller 2008] applied to each particle in the system through the following steps:

1. $\dot{\mathbf{x}}(t) = [\mathbf{x}(t) - \mathbf{x}(t - \Delta t)]/\Delta t$.
2. $\ddot{\mathbf{x}}(t) = \mathbf{F}(\mathbf{x}(t), \dot{\mathbf{x}}(t))/m$.
3. $\mathbf{x}(t + \Delta t) = 2\mathbf{x}(t) - \mathbf{x}(t - \Delta t) + \ddot{\mathbf{x}}(t)\Delta t^2$.

Here, $\mathbf{F}(t)$ is the current total force applied to the particle, m is the particle mass, $\ddot{\mathbf{x}}(t)$ is its acceleration, $\dot{\mathbf{x}}(t)$ is the velocity, $\mathbf{x}(t)$ is the current position, and Δt is the time step of the simulation (i.e., how much time the simulation is advanced for each iteration of the algorithm).

The Verlet method is very popular in real-time applications because it is simple and fourth-order accurate, meaning that the error for the position computation is $O(\Delta t^4)$. This makes the Verlet method two orders of magnitude more precise than the explicit Euler method, and at the same time, it avoids the computational cost involved in the Runge-Kutta fourth-order method. In the Verlet scheme, however, velocity is only first-order accurate; in this case, this is not really important because velocity is considered only for damping the springs.

22.3 Collision Handling

Generally, collision handling is composed of two phases, collision detection and collision response. The outcome of collision detection is the set of particles that are currently colliding with some other primitive. Collision response defines how these collisions are solved to bring the colliding particles to a legal state (i.e., not inside a collision primitive). One of the key advantages of the Verlet integration scheme is the easiness of handling collision response. The position at the next time step depends only on the current position and the position at the previous step. The velocity is then estimated by subtracting one from the other. Thus, to solve a collision, it is sufficient to modify the current position of the colliding particle to bring it into a legal state, for example, by moving it perpendicularly out toward the collision surface. The change to the velocity is then handled automatically by considering this new position. This approach is fast and stable, even though it remains valid only when the particles do not penetrate too far.

In our cloth simulation, as the state of the particle is being updated, if the collision test is positive, the particle is displaced into a valid state. For example, let's consider a stationary sphere placed into the scene. In this simple case, a collision between the sphere and a particle happens when the following condition is satisfied:

$$\|\mathbf{x}(t + \Delta t) - \mathbf{c}\| - r < 0,$$

where $\mathbf{c}$ and r are the center and the radius of the sphere, respectively. If a collision occurs, then it is handled by moving the particle into a valid state by moving its position just above the surface of the sphere. In particular, the particle should be displaced along the normal of the surface at the impact point. The position of the particle is updated according to the formula

$$\mathbf{d} = \frac{\mathbf{x}(t + \Delta t) - \mathbf{c}}{\|\mathbf{x}(t + \Delta t) - \mathbf{c}\|},$$

$$\mathbf{x}'(t + \Delta t) = \mathbf{c} + \mathbf{d}r,$$

where $\mathbf{x}'(t + \Delta t)$ is the updated position after the collision. If the particle does not penetrate too far, $\mathbf{d}$ can be considered as an acceptable approximation of the normal to the surface at the impact point.

22.4 CPU Implementation

We first describe the implementation of the algorithm for the CPU as a reference for the implementations on the GPU described in the following sections.

During the design of an algorithm for the GPU, it is critical to minimize the amount of data that travels on the main memory bus. The time spent on the bus is actually one of the primary bottlenecks that strongly penalize performance [Nvidia 2010]. The transfer bandwidth of a standard PCI-express bus is 2 to 8 GB per second. The internal bus bandwidth of a modern GPU is approximately 100 to 150 GB per second. It is very important, therefore, to minimize the amount of data that travels on the bus and keep the data on the GPU as much as possible.

In the case of cloth simulation, only the current and the previous positions of the particles are needed on the GPU. The algorithm computes directly on GPU the rest distance of the springs and which particles are connected by the springs. The state of each particle is represented by the following attributes:

1. The current position (four floating-point values).
2. The previous position (four floating-point values).
3. The current normal vector (four floating-point values).

Even though the normal vector is computed during the simulation, it is used only for rendering purposes and does not affect the simulation dynamics. Here, the normal vector of a particle is defined to be the average of the normal vectors of the triangulated faces to which the particle belongs. A different array is created for storing the current positions, previous positions, and normal vectors. As explained in later sections of this chapter, for the GPU implementation, these attributes are loaded as textures or buffers into video memory. Each array stores the attributes for all the particles. The size of each array is equal to the size of an attribute (four floating-point values) multiplied by the number of particles. For example, the position of the i-th particle $\mathbf{p}_i$ is stored in the positions array and accessed as follows:

$$\mathbf{pos}_i \leftarrow \text{vec3}(\text{in\_pos}[i * 4], \text{in\_pos}[i * 4 + 1], \text{in\_pos}[i * 4 + 2],$$
$$\text{in\_pos}[i * 4 + 3])$$

The cloth is built as a grid of $n \times n$ particles, where n is the number of particles composing one side of the grid. Regardless of the value of n, the horizontal and the vertical spatial dimensions of the grid are always normalized to the range

[0,1]. A particle is identified by its array index i, which is related to the row and the column in the grid as follows:

$$row_i = \lfloor i/n \rfloor,$$
$$col_i = i \bmod n.$$

From the row and the column of a particle, it is easy to access its neighbors by simply adding an offset to the row and the column, as shown in the examples in Figure 22.2.

The pseudocode for calculating the dynamics of the particles in an $n \times n$ grid is shown in Listing 22.1. In steps 1 and 2, the current and previous positions of the the i-th particle are loaded in the local variables $\mathbf{pos}_i^t$ and $\mathbf{pos}_i^{t-1}$, respectively, and then the current velocity $\mathbf{vel}_i^t$ is estimated in step 3. In step 4, the total force $\mathbf{force}_i$ is initialized with the gravity value. Then, the `for` loop in step 5 iterates over all the neighbors of $\mathbf{p}_i$ (steps 5.1 and 5.2), spring forces are computed (steps 5.3 to 5.5), and they are accumulated into the total force (step 5.6). Each neighbor is identified and accessed using a 2D offset ($x_{\text{offset}}, y_{\text{offset}}$) from the position of $\mathbf{p}_i$ within the grid, as shown in Figure 22.2. Finally, the dynamics are computed in step 6, and the results are written into the output buffers in steps 7 and 8.

```
for each particle pᵢ
1. posᵢᵗ ← xᵢ(t)
2. posᵢᵗ⁻¹ ← xᵢ(t − Δt)
3. velᵢᵗ = (posᵢᵗ − posᵢᵗ⁻¹)/Δt
4. forceᵢ = (0, -9.81, 0, 0)
5. for each neighbor (rowᵢ + x_offset, colᵢ + y_offset)
          if (rowᵢ + x_offset, colᵢ + y_offset) is inside the grid
               5.1. i_neigh = (rowᵢ + y_offset) * n + colᵢ + x_offset
               5.2. pos'_neigh ← x_neigh(t)
               5.3. d_rest = ‖(x_offset, y_offset)‖/n
               5.4. d_curr = ‖pos'_neigh − posᵢᵗ‖

               5.5. force_spring = − (pos'_neigh − posᵢᵗ)/‖pos'_neigh − posᵢᵗ‖ (d_curr − d_rest)·k − velᵢᵗ·b

               5.6. forceᵢ += force_spring
6. posᵢᵗ⁺¹ = 2·posᵢᵗ − posᵢᵗ⁻¹ + (forceᵢ/m)·Δt²
7. xᵢ(t) ← posᵢᵗ⁺¹
8. xᵢ(t − Δt) ← posᵢᵗ
```

Listing 22.1. Pseudocode to compute the dynamics of a single particle i belonging to the $n \times n$ grid.

22.5 GPU Implementations

The different implementations for each GPGPU computing platform (GLSL, OpenCL, and CUDA) are based on the same principles. We employ the so-called "ping-pong" technique that is particularly useful when the input of a simulation step is the outcome of the previous one, which is the case in most physically-based animations. The basic idea is rather simple. In the initialization phase, two buffers are loaded on the GPU, one buffer to store the input of the computation and the other to store the output. When the computation ends and the output buffer is filled with the results, the pointers to the two buffers are swapped such that in the following step, the previous output is considered as the current input. The results data is also stored in a vertex buffer object (VBO), which is then used to draw the current state of the cloth. In this way, the data never leaves the GPU, achieving maximal performance. This mechanism is illustrated in Figure 22.4.

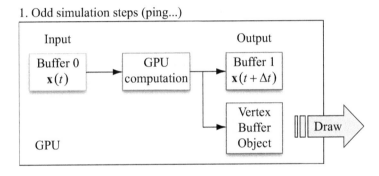

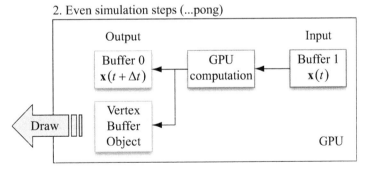

Figure 22.4. The ping-pong technique on the GPU. The output of a simulation step becomes the input of the following step. The current output buffer is mapped to a VBO for fast visualization.

22.6 GLSL Implementation

This section describes the implementation of the algorithm in GLSL 1.2. The source code for the vertex and fragment shaders is provided in the files `ver-let_cloth.vs` and `verlet_cloth.fs`, respectively, on the website. The position and velocity arrays are each stored in a different texture having $n \times n$ dimensions. In such textures, each particle corresponds to a single texel. The textures are uploaded to the GPU, and then the computation is carried out in the fragment shader, where each particle is handled in a separate thread. The updated state (i.e., positions, previous positions, and normal vectors) is written to three distinct render targets.

Frame buffer objects (FBOs) are employed for efficiently storing and accessing the input textures and the output render targets. The ping-pong technique is applied through the use of two frame buffer objects, FBO1 and FBO2. Each FBO contains three textures storing the state of the particles. These three textures are attached to their corresponding FBOs as color buffers using the following code, where `fb` is the index of the FBO and `texid[0]`, `texid[1]`, and `texid[2]` are the indices of the textures storing the current positions, the previous positions, and the normal vectors of the particles, respectively:

```
glBindFramebufferEXT(GL_FRAMEBUFFER_EXT, fbo->fb);
glFramebufferTexture2DEXT(GL_FRAMEBUFFER_EXT,
    GL_COLOR_ATTACHMENT0_EXT, GL_TEXTURE_2D, texid[0], 0);
glFramebufferTexture2DEXT(GL_FRAMEBUFFER_EXT,
    GL_COLOR_ATTACHMENT1_EXT, GL_TEXTURE_2D, texid[1], 0);
glFramebufferTexture2DEXT(GL_FRAMEBUFFER_EXT,
    GL_COLOR_ATTACHMENT2_EXT, GL_TEXTURE_2D, texid[2], 0);
```

In the initialization phase, both of the FBOs holding the initial state of the particles are uploaded to video memory. When the algorithm is run, one of the FBOs is used as input and the other one as output. The fragment shader reads the data from the input FBO and writes the results in the render targets of the output FBO (stored in the color buffers). We declare the output render targets by using the following code, where `fb_out` is the FBO that stores the output:

```
glBindFramebufferEXT(GL_FRAMEBUFFER_EXT, fb_out);
GLenum mrt[] = {GL_COLOR_ATTACHMENT0_EXT,
    GL_COLOR_ATTACHMENT1_EXT, GL_COLOR_ATTACHMENT2_EXT};
glDrawBuffers(3, mrt);
```

In the next simulation step, the pointers to the input and output FBOs are swapped so that the algorithm uses the output of the previous iteration as the current input.

The two FBOs are stored in the video memory, so there is no need to upload data from the CPU to the GPU during the simulation. This drastically reduces the amount of data bandwidth required on the PCI-express bus, improving the performance. At the end of each simulation step, however, position and normal data is read out to a pixel buffer object that is then used as a VBO for drawing purposes. The position data is stored into the VBO directly on the GPU using the following code:

```
glReadBuffer(GL_COLOR_ATTACHMENT0_EXT);
glBindBuffer(GL_PIXEL_PACK_BUFFER, vbo[POSITION_OBJECT]);
glReadPixels(0, 0, texture_size, texture_size,
    GL_RGBA, GL_FLOAT, 0);
```

First, the color buffer of the FBO where the output positions are stored is selected. Then, the positions' VBO is selected, specifying that it will be used as a pixel buffer object. Finally, the VBO is filled with the updated data directly on the GPU. Similar steps are taken to read the normals' data buffer.

22.7 CUDA Implementation

The CUDA implementation works similarly to the GLSL implementation, and the source code is provided in the files verlet_cloth.cu and verlet_cloth_kernel.cu on the website. Instead of using FBOs, this time we use memory buffers. Two pairs of buffers in video memory are uploaded into video memory, one pair for current positions and one pair for previous positions. Each pair comprises an input buffer and an output buffer. The kernel reads the input buffers, performs the computation, and writes the results in the proper output buffers. The same data is also stored in a pair of VBOs (one for the positions and one for the normals), which are then visualized. In the beginning of the next iteration, the output buffers are copied in the input buffers through the cudaMemcpyDeviceToDevice call. For example, in the case of positions, we use the following code:

```
cudaMemcpy(pPosOut, pPosIn, mem_size, cudaMemcpyDeviceToDevice);
```

It is important to note that this instruction does not cause a buffer upload from the CPU to the GPU because the buffer is already stored in video memory. The output data is shared with the VBOs by using `graphicsCudaResource` objects, as follows:

```
// Initialization, done only once.
cudaGraphicsGLRegisterBuffer(&cuda_vbo_resource, gl_vbo,
    cudaGraphicsMapFlagsWriteDiscard);

// During the algorithm execution.
cudaGraphicsMapResources(1, cuda_vbo_resource, 0);
cudaGraphicsResourceGetMappedPointer((void **) &pos,
    &num_bytes, cuda_vbo_resource);
executeCudaKernel(pos, ...);
cudaGraphicsUnmapResources(1, cuda_vbo_resource, 0);
```

In the initialization phase, we declare that we are sharing data in video memory with OpenGL VBOs through CUDA graphical resources. Then, during the execution of the algorithm kernel, we map the graphical resources to buffer pointers. The kernel computes the results and writes them in the buffer. At this point, the graphical resources are unmapped, allowing the VBOs to be used for drawing.

22.8 OpenCL Implementation

The OpenCL implementation is very similar to the GLSL and CUDA implementations, except that the data is uploaded at the beginning of each iteration of the algorithm. At the time of this writing, OpenCL has a rather young implementation that sometimes leads to poor debugging capabilities and sporadic instabilities. For example, suppose a kernel in OpenCL is declared as follows:

```
__kernel void hello(__global int *g_idata);
```

Now suppose we pass input data of some different type (e.g., a `float`) in the following way:

```
float input = 3.0F;
cfloatlSetKernelArg(ckKernel, 0, sizeof(float), (void *) &input);
clEnqueueNDRangeKernel(cqQueue, ckKernel, 1, NULL,
    &_szGlobalWorkSize, &_szLocalWorkSize, 0, 0, 0);
```

When executed, the program will fail silently without giving any error message because it expects an `int` instead of a `float`. This made the OpenCL implementation rather complicated to develop.

22.9 Results

The described method has been implemented and tested on two different machines:

■ A desktop PC with an Nvidia GeForce GTS250, 1GB VRAM and a processor Intel Core i5.
■ A laptop PC with an Nvidia Quadro FX 360M, 128MB VRAM and a processor Intel Core2 Duo.

We collected performance times for each GPU computing platform, varying the numbers of particles and springs, from a grid resolution of 32×32 (1024 particles and 11,412 springs) to 256×256 (65,536 particles and approximately 700,000 springs). Numerical results are collected in the plots in Figures 22.5 and 22.6.

From the data plotted in Figures 22.5 and 22.6, the computing superiority of the GPU compared with the CPU is evident. This is mainly due to the fact that this cloth simulation algorithm is strongly parallelizable, like most of the particle-based approaches. While the computational cost on the CPU keeps growing linearly with the number of particles, the computation time on the GPU remains relatively low because the particle dynamics are computed in parallel. On the GTS250 device, this leads to a performance gain ranging from 10 to 40 times, depending on the number of particles.

It is interesting to note that in this case, GLSL has a much better performance than CUDA does. This can be explained by considering how the memory is accessed by the GPU kernels. In the GLSL fragment program, images are employed to store particle data in texture memory, while in CUDA and OpenCL, these data is stored in the global memory of the device. Texture memory has two main advantages [Nvidia 2010]. First, it is cached, and thus, video memory is accessed only if there is a cache miss. Second, it is built in such a way as to optimize the access to 2D local data, which is the case because each particle corresponds to a pixel, and it must have access to the positions of its neighbors, which are stored in the immediately adjacent texture pixels. Furthermore, the results in GLSL are stored in the color render targets that are then directly mapped to VBOs and drawn on the screen. The data resides in video memory and does not need to be copied between different memory areas. This makes the entire process extremely fast compared with the other approaches.

The plots also highlight the lower performance of OpenCL compared with CUDA. This difference is caused by the fact that it has been rather difficult to tune the number of global and local work items due to causes requiring further

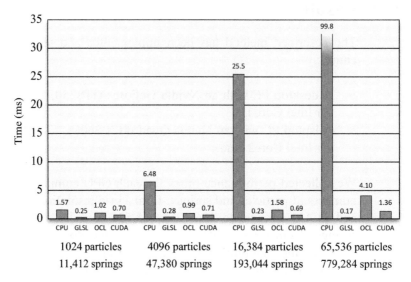

Figure 22.5. Computation times measured on different computation platforms using a GeForce GTS 250 device (16 computing units, 128 CUDA cores).

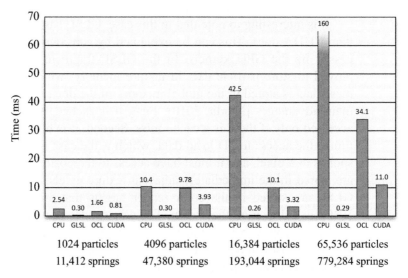

Figure 22.6. Computation times measured on different computation platforms using a Quadro FX 360M device (2 computing units, 16 CUDA cores).

investigation. OpenCL is a very young standard, and both the specification and the driver implementation are likely to change in the near future in order to avoid such instabilities.

The GLSL program works on relatively old hardware, and different from CUDA, it does not require Nvidia hardware. CUDA on the other hand, is a more flexible architecture that has been specifically devised for performing computing tasks (not only graphics, like GLSL), which is easier to debug and provides access to hardware resources, like the shared memory, allowing for a further boost to the performance. OpenCL has the same features as CUDA, but its implementation is rather naive at the moment, and it is harder to debug. However, different from CUDA, it has been devised to run on the widest range of hardware platforms (including consoles and mobile phones), not limited to Nvidia ones, and thus, it is the main candidate for becoming the reference platform for GPGPU in the near future.

The main effort when dealing with GPGPU is in the design of the algorithm. The challenging task that researchers and developers are currently facing is how to redesign algorithms that have been originally conceived to run in a serial manner for the CPU, to make them parallel and thus suitable for the GPU. The main disadvantage of particle-based methods is that they require a very large number of particles to obtain realistic results. However, it is relatively easy to parallelize algorithms handling particle systems, and the massive parallel computation capabilities of modern GPUs now makes it possible to simulate large systems at interactive rates.

22.10 Future Work

Our algorithm for cloth simulation can be improved in many ways. In the CUDA and OpenCL implementations, it would be interesting to exploit the use of shared memory, which should reduce the amount of global accesses and lead to improved performance.

For future research, we would like to investigate ways to generalize this algorithm by introducing connectivity information [Tejada 2005] that stores the indexes of the neighbors of each particle. This data can be stored in constant memory to hide as much as possible the inevitable latency that using this information would introduce. By using connectivity, it would be possible to simulate deformable, breakable objects with arbitrary shapes, not only rectangular pieces of cloth.

22.11 Demo

An implementation of the GPU cloth simulation is provided on the website, and it includes both the source code in C++ and the Windows binaries. The demo allows you to switch among the computing platforms at run time, and it includes a hierarchical profiler. Even though the source code has been developed for Windows using Visual Studio 2008, it has been written with cross-platform compatibility in mind, without using any Windows-specific commands, so it should compile and run on *nix platforms (Mac and Linux). The demo requires a machine capable of running Nvidia CUDA, and the CUDA Computing SDK 3.0 needs to have been compiled. A video is also included on the website.

Acknowledgements

The shader used for rendering the cloth is "fabric plaid" from RenderMonkey 1.82 by AMD and 3DLabs. The author is grateful to Professor Ingemar Ragnemalm for having introduced him to the fascinating world of GPGPU.

References

[Müller 2008] Matthias Müller, Jos Stam, Doug James, and Nils Thürey. "Real Time Physics." ACM SIGGRAPH 2008 Course Notes. Available at http://www.matthiasmueller.info/realtimephysics/index.html.

[Nvidia 2010] "NVIDIA CUDA Best Practices Guide," Version 3.0, 2010. Available at http://developer.download.nvidia.com/compute/cuda/3_0/toolkit/docs/NVIDIA_CUDA_BestPracticesGuide.pdf.

[Tejada 2005] Eduardo Tejada and Thomas Ertl. "Large Steps in GPU-Based Deformable Bodies Simulation." *Simulation Modelling Practice and Theory* 13:8 (November 2005), pp. 703–715.

23

A Jitter-Tolerant Rigid Body Sleep Condition

Eric Lengyel
Terathon Software

23.1 Introduction

One of the primary optimizations employed by any physics engine is the ability to put a rigid body to sleep when it has reached a resting state. A sleeping rigid body is not processed by the physics engine until some kind of event occurs, such as a collision or broken contact, necessitating that it is woken up and simulated once again. The fact that most rigid bodies are in the sleeping state at any one time is what allows a large number of simulated objects to exist in a game level without serious performance problems.

A problem faced by all physics engines is how to decide that a rigid body has actually come to rest. If we put an object to sleep without being sure that it has come to rest, then we risk accidentally freezing it in mid-simulation, which can look odd to the player. On the other hand, if we are too conservative and wait for too strict of a condition to be met before we put an object to sleep, then we can run into a situation where too many objects are being simulated unnecessarily and performance suffers.

The sleep decision problem is complicated by the fact that all physics engines exhibit some jitter no matter how good the constraint solver is. If the right sleep condition isn't chosen in the design of a physics engine, then jitter can prevent an object from ever going to sleep. Or at the very least, jitter can delay an object from entering the sleep state, and this generally causes a higher number of objects to be simulated at any given time. This chapter discusses a simple condition that can be used to determine when it is the proper time to put a rigid body to sleep, and it is highly tolerant to jitter.

23.2 The Sleep Condition

A physics engine typically puts a rigid body to sleep once some condition has been satisfied for a certain number of simulation steps. The most obvious condition to test is that the linear velocity and angular velocity of the rigid body both stay under particular thresholds. However, using such a condition is not robust because even if the object isn't really going anywhere, jitter can cause its step-to-step velocity to be higher than the threshold value.

The solution is to maintain an axis-aligned world-space bounding box for a rigid body's center of mass $\mathbf{C}$. We begin with a bounding box whose minimal and maximal corners are both $\mathbf{C}$ (so that the box's dimensions are zero), and we expand the box at each simulation step to include the new center of mass. After a certain number of steps n have been accumulated, we look at the size of the bounding box. If the largest dimension is less than a certain threshold value t, then we can put the object to sleep. The object can jitter around inside the box all it wants, but as long as it doesn't move further than t units of distance along any axis over n steps, then it still goes to sleep.

Of course, this solution isn't quite enough. The rigid body could be rotating about its center of mass without any linear translation, in which case the center of mass stays inside our bounding box, and the physics engine mistakenly decides to put it to sleep. To prevent this from happening, we can add a second bounding box away from the center of mass that represents the volume containing some other point that moves with the rigid body. The easiest point to choose would be a point one unit away from the center of mass along the object's world-space x-axis. If the matrix $\begin{bmatrix} \mathbf{x} & \mathbf{y} & \mathbf{z} \end{bmatrix}$ transforms the rigid body into world space, then our new point is given by $\mathbf{C} + \mathbf{x}$. As we accumulate a bounding box for the point $\mathbf{C}$, we also accumulate a second bounding box for the point $\mathbf{C} + \mathbf{x}$ calculated at every step. If the size of both boxes stays under our threshold t for n steps, then the object can be put to sleep.

We aren't quite done yet. The rigid body could be rotating about the axis $\mathbf{x}$ through its center of mass, in which case the points $\mathbf{C}$ and $\mathbf{C} + \mathbf{x}$ would stay inside their bounding box thresholds, but the object should not be put to sleep. We need to add one more test point that lies off of the x-axis. The most straightforward choice would be a point one unit away from the center of mass along the y-axis. So we maintain bounding boxes for the three points $\mathbf{C}$, $\mathbf{C} + \mathbf{x}$, and $\mathbf{C} + \mathbf{y}$, as shown in Figure 23.1. Using three points that are not collinear ensures that there is no motion that a rigid body can undergo that would be mistaken for a resting state.

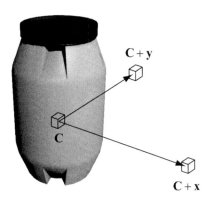

Figure 23.1. Bounding boxes are maintained for the center of mass **C** and the two points **C** + **x** and **C** + **y**, where **x** and **y** are the first two columns of the matrix transforming the rigid body into world space.

The number of steps n and the bounding box size threshold t are parameters that can be adjusted in the physics engine. Typically, if the three test points remain inside small enough bounding boxes for one second, then it's safe to put a rigid body to sleep. The size threshold for the two points away from the center of mass can be different from that used for the center of mass in order to impose angular velocity limits.

Whenever the sleep test fails because one of the bounding boxes has grown too large during a particular simulation step, the whole test should be reset. That is, the current values of **C**, **C** + **x**, and **C** + **y** should be used to reinitialize the bounding boxes, and the resting step count should be restarted at one. For a rigid body that is actively moving in some way, this reset occurs on every simulation step.

Part III

Systems Programming

Part III

24

Bit Hacks for Games

Eric Lengyel
Terathon Software

Game programmers have long been known for coming up with clever tricks that allow various short calculations to be performed more efficiently. These tricks are often applied inside tight loops of code, where even a tiny savings in CPU clock cycles can add up to a significant boost in speed overall. The techniques usually employ some kind of logical bit manipulation, or "bit twiddling," to obtain a result in a roundabout way with the goal of reducing the number of instructions, eliminating expensive instructions like divisions, or removing costly branches. This chapter describes a variety of interesting bit hacks that are likely to be applicable to game engine codebases.

Many of the techniques we describe require knowledge of the number of bits used to represent an integer value. The most efficient implementations of these techniques typically operate on integers whose size is equal to the native register width of the CPU running the code. This chapter is written for integer registers that are 32 bits wide, and that is the size assumed for the int type in C/C++. All of the techniques can be adapted to CPUs having different native register widths (most commonly, 64 bits) by simply changing occurrences of the constants 31 and 32 in the code listings and using the appropriately sized data type.

24.1 Integer Sign Manipulation

We begin with a group of techniques that can be used to extract or modify information about the sign of an integer value. These techniques, and many more throughout this chapter, rely on the assumption that a signed right shift using the >> operator preserves the sign of the input by smearing the original sign bit across the high bits opened by the shift operation. Strictly speaking, shifting a negative number to the right produces implementation-defined results according to the C++ standard[1], but all compilers likely to be encountered in game development give the expected outcome—if the bits are shifted to the right by n bit positions, then the value of the highest bit of the input is replicated to the highest n bits of the result.

Absolute Value

Most CPU instruction sets do not include an integer absolute value operation. The most straightforward way of calculating an absolute value involves comparing the input to zero and branching around a single instruction that negates the input if it happens to be less than zero. These kinds of code sequences execute with poor performance because the branch prevents good instruction scheduling and pollutes the branch history table used by the hardware for dynamic branch prediction.

A better solution makes clever use of the relationship

$$-x = \sim x + 1, \tag{24.1}$$

where the unary operator $\sim$ represents the bitwise NOT operation that inverts each bit in x. In addition to using the $\sim$ operator in C/C++, the NOT operation can be performed by taking the exclusive OR between an input value and a value whose bits are all 1s, which is the representation of the integer value -1. So we can rewrite Equation (24.1) in the form

$$-x = (x \wedge -1) - (-1), \tag{24.2}$$

where the reason for subtracting -1 at the end will become clear in a moment.

If we shift a signed integer right by 31 bits, then the result is a value that is all ones for any negative integer and all zeros for everything else. Let m be the value of x shifted right by 31 bits. Then a formula for the absolute value of x is given by

$$|x| = (x \wedge m) - m. \tag{24.3}$$

If $x < 0$, then $m = -1$ and this is equivalent to Equation (24.2). If $x \geq 0$, then $m = 0$, and the right-hand side of Equation (24.3) becomes a no-op because performing an exclusive OR with zero does nothing and subtracting zero does nothing. So now we have a simple formula that calculates the absolute value in three instructions, as shown in Listing 24.1. The negative absolute value operation is also

---

[1] See Section 5.8, Paragraph 3 of the C++ standard.

shown, and it is achieved by simply reversing the subtraction on the right-hand side of Equation (24.3).

Since there is no branching in Listing 24.1 (and the functions are inlined), the basic block sizes are larger in the calling code, and the compiler is free to schedule other instructions among the instructions used for the absolute value. This results in higher instruction throughput, and thus faster code.

```
inline int Abs(int x)
{
    int m = x >> 31;        // m = (x < 0) ? -1 : 0
    return ((x ^ m) - m);
}

inline int Nabs(int x)
{
    int m = x >> 31;        // m = (x < 0) ? -1 : 0
    return (m - (x ^ m));
}
```

Listing 24.1. These functions calculate the absolute value and negative absolute value.

Note that the absolute value function breaks down if the input value is 0x80000000. This is technically considered a negative number because the most significant bit is a one, but there is no way to represent its negation in 32 bits. The value 0x80000000 often behaves as though it were the opposite of zero, and like zero, is neither positive nor negative. Several other bit hacks discussed in this chapter also fail for this particular value, but in practice, no problems typically arise as a result.

Sign Function

The sign function $\mathrm{sgn}(x)$ is defined as

$$\mathrm{sgn}(x) = \begin{cases} 1, & \text{if } x > 0; \\ 0, & \text{if } x = 0; \\ -1, & \text{if } x < 0. \end{cases} \tag{24.4}$$

This function can be calculated efficiently without branching by realizing that for a nonzero input x, either x >> 31 is all ones or -x >> 31 is all ones, but not both. If x is zero, then both shifts also produce the value zero. This leads us to the code shown in Listing 24.2, which requires four instructions. Note that on PowerPC processors, the sign function can be evaluated in three instructions, but that sequence makes use of the carry bit, which is inaccessible in C/C++. (See [Hoxey et al. 1996] for details.)

```
inline int Sgn(int x)
{
    return ((x >> 31) - (-x >> 31));
}
```

Listing 24.2. This function calculates the sign function given by Equation (24.4).

Sign Extension

Processors typically have native instructions that can extend the sign of an 8-bit or 16-bit integer quantity to the full width of a register. For quantities of other bit sizes, a sign extension can be achieved with two shift instructions, as shown in Listing 24.3. An n-bit integer is first shifted right by $32 - n$ bits so that the value occupies the n most significant bits of a register. (Note that this destroys the bits of the original value that are shifted out, so the state of those bits can be ignored in cases when an n-bit quantity is being extracted from a larger data word.) The result is shifted right by the same number of bits, causing the sign bit to be smeared.

On some PowerPC processors, it's important that the value of n in Listing 24.3 be a compile-time constant because shifts by register values are microcoded and cause a pipeline flush.

```
template <int n> inline int ExtendSign(int x)
{
    return (x << (32 - n) >> (32 - n));
}
```

Listing 24.3. This function extends the sign of an n-bit integer to a full 32 bits.

24.2 Predicates

We have seen that the expression x >> 31 can be used to produce a value of 0 or −1 depending on whether x is less than zero. There may also be times when we want to produce a value of 0 or 1, and we might also want to produce these values based on different conditions. In general, there are six comparisons that we can make against zero, and an expression generating a value based on these comparisons is called a *predicate*.

Table 24.1 lists the six predicates and the branchless C/C++ code that can be used to generate a 0 or 1 value based on the boolean result of each comparison. Table 24.2 lists negations of the same predicates and the code that can be used to generate a mask of all 0s or all 1s (or a value of 0 or −1) based on the result of each comparison. The only difference between the code shown in Tables 24.1 and 24.2 is that the code in the first table uses unsigned shifts (a.k.a. logical shifts), and the second table uses signed shifts (a.k.a. arithmetic or algebraic shifts).

Predicate	Code	Instructions
x = (a == 0);	x = (unsigned) ~(a | -a) >> 31;	4
x = (a != 0);	x = (unsigned) (a | -a) >> 31;	3
x = (a > 0);	x = (unsigned) -a >> 31;	2
x = (a < 0);	x = (unsigned) a >> 31;	1
x = (a >= 0);	x = (unsigned) ~a >> 31;	2
x = (a <= 0);	x = (unsigned) (a - 1) >> 31;	2

Table 24.1. For each predicate, the code generates the value 1 if the condition is true and generates the value 0 if the condition is false. The type of a and x is signed integer.

Predicate	Code	Instructions
x = -(a == 0);	x = ~(a | -a) >> 31;	4
x = -(a != 0);	x = (a | -a) >> 31;	3
x = -(a > 0);	x = -a >> 31;	2
x = -(a < 0);	x = a >> 31;	1
x = -(a >= 0);	x = ~a >> 31;	2
x = -(a <= 0);	x = (a - 1) >> 31;	2

Table 24.2. For each predicate, the code generates the value −1 if the condition is true and generates the value 0 if the condition is false. The type of a and x is signed integer.

On PowerPC processors, the `cntlzw` (count leading zeros word) instruction can be used to evaluate the first expression in Table 24.1, (a == 0), by calculating `cntlzw(a) >> 5`. This works because the `cntlzw` instruction produces the value 32 when its input is zero and a lower value for any other input. When shifted right five bits, the value 32 becomes 1, and all other values are completely shifted out to produce 0. A similar instruction called BSR (bit scan reverse) exists on x86 processors, but it produces undefined results when the input is zero, so it cannot achieve the same result without a branch to handle the zero case.

Predicates can be used to perform a variety of conditional operations. The expressions shown in Table 24.1 are typically used to change some other value by one (or a power of two with the proper left shift), and the expressions shown in Table 24.2 are typically used as masks that conditionally preserve or clear some other value. We look at several examples in the remainder of this section.

Conditional Increment and Decrement

To perform conditional increment or decrement operations, the expressions shown in Table 24.1 can simply be added to or subtracted from another value, respectively. For example, the conditional statement

```
if (a >= 0) x++;
```

can be replaced by the following non-branching code:

```
x += (unsigned) ~a >> 31;
```

Conditional Addition and Subtraction

Conditional addition and subtraction can be performed by using the expressions shown in Table 24.2 to mask the operations. For example, the conditional statement

```
if (a >= 0) x += y;
```

can be replaced by the following non-branching code:

```
x += y & (~a >> 31);
```

Increment or Decrement Modulo N

The mask for the predicate (a < 0) can be used to implement increment and decrement operations modulo a number n. Incrementing modulo 3 is particularly common in game programming because it's used to iterate over the vertices of a triangle or to iterate over the columns of a 3×3 matrix from an arbitrary starting index in the range $[0, 2]$.

To increment a number modulo n, we can subtract $n - 1$ from it and compare against zero. If the result is negative, then we keep the new value; otherwise, we wrap around to zero. For the decrement operation, we subtract one and then compare against zero. If the result is negative, then we add the modulus n. These operations are shown in Listing 24.4, and both generate four instructions when n is a compile-time constant.

```
template <int n> inline int IncMod(int x)
{
    return ((x + 1) & ((x - (n - 1)) >> 31));
}

template <int n> inline int DecMod(int x)
{
    x--;
    return (x + ((x >> 31) & n));
}
```

Listing 24.4. These functions increment and decrement the input value modulo n.

Clamping to Zero

Another use of masks is clamping against zero. The minimum and maximum functions shown in Listing 24.5 take a single input and clamp to a minimum of zero or a maximum of zero. On processors that have a logical AND with complement instruction, like the PowerPC, both of these functions generate only two instructions.

```
inline int MinZero(int x)
{
    return (x & (x >> 31));
}

inline int MaxZero(int x)
{
    return (x & ~(x >> 31));
}
```

Listing 24.5. These functions take the minimum and maximum of the input with zero.

Minimum and Maximum

Branchless minimum and maximum operations are difficult to achieve when they must work for the entire range of 32-bit integers. (However, see [1] for some examples that use special instructions.) They become much easier when we can assume that the difference between the two input operands doesn't underflow or overflow when one is subtracted from the other. Another way of putting this is to say that the input operands always have two bits of sign or that they are always in the range $\left[-2^{30}, 2^{30} - 1\right]$. When this is the case, we can compare the difference to zero in order to produce a mask used to choose the minimum or maximum value. The code is shown in Listing 24.6. Both functions generate four instructions if a logical AND with complement is available; otherwise, the `Min()` function generates five instructions.

24.3 Miscellaneous Tricks

This section presents several miscellaneous tricks that can be used to optimize code. Some of the tricks are generic and can be applied to many different situations, while others are meant for specific tasks.

Clear the Least Significant 1 Bit

The least significant 1 bit of any value x can be cleared by logically ANDing it with $x - 1$. This property can be used to count the number of 1 bits in a value by repeatedly clearing the least significant 1 bit until the value becomes zero. (See [Anderson 2005] for more efficient methods, however.)

```
inline int Min(int x, int y)
{
    int a = x - y;
    return (x - (a & ~(a >> 31)));
}

inline int Max(int x, int y)
{
    int a = x - y;
    return (x - (a & (a >> 31)));
}
```

Listing 24.6. These functions return the minimum and maximum of a pair of integers when we can assume two bits of sign.

Test for Power of Two

If we clear the least significant 1 bit and find that the result is zero, then we know that the original value was either zero or was a power of two because only a single bit was set. Functions that test whether a value is a power of two, with and without returning a positive result for zero, are shown in Listing 24.7.

```
inline bool PowerOfTwo(int x)
{
    int y = x - 1;    // y is negative only if x == 0.
    return ((x & y) - (y >> 31) == 0);
}

inline bool PowerOfTwoOrZero(int x)
{
    return ((x & (x - 1)) == 0);
}
```

Listing 24.7. These functions test whether a value is a power of two.

Test for Power of Two Minus One

The bits to the right of the least significant 0 bit of any value x can be cleared by logically ANDing it with $x+1$. If the result is zero, then that means that the value was composed of a contiguous string of n 1 bits with no trailing 0 bits, which is the representation of $2^n - 1$. Note that this test gives a positive result for zero, which is correct because zero is one less than 2^0.

Determine Whether a Voxel Contains Triangles

In the marching cubes algorithm, an 8-bit code is determined for every cell in a voxel grid, and this code maps to a set of triangulation cases that tell how many vertices and triangles are necessary to extract the isosurface within the cell. The codes 0x00 and 0xFF correspond to cells containing no triangles, and such cells are skipped during the mesh generation process. We would like to avoid making two comparisons and instead make only one so that empty cells are skipped more efficiently.

Suppose that voxel values are signed 8-bit integers. We form the case code by shifting the sign bits from the voxels at each of the eight cell corners into a different position in an 8-bit byte, as shown in Listing 24.8. We then exclusive OR the case code with any one of the voxel values shifted right seven bits. The result is zero for exactly the cases 0x00 and 0xFF, and it's nonzero for everything else.

```
unsigned long caseCode   = ((corner[0] >> 7) & 0x01)
                         | ((corner[1] >> 6) & 0x02)
                         | ((corner[2] >> 5) & 0x04)
                         | ((corner[3] >> 4) & 0x08)
                         | ((corner[4] >> 3) & 0x10)
                         | ((corner[5] >> 2) & 0x20)
                         | ((corner[6] >> 1) & 0x40)
                         | (corner[7] & 0x80);

if ((caseCode ^ ((corner[7] >> 7) & 0xFF)) != 0)
{
    // Cell has a nontrivial triangulation.
}
```

Listing 24.8. The case code for a cell is constructed by shifting the sign bits from the eight corner voxel values into specific bit positions. One of the voxel values is shifted seven bits right to produce a mask of all 0s or all 1s, and it is then exclusive ORed with the case code to determine whether a cell contains triangles.

Determine the Index of the Greatest Value in a Set of Three

Given a set of three values (which could be floating-point), we sometimes need to determine the index of the largest value in the set, in the range $[0,2]$. In particular, this arises when finding the support point for a triangle in a given direction for the Gilbert-Johnson-Keerthi (GJK) algorithm. It's easy to perform a few comparisons and return different results based on the outcome, but we would like to eliminate some of the branches involved in doing that.

For a set of values $\{v_0, v_1, v_2\}$, Table 24.3 enumerates the six possible combinations of the truth values for the comparisons $v_1 > v_0$, $v_2 > v_0$, and $v_2 > v_1$, which we label as b_0, b_1, and b_2, respectively. As it turns out, the sum of $b_0 | b_1$ and $b_1 \& b_2$ produces the correct index for all possible cases. This leads us to the code shown in Listing 24.9.

Case	$b_0 = (v_1 > v_0)$	$b_1 = (v_2 > v_0)$	$b_2 = (v_2 > v_1)$	$b_0 \mid b_1$	$b_0 \& b_1$	Sum
v_0 largest	0	0	0	0	0	0
	0	0	1	0	0	0
v_1 largest	1	0	0	1	0	1
	1	1	0	1	0	1
v_2 largest	0	1	1	1	1	2
	1	1	1	1	1	2
Impossible	0	1	0	1	0	1
	1	0	1	1	0	1

Table 24.3. This table lists all possible combinations for the truth values b_0, b_1, and b_2 relating the values v_0, v_1, and v_2. The sum of $b_0 \mid b_1$ and $b_1 \& b_2$ gives the index of the largest value.

```
template <typename T> int GetGreatestValueIndex(const T *value)
{
    bool b0 = (value[1] > value[0]);
    bool b1 = (value[2] > value[0]);
    bool b2 = (value[2] > value[1]);
    return ((b0 | b1) + (b1 & b2));
}
```

Listing 24.9. This function returns the index, in the range $[0, 2]$, corresponding to the largest value in a set of three.

24.4 Logic Formulas

We end this chapter with Table 24.4, which lists several simple logic formulas and their effect on the binary representation of a signed integer. (See also [Ericson 2008].) With the exception of the last entry in the table, all of the formulas can be calculated using two instructions on the PowerPC processor due to the availability of the andc, orc, and eqv instructions. Other processors may require up to three instructions.

Formula	Operation / Effect	Notes
`x & (x - 1)`	Clear lowest 1 bit.	If result is 0, then x is 2^n.
`x | (x + 1)`	Set lowest 0 bit.	
`x | (x - 1)`	Set all bits to right of lowest 1 bit.	
`x & (x + 1)`	Clear all bits to right of lowest 0 bit.	If result is 0, then x is $2^n - 1$.
`x & -x`	Extract lowest 1 bit.	
`~x & (x + 1)`	Extract lowest 0 bit (as a 1 bit).	
`~x | (x - 1)`	Create mask for bits other than lowest 1 bit.	
`x | ~(x + 1)`	Create mask for bits other than lowest 0 bit.	
`x | -x`	Create mask for bits left of lowest 1 bit, inclusive.	
`x ^ -x`	Create mask for bits left of lowest 1 bit, exclusive.	
`~x | (x + 1)`	Create mask for bits left of lowest 0 bit, inclusive.	
`~x ^ (x + 1)`	Create mask for bits left of lowest 0 bit, exclusive.	Also $x \equiv (x + 1)$.
`x ^ (x - 1)`	Create mask for bits right of lowest 1 bit, inclusive.	0 becomes -1.
`~x & (x - 1)`	Create mask for bits right of lowest 1 bit, exclusive.	0 becomes -1.
`x ^ (x + 1)`	Create mask for bits right of lowest 0 bit, inclusive.	-1 remains -1.
`x & (~x - 1)`	Create mask for bits right of lowest 0 bit, exclusive.	-1 remains -1.

Table 24.4. Logic formulas and their effect on the binary representation of a signed integer.

References

[Anderson 2005] Sean Eron Anderson. "Bit Twiddling Hacks." 2005. Available at http://graphics.stanford.edu/~seander/bithacks.html.

[Ericson 2008] Christer Ericson. "Advanced Bit Manipulation-fu." realtimecollision detection.net - the blog, August 24, 2008. Available at http://realtimecollision detection.net/blog/?p=78.

[Hoxey et al. 1996] Steve Hoxey, Faraydon Karim, Bill Hay, and Hank Warren, eds. *The PowerPC Compiler Writer's Guide*. Palo Alto, CA: Warthman Associates, 1996.

25

Introspection for C++ Game Engines

Jon Watte
IMVU, Inc.

25.1 Introduction

This gem describes a mechanism for adding general-purpose introspection of data types to a C++ program, using a minimum of macros, mark-up, or repetition. Introspection is the capability of a computer program to look at its own data and make modifications to it. Any high-level language of note has a rich introspection API, generally as part of an even larger reflection API, but users of C++ have had to make do with the built-in class type_info ever since the early days. A conversation with the introspection system of a language such as C#, Java, or Python might look something like this:

> "Hello, introspection system, I'd like you to tell me a bit about this here piece of data I have!"

> "Why, certainly, game program! It's a data structure you'd like to call a TreasureChest."

> "That's useful to know, but what I really want to know is whether it has a property called 'Position'?"

> "Yes, it does! It's of type float3, containing the position in the world."

> "That's great! Actually, now that I think about it, the designer just clicked on it, and wants to edit all the properties. What are they?"

> "Well, we've got 'Name', which is a string, and 'Contents', which is a list of references to Object templates, and 'Model', which is a string used to reference the 3D mesh used to render the object, and …"

In code, it looks something more like Listing 25.1 (using C#/.NET).

```
string typeName = theObject.GetType().Name;
PropertyInfo info = theObject.GetType().GetProperty("Position");
/* ... */
foreach (PropertyInfo pi in theObject.GetType().GetProperties())
{
    theEditor.AddProperty(theObject, pi.Name,
        new GetterSetter(pi.ReflectedType));
}
```

Listing 25.1. Properties in a dynamic language.

So, what can we do in C++? Using only the standard language and library, we can't do much. You can find out the name of a type (using `typeid`), and the size of a type (using `sizeof`), and that's about it. You can also find out whether a given object instance derives from a given base class (using the horribly expensive `dynamic_cast<>`), but only if the type has a virtual table, and you can't iterate over the set of base classes that it derives from, except perhaps by testing against all possible base classes. Despite these draw-backs, C++ game programmers still need to deal with data, display it in editors, save it to files, send it over networks, wire compatible properties together in object/component systems, and do all the other data-driven game development magic that a modern game engine must provide.

Trying to solve this problem, many engines end up with a number of different mechanisms to describe the data in a given object, entity, or data structure. For example, you may end up having to write code that's something like what is shown in Listing 25.2.

```
class TreasureChest : public GameObject
{
    private:

        /* properties */
        string          name;
        float3          position;
        list<ObjectRef> contents;
        string          model;

    public:
```

```
/* saving */
void WriteOut(Archive &ar)
{
    ar.beginObject("TreasureChest");
    GameObject::WriteOut(ar);
    ar.write(name);
    ar.write(position);
    ar.write(contents.size());

    for (list<ObjectRef>::iterator ptr(contents.begin()),
        end(contents.end()); ptr != end; ++ptr)
    {
        ar.write(*ptr);
    }

    ar.write(model);
    ar.endObject();
}

/* loading */
void ReadIn(Archive &ar)
{
    ar.beginObject("TreasureChest");
    GameObject::ReadIn(ar);
    ar.read(name);
    ar.read(position);
    size_t size;
    ar.read(size);

    while (size-- > 0)
    {
        contents.push_back(ObjectRef());
        ar.read(contents.back());
    }

    ar.read(model);
    ar.endObject();
}

#if EDITOR
```

```
        /* editing */
        void AddToEditor(Editor &ed)
        {
            ed.beginGroup("TreasureChest");
            GameObject::AddToEditor(ed);
            ed.addString(name, "name", "The name of the object");
            ed.addPosition(position, "position",
                "Where the object is in the world");
            ed.addObjectRefCollection(contents, "contents",
                "What's in this chest");
            ed.addFilename(model, "model",
                "What the chest looks like", "*.mdl");
            ed.endGroup();
        }

    #endif
}
```

Listing 25.2. `TreasureChest` introspection, the verbose way.

This approach has several problems, however. For example, each addition of a new property requires adding it in several parts of the code. You have to remember the order in which properties are serialized, and you generally have to write even more support code to support cloning, templating, and other common operations used by most game engines and editors. More than one game has shipped with bugs caused by getting one of these many manual details wrong. If you apply the technique in this gem, you will be able to avoid this whole class of bugs—as well as avoid a lot of boring, error-prone typing. Next to having the language run-time library do it for you, this is the best you can get!

25.2 The Demo

The code that comes with this gem implements a simple network protocol for a text-based chat server. A user can join a chat, get a list of other users in the chat, and send and receive chat text. The chat server itself stores and edits some information about each user, such as an email address and a password. While simple, this is enough to show how to apply the mechanism of this gem to solve a variety of problems in a real program.

There are four related programs:

- A test program, which exercises the API mechanisms one at a time to let you easily verify the functionality of the different pieces.
- A server program, which lets you host a simple chat server (from the command line).
- A client program, which lets you connect to a running chat server and exchange chat messages with other users (again, from the command line).
- An editor program, which lets you edit the list of users used by the server program, using the introspection mechanisms outlined in this chapter.

These are baked into two executables: the `introspection` test program, to verify that the API works as intended, and the `simplechat` program, to act as a client, server, or user list editor for the simple chat system.

To build the programs, either use the included solution and project files for Microsoft Visual Studio 2010 (tested on Windows 7) or use the included GNU make file for GCC (tested on Ubuntu Linux 10.04). Run the sample programs from the command line.

25.3 The Gem

The implementation of this gem lives in the files `introspection.h`, `introspection.cpp`, and `protocol.cpp`. While the implementation makes use of type inference and other template metaprogramming tricks, the API from a user's perspective is designed to be simple.

Decorate each data structure you want to introspect using an `INTROSPECTION()` macro, as illustrated by the example shown in Listing 25.3. This macro takes two arguments: the name of the data structure itself and a list of members that you want to introspect. Each introspection member is described using a separate macro that lists its name and a description of the role of the member. This can be as simple as a plain C string (to display in an editor interface, for example) or as complex as a rules-enforcing and network-encoding object that makes sure floating-point values are within a given range and transmitted using a specified number of bits and other variant behavior.

While you still have to list each member twice, once for declaring it as an actual member and once for declaring it introspectable, this mechanism cuts down on the amount of boilerplate code you need to write. It also keeps the declaration of introspection in the header, right near the declaration of the members, to reduce the likelihood that someone adds a member without adding the appro-

```
#include <introspection/introspection.h>

struct UserInfo
{
    std::string        name;
    std::string        email;
    std::string        password;
    int                shoe_size;

    INTROSPECTION
    (
        UserInfo,
        MEMBER(name, "user name")
        MEMBER(email, "e-mail address")
        MEMBER(password, "user password")
        MEMBER(shoe_size,
            introspection::int_range("shoe size (European)", 30, 50))
    );
};
```

Listing 25.3. Introspection example (the simple way).

priate introspection information. Compared to the previous example, this is a significant improvement in both maintainability and ease of use!

The new example uses the standard C++ library string class for storage of strings. This is because, when dealing with strings of indeterminate size (such as those found in files, network packets, or graphical user interface editors), memory allocation and deallocation can otherwise become a problem. Because every C++ compiler comes with a decent implementation of the string class, as well as container classes like set, list, and vector, I have chosen to use those implementations in this gem rather than build custom, game-specific containers. If you have special needs, such as fitting into a tight console memory space with a fixed-address memory allocator, adapting the implementation to other string and container classes is quite possible—read the implementation in introspection.h for more details.

To use the code as is in your own game, it's easiest to copy the files introspection.cpp, introspection.h, not_win32.h, and protocol.cpp into your own project where they can be easily found. A more structured approach would be to build a library out of the .cpp files and set the include files in the project settings (or the makefile) to reference the parent of the introspection directory. Note that include files in the samples are included using angle brackets, naming the introspection directory <introspection/introspection.h>.

Given the above declaration, you can now do a number of interesting things to any instance of struct UserInfo. Most importantly, you can get a list of the members of the structure, as well as their type and offset within the structure, programmatically. The function test_introspection() in the sample main.cpp file shows how to do this and is illustrated in Listing 25.4.

```
void test_introspection()
{
    std::stringstream    ss;

    const type_info_base& tib = UserInfo::member_info();
    for (member_t::iterator ptr(tib.begin()), end(tib.end());
            ptr != end; ++ptr)
    {
        ss << "member: " << (*ptr).name() <<
            " desc: " << (*ptr).info().desc() <<
            " offset: " << (*ptr).access().offset() <<
            " size: " << (*ptr).access().size();

        if ((*ptr).access().compound())
        {
            ss << " [compound]";
        }

        if ((*ptr).access().collection())
        {
            ss << " {collection}";
        }

        ss << std::endl;
        std::cout << ss.str();
        ss.str("");
    }
}
```

Listing 25.4. Implementation of the test_introspection() function.

The name of the type that contains information about each compound type is `type_info_base`. It contains standard-template-library-style iterator accessors `begin()` and `end()` to iterate over the members of the type, each of which is described using a `member_t` instance. Additionally, `type_info_base` contains an `access()` accessor for the `member_access_base` type, which implements operations on the compound type itself, such as serializing it to and from a binary stream, converting it to and from a text representation, and creating and destroying instances in raw memory. Each type (including the basic types like `int` and `float`) has a corresponding `member_access_base`, so this structure tells you whether a type is compound (such as `struct UserInfo`) and whether it is a collection (such as a list or a vector).

25.4 Lifting the Veil

Examining every line of the implementation of the gem in detail is beyond the scope of this chapter, but there are a few important things that deserve particular attention.

Syntactically, the simple INTROSPECTION (*name, member member* ...) macro takes only two arguments. Instead of listing each member using a comma, each member is listed as a separate MEMBER() macro, which in turn expands to contain the comma if the implementation requires it. The current implementation expands the INTROSPECTION() macro to a static inline member function that returns a static local instance of the member description type, and it initializes that instance using the expansion of the member macros. The implementation of the macros is shown in Listing 25.5.

```
#define INTROSPECTION(type, members) \
    typedef type self_t; \
    static inline const introspection::type_info_base& member_info() \
    { \
        static introspection::member_t data[] = \
        { \
            members \
        }; \
        \
        static introspection::type_info_t<type> info( \
            data, sizeof(data) / sizeof(data[0]), \
            introspection::struct_access_t<type>::instance()); \
        \
```

```
        return (info); \
    }

#define MEMBER(name, desc) \
    introspection::member_instance(&self_t::name, #name, desc),
```

Listing 25.5. The INTROSPECTION macro implementation.

This keeps the namespace of the introspected type (such as UserInfo) clean, only introducing the typedef self_t and the member_info() member function. Access to each member of the structure is done using pointer-to-member syntax. However, the actual pointer to member does not need to be dereferenced because template metaprogramming can pick it apart and turn it into an offset-and-typecast construct, to save run-time CPU cycles.

25.5 To and From a Network

When you receive a piece of data from a network, you need to instantiate the right data type, based on what the received packet is. Once the data type is dispatched, you have to correctly destroy the instance again, freeing up any additional memory used by members of types such as string or list. The introspection mechanism of this gem builds this into the member_access_base type, giving you the size() function to let you know how much memory to allocate, and create() and destroy() functions to manage the lifetime of the structure instance.

Additionally, the gem provides the PROTOCOL() and PDU() macros to allow definition of a suite of related data types. Each data type added with the PDU() macro is given a small integer ID, which you can use on the network to identify which kind of packet is being sent. Given the PROTOCOL instance, you can map a type to an ID and an ID to a type using the protocol_t type members code() (to get the integer code for a type) and type() (to get the type_info_base for a code).

On top of this, you can build a type-safe protocol packet dispatcher, where each dispatcher is selected by the appropriate type. End to end, the networking API then uses the actual structure type to select which code to prefix the packet with and serialize the structure to a sequence of bytes using the serialization from the INTROSPECTION() macro. It then sends this data to the other end, prefixed by the size of the packet. The receiving end first decodes the integer of the packet type and then selects the appropriate unpacker type instance based on the code.

Using the `size()` and `create()` functions, it inflates an instance of the appropriate structure into an allocated array of bytes and constructs it using the appropriate C++ constructor, after which it passes the instance through a dispatcher selected by the integer code (type) of the packet. When the dispatcher returns, the `destroy()` function calls the appropriate C++ destructor, and the memory can be returned to the system as an array of bytes again.

This is illustrated in more detail in the `client` and `server` programs, where it is implemented in the `protocol_t::encode()` and `protocol_t::decode()` functions, respectively, using the memory stream functions of the `simple_stream` class. A good example use of this mechanism is the `send_a_message()` function from the `client.cpp` file, as shown in Listing 25.6.

```cpp
static bool send_a_message(const char *line)
{
    SaySomethingPacket    ssp;
    simple_stream         ss;

    ssp.message = line;

    // Make space for the frame size field (short).
    ss.write_bytes(2, "\0");
    my_proto.encode(ssp, ss);
    size_t plen = ss.position() - 2;
    unsigned char *pdata = (unsigned char *) ss.unsafe_data();

    // Generate a big-endian short for number of bytes count.
    pdata[0] = (plen >> 8) & 0xFF;
    pdata[1] = plen & 0xFF;
    int l = send(sockfd, (const char *) pdata, plen + 2, 0);
    if (l < 0)
    {
        // WSAGetLastError() may have been cleared by the other thread.
        fprintf(stderr, "send error: %d\n", WSAGetLastError());
    }

    return (l == plen + 2);
}
```

Listing 25.6. Sending a C++ data structure using automatic marshaling.

25.6 In Closing

Finally, I have set up a forum for discussion about this gem, as well as any errata and code update releases, that you can find at my website: http://www. enchantedage.com/geg2-introspection. I hope you find that this gem saves you a lot of error-prone typing, and I hope to hear your experiences and feedback in the forum!

26

A Highly Optimized Portable
Memory Manager

Jason Hughes
Steel Penny Games, Inc.

26.1 Introduction

Every game has a memory manager of some sort. On PCs, this tends to be Microsoft's C run-time library memory manager. On the various consoles, it's likely to either be the platform-specific memory manager that was written by the hardware vendor or the one provided by the compiler company in its run-time library. Do they all work the same way? No. Do they exhibit the same performance characteristics? Absolutely no. Some allow the heap to become fragmented very quickly, while others may be very slow for small allocations or when the number of allocations becomes quite large, and still others may have a high per-allocation overhead that invisibly eats away at your memory.

Memory allocation is a fundamental operation, thus, it has to satisfy a wide number of use cases robustly and efficiently. This is a serious technical challenge. Even a good implementation can harm a game's performance if exercised in just the wrong way. A naive implementation can utterly cripple performance or cause crashes due to artificial low-memory situations (e.g., fragmentation or overhead). The good news is that most of the provided memory managers are relatively efficient and work well enough for simple cases with few allocations.

After enough experiences where cross-platform stability and performance came down strictly to the memory manager, however, you may be tempted to implement your own that is scalable and easy to tune for best performance. These days, with so many platforms to support, it's a mark of quality for an engine to run well across all machines.

Alternatives to Rolling Your Own

Surely someone has already written a good heap. Doug Lea's dlmalloc is one of the best and is typically the memory manager of choice on Linux. There are many derivatives on the internet that claim some improvements, but there are many flavors to choose from. dlmalloc is a good choice primarily because it's very stable and well-tested, it's public domain code that is freely available, and it runs on 32-bit and 64-bit systems of either endianness. Very compelling.

The main gripes with dlmalloc is that it is a mess to look at, it's horrible to debug in case of heap corruption, and it's easily corrupted in case of a buffer overrun. As for usability, there's no way to create one-off heaps with it that have been optimized for special use cases, where you can predict sizes and counts of allocations. dlmalloc also has a fairly high per-allocation overhead that chews up memory. As general-purpose heaps go, it's excellent, but for games, we sometimes need custom solutions with more bells and whistles.

Desirable Properties

After writing about eight or nine different memory managers, a list of priorities emerges from recognizing the importance of certain attributes that make a memory manager good. Here are a few:

1. *Must not thrash the CPU cache.* High performance comes with limiting the number of bytes that have to interact with RAM. Whatever is touched should fit within a minimum number of cache lines, both to reduce memory latency and to limit the amount of cache disruption to the rest of the program.
2. *No searching for free space.* The naive implementation of a memory manager is a linked list of blocks that are marked free or empty. Scanning this list is hugely expensive and slow under all circumstances. Good memory managers have lookup tables of some sort that point to places where free blocks of memory of various sizes are.
3. *Minimum overhead per allocation.* Reducing overhead in a memory manager is almost like adding compression to memory—you can fit more data in the same physical space.
4. *Should be easy to debug.* Sometimes memory problems happen, and it's important to consider the difficulty in tracking down such issues. Sometimes this means temporarily adding features to the memory manager. Ideally, the debug build should do some basic instrumentation as it runs that determines whether memory has been trampled without slowing down the system.
5. *Should resist corruption.* Most memory managers are organized such that blocks of program data are sandwiched between heap tracking information.

This is particularly fragile because buffer overruns tend to be utterly fatal to the system, but not immediately. The system keeps running until the memory manager has to allocate or free memory in that region and attempts to use a pointer that is garbage. However, determining what exactly is smashing the heap is a challenge, but one made less critical if the memory manager itself doesn't crash in response to a programming error. Ideally, program data should only be able to trample other program data, and heap tracking information is separated and thus protected.

6. *Should not fragment easily.* Fragmentation leads to situations where a small allocation in the middle of memory prevents a large allocation from succeeding, even though the total number of available bytes (if you add the separate free chunks together) is more than enough to accommodate the allocation. It's also a common failure of memory managers to spend a lot of run-time cycles reducing existing fragmentation or actively attempting to prevent it. The wrong approach is creating complex policies that define where allocations should go, in what order allocations should be placed in the heap, or how to coalesce adjacent free blocks. Some systems even go to the extent of relocating memory to manually defragment the heap, similar to Java's memory model. In the end, preventing fragmentation should be a design criterion that is inherent in the system, not paid for at run time.

26.2 Overview

The following design is one such system that satisfies the minimum criteria, as specified in the introduction, though there are many others possible. There are certain trade-offs that are often made during large design processes that ultimately shape the final software product. Although we describe a specific instance of the memory manager, the version found on the website is actually a template-based solution that allows reconfiguration of many parameters so you can easily experiment with what is right for your specific situation. In fact, it would be easy to capture all allocations and deletions in a log file with your current memory manager, then replay the operations in a unit test setting to measure exactly what performance and memory fragmentation would be under realistic heap conditions.

Fighting Fragmentation Is Job #1

From experience, the greatest problem that needs to be addressed by a games memory manager is fragmentation. There are two types of fragmentation: internal and external.

- *Internal fragmentation* occurs when a single allocation is requested for M bytes, but the memory manager cannot allocate exactly M bytes, so it rounds up by an extra N bytes, so $M + N$ bytes are reserved for the allocation. Internal fragmentation can be considered a factor of systemic overhead, and is tolerable if kept to a minimum, because all that unusable memory will be fully reclaimed when the block is freed.

- *External fragmentation* occurs when memory is known to be free and available by the memory manager but has been subdivided by allocations so that it is not contiguous and, thus, cannot be allocated in a single large block. The degree of fragmentation can be measured in many ways, but developers are traditionally concerned with the "big block," that is, the single largest allocation possible at any given time. The ratio of the big block to total available memory is another meaningful way to express how fragmented memory is. As that ratio tends toward zero, the number of free regions increases while the sizes of free spaces decrease, and the largest single allocation possible becomes a fraction of total available memory.

External fragmentation, henceforth referred to simply as fragmentation, is the source of many apparent stability problems in otherwise well-written programs. For games, this is particularly challenging because in this industry, memory tends

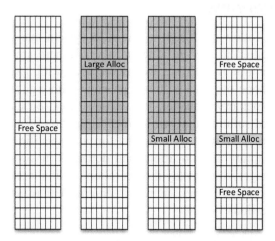

Figure 26.1. From left to right: the heap is free and unfragmented, one large allocation is made, one small allocation is made, the large allocation is freed. In the rightmost diagram, the heap is partitioned in half and exhibits an approximately 0.5 fragmentation ratio.

to be considerably more limited, and players are more likely to play the game without resetting for hours or even days on end. Thus, the dreaded "uptime crash" is almost unavoidable without special precautions, careful planning, and perhaps a custom memory management strategy that attempts to avoid fragmentation automatically.

For further clarification of how fragmentation can occur within a program, see Figure 26.1. Fragmentation can be demonstrated in three simple operations: allocate twice and free once. If one allocation is quite large and is followed by a small allocation, then once you release the large block back to the system, the memory manager now has an approximate 0.5 fragmentation ratio. If the next allocation is slightly larger than 50 percent of the total available memory, it will fail.

Most decent memory managers have allocation strategies that react differently based on the size of the allocation. Often, there is a specific handler for tiny allocations (under 256 bytes, for instance) as well as a general allocation method for everything larger. The idea behind this is primarily to reduce overhead in allocating very small blocks of memory that tend to represent the lion's share of allocations in most C/C++ programs. One happy consequence is that it prevents that group of allocations from possibly splitting the big block in half and causing fragmentation by preventing them from coming from the same memory pool.

By extending this understanding to the design of a memory manager, it is possible to reduce external fragmentation, at some small expense of internal fragmentation. To illuminate by exaggeration, if you round up all of your allocations to the largest size you'll ever allocate, then no external fragmentation is important because no allocation will ever fail due to a fragmented heap, and any allocation that gets freed can always fit any other allocation that may come in the future. Of course, all of your memory would be exhausted before that could happen! Preposterous as it is, scaling back the idea and applying it to smaller allocations, where rounding up the size is less significant, makes the concept viable. The trick is to make it fast and limit the amount of internal fragmentation (i.e., wasted space) the strategy produces.

Our approach is to make a very fast small block allocator (SBA) for allocations under 256 bytes, a reasonably fast medium block allocator (MBA) for allocations that are larger than 256 bytes but smaller than a large allocation, and a large block allocator (LBA) for any allocation of at least 4096 bytes. As mentioned above, the code on the website is templatized, so these numbers can be modified trivially for your purposes, and when set to equal sizes, you can completely remove the SBA, the MBA, or both.

Paged Allocation

It is fine to suggest different allocation strategies for different-sized memory requests, but that memory has to come from somewhere. The simplest method is to preallocate a fixed number of blocks for each allocation size and hope you don't exceed that limit. Having done this on shipped titles, we honestly don't recommend it. However, it has one benefit in that you can determine the size of an allocation based on the address alone (because it is either in the small, medium, or large region of memory). Still, it is far better to have a flexible solution that doesn't require tweaking for every project having slightly different memory allocation characteristics.

The solution is to preallocate in large batches, or pages, and string these pages together to serve as small- and medium-sized memory heaps. Ideally, these are quite large, granular pages of equal sizes for both SBA and MBA pages, or MBA pages are at least an even multiple of the page size used for SBA. This is preferable, as mentioned above, due to the inherent resistance to fragmentation when allocations are all the same size. By making page sizes large, there are fewer of them allocated, leading to fewer chances for failure and less overall CPU time spent managing pages.

Inevitably, a single page does not hold all of the allocations the game needs for a specific-sized allocation, so pages must be linked together by some data structure. We selected a linked list for pages to make the traversal cheap and to make it easy to extract or insert pages with minimal cache line misses. Searching a linked list is slow, so we come up with ways to prevent the search entirely.

Easing Portability with the OSAPI Class

The exact policy for how and where pages come from is going to vary per platform, so the memory manager template is parameterized by a class called OSAPI. OSAPI has functions to allocate and free SBA and MBA pages, allocate and free large blocks, and identify whether a pointer was allocated by an SBA, MBA, or LBA call. It also specifies the small and medium page sizes that are created when new pages are requested.

By making this a separate class, you can experiment with various page management methods without requiring modifications to the rest of the memory manager. Considering the OSAPI implementation is likely to be almost identical across platforms except for how memory is allocated, the separation may seem unnecessary, but in the event that one platform needs to pay very special attention to where memory comes from for certain kinds of data, this class makes just such an implementation simple. For example, MEM1 is faster than MEM2 on the

Nintendo Wii. It is reasonable to force all small and medium allocations to be in MEM1 by providing all pages from MEM1, and meanwhile force all large allocations to be in MEM2, which is fine because the CPU is unlikely to need direct access to large allocations.

26.3 Small Block Allocator

The SBA is implemented as a full page of equally sized blocks, minus a small section of data at the end of the page called the `Tail`, which includes a bit mask indicating which blocks are free, pointers to the previous and next pages, a count of free blocks in the page (for verification purposes only), and the number of bytes in a single block. Since the number of blocks in an SBA page depends on both the number of bytes in a block and the size of the page, this is a variable-sized structure and is thus slightly more complicated to deal with than your typical C structure. Figure 26.2 shows the memory layout for a 4-kB page. The entire `Tail` and bit mask fits entirely within a single 32-byte block, which is also smaller than the L1 and L2 cache line size for modern processors.

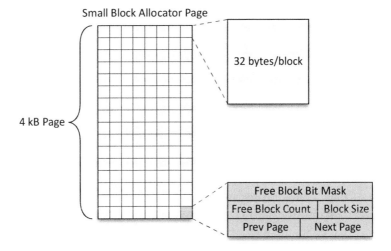

Figure 26.2. This is the memory layout for an example SBA page that totals 4 kB segmented into 32-byte allocations. Notice the tail is the last block and is quite small relative to the amount of memory managed.

Due to the added complexity of handling variable-sized structures at the Tail, the code is written in such a way that the fixed-sized portion of the structure is stored in the Tail, and the bit mask is arranged such that it ends just before the start of the Tail.

Why put the data at the end of the page rather than at the start? It's done partly out of habit and partly because placing a variable structure at the start of the page would most likely require some decision making about whether the first block should occur on the next byte, or on the next 32-byte address, or aligned based on the last block in the page. The choices are likely to be platform-specific, too. Since this is meant to be a simple, portable system that operates identically as much as possible on all platforms, the decision to move the metadata to the end of the page removes all doubt and simplifies the code somewhat. The main risk is that a very misbehaving program could smash the entire page by overrunning its memory and destroying the tracking information at the end of the page. True, but this is one reason why the free block count exists in the Tail structure—to sanity-check the validity of the block whenever problems arise.

How Alloc() Works

Once a page is found that provides memory allocations of the appropriate size, the job of the SBA is straightforward. First, we use the page size to determine the location of the Tail structure and use the block size to locate the start of the free block bit mask. Then, we scan through the entire bit mask looking for a bit set to one. A bit set to one indicates a free block at that index. Once found, the bit is set to zero, the free count in the Tail structure is decremented for accounting purposes, and a pointer is returned to the caller that points to the newly allocated block of memory.

Notice that we do not describe a failure case for allocation. Of course, we could devise a system that allows for depleted pages to coexist in the same list as pages with blocks remaining, but this would be wasteful. Completely depleted pages mixed in with those that have allocations available would always degenerate into a linked-list search through pages, ending either in an allocation or a new page being created in the list. A little experimentation with a profiler showed 75 percent of all CPU time was spent waiting for memory cache misses on each page. So, the memory manager must also handle segregating depleted pages from those that have blocks remaining. Thus, every time an allocation is requested, it is always found on the very first page it checks. The result of an allocation can cause that page to be put in the depleted page list or a new page to be created should the free block page list be empty.

How `Free()` Works

Deleting a block of memory is a thornier problem. All the system knows is that, based on the address falling within an SBA page, the SBA must free this pointer. To determine this, it must also have found the page it came from. Knowing that all pages in the SBA are the same size, the `Tail` structure is always at the same offset from the start of a page. However, the start of the bit mask varies depending on the number of blocks in the page. This can be calculated from the number of bytes per block, which we stored in the `Tail` structure for this purpose.

Based on the block size and the pointer, the index into the free block bit mask that represents this memory is simply (`ptr - pageStart`) / `blockSize`. We set the bit to one, indicating it is free for allocation. Notice that there is no need to explicitly handle adjacent free block coalescing because every block in the page is the exact size that is allocated.

Finally, the results of a free operation on a page that was completely depleted of free blocks, is to reinsert the page into the list of available memory pages.

Performance Analysis

The two major operations, allocation and deallocation, are very fast. Deallocation is a constant-time $O(1)$ operation and has no loops once the SBA page has been located. Locating the page is a separate operation, described in Section 26.6. Allocation requires scanning for set bits in a bit mask, on average $n/2$, where a page is divided into n blocks. A naive C implementation of scanning for set bits using bytes can be this slow, however, we can trivially test 32, 64, or even 128 bits at a time by loading the mask into a register and comparing against zero, generally in just two instructions. Intel CPUs can use the bit scan forward (`BSF`) instruction to check for the first set bit in either a 32- or 64-bit register in just a few cycles. In addition, for most practical page and block sizes, the entire `Tail` structure and free block bit mask fits within a single L1 cache line, ensuring the fastest possible allocation performance.

As a specific example, given a 4-kB page with 32 bytes per block, there are 127 allocatable blocks per page plus one reserved block for the `Tail` structure. This is 127 bits for the bit mask, which fits in only four 32-bit words. Worst case, obtaining an allocation requires four 4-byte requests from memory, but on average finds a free block in two requests. The first load causes a delay as a cache line is flushed and another is pulled into the L2 and L1 caches, but the second load requires only one or two cycles on typical hardware.

Memory Overhead

The SBA scheme described is very memory-efficient. For a given page, there is a fixed amount of overhead due to the `Tail` structure of exactly 12 bytes. There is also a variable amount of overhead for the bit mask, which is always rounded up to the next four-byte boundary, and occupies one bit per block in the page. For the 4-kB page with 32 bytes per block above, this works out to 16 bytes for the bit mask. So the total amount of management overhead within a single SBA page is less than two bits per allocation. In practical settings, pages are probably larger than 4 kB, resulting in overhead that rapidly approaches one bit per allocation. Contrast this with dlmalloc, which has a relatively efficient 16 bytes of overhead per allocation.

26.4 Medium Block Allocator

The MBA serves a slightly different purpose than the SBA, but does so through similar techniques. The MBA works by always rounding up allocations to a fixed block size so that fragmentation is more manageable. Similar to the SBA, the MBA expects to be handed pages of some fixed size, requiring a `Tail` structure at the end of each page to manage multiple memory pages. Also, a bit mask is used to mark individual blocks as free or available. The similarities end there.

Since the MBA must service many different allocation lengths, it is infeasible to create separate large pages just for each granular jump in size. This is particularly true since the page size is fixed—as the medium allocations increase in size, a single page might not hold more than one or two allocations of maximum length, and any remainder is wasted. Thus, a new technique is required that allows many allocations of varying length to share a single page. Further, all pages need to have the same block size, or allocation granularity, as a method of improving performance as well as simplifying the code that handles allocation and deallocation. It would be difficult, if not impossible, to predict the ideal block size for a page since allocation sizes vary, so it works out better to have the user just supply one that is globally applied for MBA pages.

The logical first step is to allow sequential blocks to be considered as a single allocation. The allocation function can be modified to scan the free block bit mask to find a sequence of set bits long enough to enclose the allocation or step to the next page. Should no page have enough blocks, a new page can be allocated. However, this is a slow operation due to the number of bit scans and potential number of page traversals, sometimes resulting in a long linked list traversal that ultimately fails.

A major improvement is to record the largest block that can be allocated in the `Tail` structure of the page, along with the bit index of where that block is located. Armed with this information, the memory manager can simply keep a table of page lists, ordered by big block size, so that allocation is always attempted on a page that is known to have enough space. Thus, no searching is required once a page is located, and pages can be located very quickly. More details are provided on this later.

The second wrinkle is that allocation can work with the data as specified, but we cannot free a block of memory without knowing how many blocks are in each allocation. And without knowing how long the allocation is, there's no way to embed a length variable at the end of each allocation. Most competing memory managers embed the length variable just before the pointer address in memory, but we reject that option since it means wasting the first block in any given page (which would precede any allocation on the page). That's much too wasteful.

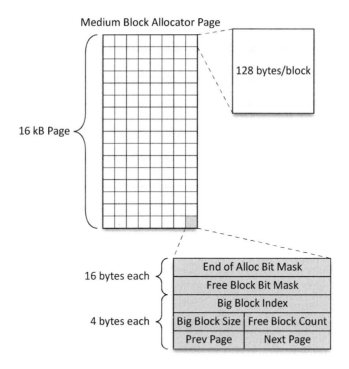

Figure 26.3. This is the memory layout of an MBA page. This particular example is a 16 kB page with 128-byte blocks. All of the management metadata fits within a single block at the end of the page.

Instead, we could store a value in an array at the end of the page, similar to how the free block bit mask is stored, although that is quite expensive as well. Considering most of the allocations are larger than one block, many of the entries in this array would be unused at any given time. Even if we stored the length per block in one byte, it would be far too wasteful!

The solution: store a single bit in a bit mask that identifies whether a block is at the end of an allocation. The pointer passed in to the `Free()` function identifies the start of the block already. Using these two pieces of information, we can extrapolate the length of an allocation simply by scanning the bit mask forward looking for a set bit. As a matter of standards, in the code provided, unallocated blocks have their end-of-allocation bit set to zero, but it won't ever be checked so it shouldn't matter. See Figure 26.3 for a more visual explanation.

How `Alloc()` Works

Once a page is found, the MBA simply looks up the large block index in the `Tail` structure, marks the range of bits as allocated in the free block bit mask, and clears all the bits except the last one in the end-of-allocation bit mask. This is enough information to allow the `Free()` function to determine the length of the original allocation given only the pointer address.

Now that the big block in the page has been at least partially allocated, we have to find the new big block in the page. This entails scanning the free block bit mask and finding the longest run of set bits, remembering the index and size of the longest one. Once this is done, it is stored in the `Tail` structure in the page for future use.

The pointer to the allocation is returned to the memory manager. The memory manager then must grab the big block size from the page and rearrange that page to a different list based on the size of its big block, so future allocations do not need to skip over insufficient memory pages.

How `Free()` Works

Once the page manager has identified the page, the MBA begins by scanning the end-of-allocation bit mask at the end of the page and counting 0 bits until a 1 bit is found. This marks the end of the allocation and implicitly defines the length of the allocation. We set the free block bit mask to one for all the blocks between the start and end of the allocation, and clear the end-of-allocation bit to zero (though this is not strictly necessary).

Since the memory region that was just released plus any adjacent free blocks might sum to a larger big block, we update the big block size, just as was done in the `Alloc()` function. Again, the memory manager requests this updated information from the page and reorders its internal structures based on the big block size in this page to optimize the next allocation.

Performance Analysis

Allocation and deallocation are very fast. The act of taking possession of the requested memory once a page is found is simply a few pointer additions and a couple of memory dereferences. Setting a few bits in the free block bit mask is a relatively quick $n/32$ operation when using 32-bit bitset operations, where n is the number of blocks that the allocation spans. `Free()` performs the same operations as `Alloc()`, except that it reads bits from the free block bit mask rather than sets them. However, searching for the big block takes $m/32$ memory accesses, where m is the number of blocks in the page. Since m is a constant for a given page and block size, technically both of these analyses are bound by a constant time ($O(1)$), meaning the number of cycles that is required can be computed at compile time. Practically speaking, it is very fast, requiring only one L1 cache line to be loaded if page and block sizes are carefully chosen for your architecture.

Here's a specific example: given a 16-kB page with 128 bytes per block, there are 127 allocatable blocks per page plus one reserved block for the `Tail` structure. This is 127 bits for the bit mask, which fits in only four 32-bit words. A good target is to make all MBA pages at least two times larger than the maximum medium block allocation size to improve memory performance when allocating close to the maximum. In the worst case, a single allocation might be half the size of a page, thus writing 64 bits, 32 bits at a time. The first load causes a delay as a cache line is flushed and another is pulled into the L2 and L1 caches, but the second load takes only one or two cycles on typical hardware.

Memory Overhead

The MBA scheme is almost as efficient as the SBA approach, given that the memory requirements are very similar with an extra bit mask and a couple more words stored in the `Tail` structure. For any page, the `Tail` structure is 20 bytes in size. There is a fixed amount of overhead for the two bit masks, which is two bits per block, aligned and rounded up to the next 32-bit word. For a 16-kB page with 128 bytes per block, this is 32 bytes for bit masks. Management overhead for this example is under three bits per block, but since allocations tend to span many blocks, this could add up. The largest reasonable allocation in this example page size would be 8 kB (half the page size), which covers 64 blocks. At two bits per block, this worst-case allocation overhead is 16 bytes. dlmalloc appears to have 16 bytes of overhead per allocation, regardless of its size, and those bytes are spread out in memory causing random cache misses. In our system, the cache lines involved in storing memory management metadata are far more likely to be accessed again in the future.

26.5 Large Block Allocator

There are many techniques for making memory management faster, but none so easy as handling a few large allocations. In general, dealing with a few large allocations is the simplest of all possible cases and, thus, is really not worth putting effort into. For this, feel free to substitute any allocation strategy that is suitably efficient. dlmalloc can be configured to have an mspace heap constrained within the address range from which your large blocks are allocated. This is quite sufficient.

If you plan to write your own LBA, a simple linked list of free blocks and a separate linked list of allocated blocks would work fine. Best-fit strategies tend to fragment more over time, but if your memory use tends to be generational (e.g., many allocations have matching lifetimes and are freed at the same time), almost any method works fine. For very specific cases, you might explicitly carve out memory for all large allocations or even use a simple stack allocator, where the entire heap is reset with a single pointer being reset to the top of the stack. Use what makes sense for your team and project.

The one point that should be made is that the system should have some way of deciding whether a pointer is an LBA pointer or whether it belongs to the SBA or MBA. The simplest way is to check the SBA and MBA first, although this may become a performance problem if those operations are slow. Alternatively, keeping a map, a hash table, or even a simple array of allocations may be fine. There shouldn't be too many large allocations, and unless your project allocates and deallocates them frequently, any slowdown here is acceptable.

26.6 Page Management in the Memory Manager

One of the early performance bottlenecks of our system was the handling and searching of pages inside the memory manager. Once each allocation strategy is fast inside a single page, it is all the more critical to make sure the memory manager itself can quickly figure out from what page to allocate memory. Finding the right page to free memory from is really an operation that is performed by the

OSAPI because it may have special knowledge of determining that based on the page handling implementation.

SBA Pages

Page management for small block pages is relatively straightforward. Since each page has a specific block size that it can allocate, the memory manager simply keeps two tables of linked lists of pages. One table contains pages that have absolutely no free space in them (and are thus useless when trying to allocate memory). The other table contains pages that have at least one free block available. As shown in Figure 26.4, each list contains pages that have equal block sizes, so whenever a page fills up, it can be moved to the full table, and the next page in the available list will service the next request. Any time a block is freed from a page in the full list, that page is moved back to the available list for that block size. Because of this, no searching for an allocation is ever required!

It should be noted that the current implementation of the page management feature for SBA always requires allocations be as tightly fitting as possible. As a result, any time there are no pages of exactly the correct block size, a new page must be allocated for that block size and added to the available list. In low-memory conditions, however, this could cause an early failure to allocate because larger block sizes might be available but not considered for allocation. You may wish to implement this as a fallback in case of page allocation failure to extend the life of your game in the event of memory exhaustion. Alternatively, you might always return a piece of memory that is already in an available page rather than allocating a new page when a block size runs out of pages, which may help utilize memory more effectively (although it is quite wasteful) because it will

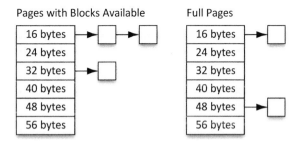

Figure 26.4. The SBA always allocates a single sized object from each list of pages. So to prevent searching, full pages are removed from the list from which we allocate.

wait until the absolute last allocation is used before requesting a new page. Or use some hybrid approach in which the next three or four block sizes are considered valid if the correct one is full.

MBA Pages

Medium block page management is more complicated only because the lists are constructed based on the size of the largest available block in each page. Consequently, for nearly every allocation or deallocation, the affected page most likely has its list pointers adjusted. Unlike SBA, it is entirely possible that no big block matches the requested allocation size exactly (see Figure 26.5 for an example of a sparse page list). In fact, a brand-new page would show the big block being the full size of the page (minus the metadata overhead in the `Tail` structure), so requesting a new page is also not likely to add a page in the list to which the allocation first indexes. So, whenever a specific-sized list is empty, the page manager must scan the larger lists looking for a page, guaranteeing that the allocation function succeeds. However, since list pointers are four bytes each in a 32-bit machine, an array of pointers actually requires multiple cache lines and far too many fetches from memory for good performance. Instead, we implement the lookup first as a bit mask that uses 0 to indicate a null pointer and 1 to indicate any non-null pointer. In this way, a single cache line can track the availability of hundreds of lists with just a few memory fetches.

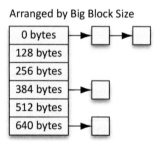

Figure 26.5. This MBA page table shows two pages that are completely full, one with 384 bytes in a single block, and another with 640 bytes in a single block. It makes no statement about how many blocks of that size might exist or how many smaller ones exist.

26.7 `OSAPI` Ideas

All interactions that the memory manager has with the underlying operating system are tucked away in the `OSAPI`. The `OSAPI` has to be able to tell immediately what kind of page an allocation comes from and be able to return the pointer to the start of that page. This has to be very fast, or it becomes the bottleneck for all other memory operations. It also must be able to allocate new pages and delete pages that are empty and no longer needed.

There are many possible approaches to implementing the `OSAPI`. Our research on this has not yet reached a final best solution, and indeed, we believe that the right solution depends on the project and platform. Even so, there are a few ways to implement the `OSAPI`'s page tracking, so we discuss some here.

The `OSAPIForWindows` is implemented such that all SBA and MBA pages are tracked in a `std::set`, which is implemented as a red-black tree. While theoretically fast, each node requires its own allocation, which can both fragment the underlying heap and be slow. Pages are allocated using the `VirtualAlloc()` function, which has some implications for ideal page size, especially on 32-bit systems where the amount of address space mapped should be equal to the physical memory being mapped. Performance is not great, but it does work well as a rudimentary example of how to implement the features required by the interface. It also suffers due to the many calls to `VirtualAlloc()` and `VirtualFree()`, causing thunks into the kernel to allocate memory. With this technique, operating system functions show up on performance profiles—never a good sign.

The `OSAPIFastForWindows` is an improved version that preallocates a big chunk of memory and assigns pages from this chunk. To maximize flexible use of this space, SBA and MBA pages are set to the same size and are allowed to change from small- to medium-sized pages while the page is empty. This is accomplished simply by tracking the type of page with an array of bits and tracking whether the page is in use with another array of bits. To limit time spent searching for free pages, a simple start index is stored in the class that remembers the last index to a freed page. It is not required to be accurate, but it does give a good place to start searching and usually returns a hit on the first attempt to allocate a page. However, this implementation has the downside of being rather static, and if there are times when too many large allocations are made, or too many small or medium allocations are made, the system could catastrophically fail.

Future systems could be written to have completely different allocation strategies. For instance, it could be smart about allocating pages from lower memory addresses and allocating large blocks from higher memory addresses until the two collide somewhere in the middle.

It is also intended that the OSAPI be platform-specific, tuned to the exact amount of memory that is used in the system. There is no real benefit to having a generic page handling system if there is no flexibility in the amount of memory available to the game on a specific piece of hardware. A custom page handler, specifically one that knows the address range that could possibly exist for allocations and can reduce the identification of pages to a shift and bit mask operation to create an index into a hash table is ideal. For instance, the Nintendo Wii can only return certain addresses for MEM1 and MEM2, so if the LBA is known to allocate exclusively from MEM2, it is easy to identify. Similarly, if the SBA and MBA allocate from MEM1, we can create a simple bit array that maps exactly the memory addresses used on the console and know which kind of page each pointer belongs to simply by rounding off the address and looking it up.

27

Simple Remote Heaps

Jason Hughes
Steel Penny Games, Inc.

27.1 Introduction

A *remote heap* is an old tool in the memory management toolbox. It is appropriate to consider using a remote heap whenever a memory architecture is segmented, such as on the Wii, where MEM1 is much faster than MEM2 and can benefit from having all the allocation and deletion work done with lower latency. A remote heap is also useful when a processor has to handle feeding other custom processors with their own local memory, such as pushing audio data to a sound chip, when the only way to touch that memory is via bulk DMA calls that have significant latency. Similarly, on the PlayStation 3, the PPU may want to treat the memory associated with each SPU as a remote heap so that it can play traffic cop, with data whizzing back and forth, without needing to have direct access to their individual memories. Conceptually, remote heaps could also be used in a unified memory architecture using multiple processors, such as the Xbox 360, where each heap is mutually distinct and partitions memory for each CPU to work on, with a guarantee that no other processor will cause mutex blocking at the central memory manager.

Considerations

There are limitations to every technique. Unfortunately, remote heaps suffer from a significant overhead in performance and local memory footprint, which can be traced primarily back to the fact that the only data that a typical program has to refer to an allocation is its address. In a standard heap, metadata describing each allocation is stored a few bytes before each allocated block of memory, so operations such as Free() tend to do some quick pointer arithmetic to do their work.

This is not as straightforward with a remote heap because that metadata has to be associated with an address, without the benefit of storing data at that address. Designing a good remote heap means flirting with hash tables, red-black trees, or similar data structures.

Unfortunately, hash tables of addresses are terrible when searching for memory to allocate (sequential allocations would be randomly scattered through memory and have $O(n)$ allocation time, where n scales with the hash table rather than with the entries used) and equally bad for merging adjacent free blocks.

Red-black trees of addresses (such as the standard template library `map`) are memory hungry and somewhat tedious to set up with allocators from fixed memory pools but do work very well when freeing blocks in $O(\log n)$ time, where n is the number of current allocations, and merging adjacent free blocks is a constant-time operation (neglecting the reordering of the tree when a node is deleted). Unfortunately, they have $O(n)$ running time for allocation because best-fit and first-fit strategies must scan all blocks to find free space.

One solution is to carefully pack bits so that the $O(n)$ searches are as fast as possible. Here, we accept the worst-case performance and try to make the best of it. There are some benefits to this method, which is why we present it. Sometimes it is ideal. In general, though, it lacks performance. We call this the "bitwise remote heap" approach because each unit of memory is represented as a single bit.

The other solution is to realize that a single address-based data structure is not ideal for all functions. However, keeping two structures in sync tends to force each data structure to have the worst-case performance of the other. Our solution is to allow them to fall out of sync and correct mistakes as they are detected. We call this the "blockwise remote heap" because it tracks individual allocations as single units.

27.2 Bitwise Remote Heap

Overview

The bitwise remote heap is very similar to the Medium Page Allocation scheme described in the previous chapter. The basic idea is to manage a block of memory (or rather, a range of addresses) in equally sized chunks by associating each chunk's availability with a single bit in a packed bitmap. If the bit is true, the block is currently free. If the bit is false, it is part of an allocation already. This convention may seem strange, but it was chosen specifically because Intel CPUs can find the next set bit in a single instruction,[1] and we're always looking for free blocks. On other processors, the opposite might be true.

One downside of this approach is that all allocations must be rounded to an even granularity of the chunk size, which can be wasteful if this size is set too large. Another disadvantage is that, when finding an allocation, the implementation may need to scan all of the bits in the bitmap, only to fail. This failure grows slower as the number of bits in the bitmap increases.

Consequently, this kind of implementation is ideal when relatively few allocations are required, they can be easily rounded to a specific size, and the amount of memory managed is relatively small. This sounds like a lot of restrictions to place on a remote heap, but in reality, they are satisfied quite frequently. Most remote heaps dealing with custom hardware, such as audio processors or specialized processors (vector units, GPUs, etc.), have alignment requirements anyway, and they often have a relatively small block of memory that needs to be managed.

Algorithm

The bitwise remote heap has only two important functions:

- `Alloc()`. The obvious method for finding a free block of memory in a bitmap is to scan for a sequence of bits that are set. This can be accomplished by loading 32 bits at a time, scanning for the first 1 bit, noting that location, then scanning for the first 0 bit after it. Often this requires spanning multiple 32-bit values in sequence, so an optimized routine is required. Also, many processor architectures have assembly language mnemonics that do this kind of search in a single instruction for further performance enhancement. Clearly, this operation has worst-case $n/32$ memory fetches, where n is the number of bits in the heap bitmap. Once the range is found, the bits in the range are set to zero. A single bit is set in an end-of-allocation bitmap that tells what the last block in the allocation is. Think of it like null-terminating a string.
- `Free()`. When deleting a chunk of memory, we simply turn a range of 0 bits into 1 bits, scanning forward until we find the allocation terminator bit in the end-of-allocation bitmap. Once found, we clear the terminating bit as well. Coalescing free blocks is completely unnecessary because adjacent free bits implicitly represent contiguous free memory. Thus, there is relatively little management overhead in deallocation.

---

[1] See the bit scan forward (BSF) instruction in *Intel 64 and IA-32 Architectures Software Developer's Manual Volume 2A: Instruction Set Reference, A-M*. Many details on bit scan operations are also available at http://chessprogramming.wikispaces.com/ BitScan.

Certain small changes can dramatically improve performance. One major bottleneck is scanning thousands of bits to look for memory. The simple trick of remembering a good place to start searching can cut the search time significantly when making repeated calls to `Alloc()`. When doing this, every allocation that succeeds remembers the very next bit as the starting point, since large blocks of memory tend to be partitioned into smaller ones, and the next piece of free memory is likely to be next. Also, whenever a block of memory is released in `Free()`, we reset the starting point if the address is lower than the current starting point. This is very similar to a first-fit allocation strategy, while skipping a lot of searching. However, this method can fail when an allocation occurs close to the end of memory and most of the free blocks are at the start, so an additional loop must check for allocations up to the starting point in the event of an allocation failure. This tiny change is extremely important for good performance. See Table 27.2 at the end of the article for a comparison.

A further improvement, which we have not personally implemented, would be to remove the $O(n)$ search entirely. This could be accomplished by keeping a hierarchical table of bitmaps, where each bit represents the free status of several bits at the lower level. Along with such a system comes inaccuracies in determining whether a block is available for an allocation, meaning more `Alloc()` failures, but it fails more quickly. Sometimes that's a good trade-off.

Finally, a simple, cheap improvement would be to create a table of free block addresses that is sorted by the size of the free block. This makes allocation extremely quick because any allocation request is simply a query against this table for any block of at least the requested size, running up the table to larger blocks if no blocks of the right size are available. However, deallocation becomes slower because now, bits must be scanned in both directions to effectively coalesce blocks. The bookkeeping required to maintain a useful table is slightly complicated and might become a significant overhead if many small fragments begin filling up the table. Thus, we would not bother keeping small free chunks in the table, since searching is likely to provide a valid block, but anything over some threshold that is less likely to find space would be a good candidate for keeping in the table. This probably results in $O(n)$ deallocation, but on a heavily loaded heap, it would be negligible in practice. So, again, it might be a good trade-off.

27.3 Blockwise Remote Heap

Overview

The blockwise remote heap is quite similar in implementation to a standard heap, except that it does not benefit from storing extra data immediately prior to the address in user-memory. Since we cannot use the address to quickly find metadata about allocations, the naive implementation leaves us with a linked list of tracking blocks, ordered by address. A naive implementation forces all allocations to scan the heap's tracking linked list before allocating or freeing any memory. A naive implementation has a slow $O(n)$ running time, where n is the number of allocations in the heap.

The implementation on the website does not attempt to make `Free()` as fast as possible but, rather, does a simple linked list search. This could be improved by creating a second data structure that keeps the tracking block pointer associated with each allocated address, perhaps in a hash table or balanced tree, but that is an exercise left to the reader. The memory requirements for such a structure are substantial and may be entirely unnecessary for applications where `Free()` is rarely or never called, but rather, the entire heap is dropped all at once.

However, since fast allocation is actually a complicated problem, we present a solution that is nearly constant-time for the majority of allocations and is linear for the rest. We do this by maintaining a bookmark table that essentially points to the first free block of each power-of-two–sized block of free memory. It remembers the last place the code found a free block of each size. Once we allocate a block, that entry in the table may no longer be accurate. Updating the entry may require a full traversal of the heap, a very slow operation, so we allow it to become stale. Instead, during allocation and free calls, we store any blocks we come across in the table at the appropriate (rounded down to the next) power of two. While there is no guarantee that the table is updated, in practice it tends to be close, at no additional performance cost. During an allocation, if a table entry appears invalid, we can always check the next-higher power of two, or the next, until one is found to be valid. In most cases, this works very well. In empirical tests, 65 percent of the allocations have positive hits on the cached references in the bookmark table, which means 65 percent of the long searches for free memory were avoided.

The tracking data for allocations is stored in a pool with a threaded free-list, making the location of a valid metadata block during allocation and deletion an $O(1)$ operation. Threaded free-lists act like a stack, since we simply want a blank node to write to when allocating (pop) and want to return a node to the unused pool when freeing (push). As in any other standard pool implementation, the used and unused structures occupy the same physical memory at different times, and we just cast to one or the other, depending on the current state of that block. To aid in debugging, as well as to facilitate lazy bookmarking, unused metadata nodes are marked with a sentinel value that cannot appear in an allocation, so we can repair the bookmark table when it gets out of date.

```
struct AllocatedBlock
{
    unsigned int mAddress;          // This is where the block starts.
    unsigned int mAllocBytes;       // Allocated bytes are at the BEGINNING
                                    // of the allocation.
    unsigned int mFreeBytes;        // Free bytes are at the END.
    unsigned int mNextBlockIndex;   // This is a linked list in
                                    // address-order and makes
                                    // allocation and merging quick.
};
```

Listing 27.1. An `AllocatedBlock` represents an allocated chunk of memory in the remote address space and some number of unallocated bytes that immediately follow it.

Many heap implementations treat free and used blocks with separate tracking data types and store a flag or sentinel to know which metadata is which. I always simplify mine by combining tracking data for used and free blocks into a single structure. This has two benefits: first, the memory requirements are extremely minimal to do so, and second, half the number of linked list iterations are required to walk the entire heap. If you think about it, just about every allocation will have at least a few bytes of free space after it (for hardware alignment reasons), and the heap has to decide whether to assign each tiny piece of memory its own node in the heap or consider that memory to be part of the allocation. These decisions need not be made at all if there is no distinction made, and thus, no cost added.

Even when there is no free memory between two allocations, this is no less efficient. It does have a minor downside in that the memory requirements for each metadata node are slightly higher, but not significantly so. See Listing 27.1 for an example of a metadata node.

In a standard heap, a metadata block, such as `AllocatedBlock`, is stored just prior to the actual address of the allocation. In a remote heap, we can't do that. Instead, we store these blocks in a typical allocation pool, where some of the blocks are currently in use, and some are not (see the `UnusedBlock` declaration in Listing 27.2). Note that both structures have "next" pointers. An `Allocated-Block` only points to other `AllocatedBlock` structures, so scans of the heap are limited to real allocations. The `UnusedBlock` only points to other `UnusedBlock` structures and is effectively a threaded free list.

Recall that the lazy power-of-two table that helps with allocation keeps pointers into this pool to know where free spaces are. Under normal circumstanc-

```
struct UnusedBlock
{
    // This is a threaded free list inside the mAllocations pool.
    // Doing this allows for O(1) location of AllocatedBlock
    // objects in the pool.
    unsigned int mNextUnusedBlockIndex;

    // If unused, this is set to sentinel. This saves us effort
    // of managing the hash table.
    unsigned int mSentinel;
};
```

Listing 27.2. These are placeholders in the local memory pool of `AllocationBlock`s, the metadata for the remote heap.

es, a pool structure would never link to an unused node, but with a lazy bookmark table, we can and do. We must, therefore, have a way to detect an unused node and handle it gracefully. This is the reason for the sentinel value. Whenever we look for an allocation using the bookmark table, we can tell if the block is still allocated by casting the block to an `UnusedBlock` and checking the `mSentinel` variable. If it is `UINT_MAX`, clearly the table is out of date, and we should look elsewhere. Otherwise, we cast to an `AllocatedBlock` and see if the number of free bytes is sufficient to satisfy the new allocation.

One other trick that simplifies the code is the use of a dummy block in the linked list of allocations. Since the initial block has all the free memory and no allocations, we need to have some way to represent that for the `Free()` function. Rather than write a hairy mess of special-case code for the first allocation, we just initialize one node to have the minimum allocated block and all the remaining free bytes and stick it in the list of allocations. This node always stays at the head of the list, and consequently, all other allocations come after it. Note, however, we never count that node as a user allocation, so the number of real allocations available to the user is one fewer than what is actually present.

Algorithm

The blockwise remote heap also has only two interesting functions:

- `Alloc()`. First, we compute what power of two can contain the allocation, use that as an index into the bookmark table, and find the metadata block for the associated slot in the table. If we find a valid block with enough free space, we're set. If not, we iteratively search the next-highest power of two until we run out of table entries. (This scan is why fragmentation is so high, because we're more likely to cut a piece out of the big block rather than search the heap looking for a tighter fit someplace else.) If no large block is verified, we search the entire heap from the start of the memory address space looking for a node with enough free bytes to afford this new allocation. Then, the new allocation information is stored in local memory in a metadata node. The allocation of a metadata block is quite simple. We pop an `Allo-catedBlock` off the metadata pool (which is always the first `UnusedBlock`), fill out the structure with information about our new allocation, reduce the free bytes of the source block to zero and assign them to the new block, and link the new block into the list after the source block, returning the remote memory pointer to the caller. Since we have a dummy block in the linked list that is always at the head, we never need to worry about updating the head pointer of this list.

- `Free()`. The operation that hurts performance most is searching the heap for a memory address (or free block of sufficient size). `Free()` has to search to figure out where an address is in the heap. This is quite slow and dominates the running time of the heap implementation. While searching, we keep the pointer to the previous node so we can collapse all the memory into it once the address is found. The merging operation simply adds the free and allocated bytes to the previous node, links around the deleted node, and releases the metadata back to the pool. Consequently, freeing data is an $O(n)$ operation, where n is the number of live allocations.

The bookmark table is updated after every allocation and deletion so that any time a piece of free memory is available, the appropriate power-of-two indexed slot is given the address to its metadata node. There is no sense in checking hundreds or thousands of nodes if they are in the bookmark table during every operation, so updates are limited to overwriting what is currently stored there, not cleaning out the table when sizes change. As a result, the table can sometimes point to metadata nodes that are smaller or larger than expected, or have even been completely deleted. So, anywhere the code consults the bookmark table, it verifies the data is good and corrects the data (by marking it empty) if not.

For better performance, an improvement could be made by using a doubly linked list of metadata nodes and a hash table or tree structure that maps addresses to metadata pointers. At some expense to memory, you can get as good as $O(1)$ running time by doing this. We have not done this, but it is a straightforward extension.

27.4 Testing Results

The testing methodology is intentionally brutal to show worst-case behavior. The test unit performs 1,000,000 allocations and deletions in a random but identically reproducible order, with an average of five allocations for every four deletions, eventually filling memory and causing memory allocation failures. The memory block size managed by the heap is 8 MB. Given the differences in each heap's allocation strategy, the number and size of allocation failures would grow at different rates based on the various parameters to each heap.

Memory Efficiency

The results in Table 27.1 show that the bitwise method is very good at packing the maximum amount of data into a remote heap, regardless of the granularity. Comparatively worse, the blockwise method is not as economical on memory use (i.e., it results in fewer successful allocations) and scales even more poorly as the number of managed allocations drops, most likely because the metadata table is full.

The allocation count reservations for the blockwise tests are chosen such that the amount of local memory used is roughly equal between the two heap types. We chose this metric not only because it's a valid comparison but also because we found that there is a hard limit above which adding more allocation space to the blockwise heap yields no benefit whatsoever, and indeed, begins to lose its CPU performance edge. The second key point is that the bitwise method is significantly more compact in memory, if memory is tight.

Heap Type	Parameter	Allocation Failures (lower is better)	Local Memory Usage
Bitwise	8-byte granularity	25268	64 KB
Bitwise	32-byte granularity	25291	16 KB
Bitwise	128-byte granularity	25425	4 KB
Blockwise	4096 metadata blocks	25898	64 KB
Blockwise	1024 metadata blocks	28802	16 KB
Blockwise	256 metadata blocks	30338	4 KB

Table 27.1. Memory allocation efficiency comparison between remote heap types.

Performance

Table 27.2 shows how drastically different the performance is between bitwise remote heaps and traditional metadata remote heaps. The two bitwise tests differ solely on the basis of whether the cached starting point was used. These test results clearly show that avoiding a complete search by caching the last free block's location is at least 50 percent better and improves performance further as the allocation bitmap gets longer. Some CPU architectures are much faster or much slower, depending on how the bit scan operation is handled. Moving to a wider data path, such as 128-bit MMX instructions, or using 64-bit registers, could potentially double or quadruple the performance, making bitwise an excellent choice. Hierarchical bitmasks could also take numerous $O(n)$ operations and make them $O(\log n)$.

With these performance characteristics in mind, it does appear that a blockwise heap is far superior in terms of performance; however, be aware that the number of allocations that the blockwise heap can fit is reduced significantly as the allocation count is reduced. The time required to allocate with the blockwise heap is relatively constant, but the current implementation causes deallocation to scale linearly with the number of allocations in the heap, hence the linear relation between allocation count and speed.

Heap Type	Parameter	Time, in Seconds (lower is better)	Local Memory Usage
Bitwise (no cache)	8-byte granularity	41.31	64 KB
Bitwise (no cache)	32-byte granularity	24.54	16 KB
Bitwise (no cache)	128-byte granularity	13.37	4 KB
Bitwise (cache)	8-byte granularity	19.90	64 KB
Bitwise (cache)	32-byte granularity	13.15	16 KB
Bitwise (cache)	128-byte granularity	8.92	4 KB
Blockwise	4096 metadata blocks	5.62	64 KB
Blockwise	1024 metadata blocks	1.60	16 KB
Blockwise	256 metadata blocks	0.56	4 KB

Table 27.2. Raw CPU performance tells a very different story.

28

A Cache-Aware Hybrid Sorter

Manny Ko

DreamWorks Animation

Sorting is one of the most basic building blocks of many algorithms. In graphics, a sort is commonly used for depth-sorting for transparency [Patney et al. 2010] or to get better Z-cull performance. It is a key part of collision detection [Lin 2000]. Dynamic state sorting is critical for minimizing state changes in a scene graph renderer. Recently, Garanzha and Loop [2010] demonstrated that it is highly profitable to buffer and sort rays within a ray tracer to extract better coherency, which is key to high GPU performance. Ray sorting is one example of a well-known practice in scientific computing, where parallel sorts are used to handle irregular communication patterns and workloads.

Well, can't we just use the standard template library (STL) sort? We can, but we can also do better. How about up to six times better? Quicksort is probably the best comparison-based sort and, on average, works well. However, its worst-case behavior can be $O(n^2)$, and its memory access pattern is not very cache-friendly. Radix sort is the only practical $O(n)$ sort out there (see the appendix for a quick overview of radix sort). Its memory access pattern during the first pass, where we are building counts, is very cache-friendly. However, the final output phase uses random scatter writes. Is there a way for us to use radix sort but minimize its weaknesses?

Modern parallel external sort (e.g., AlphaSort [Nyberg et al. 1995]) almost always uses a two-pass approach of in-memory sort followed by a merge. Each item only has to be read from disk twice. More importantly, the merge phase is very I/O friendly since the access is purely sequential. Substitute "disk" with "main memory" and "memory" with "cache," and the same considerations apply—we want to minimize reads from main memory and also love the sequential access pattern of the merge phase.

Hence, if we partition the input into substreams that fit into the cache and sort each of them with radix sort, then the scatter writes now hit the cache, and our main concern is addressed. One can substitute shared memory for cache in the above statement and apply it to a GPU-based sort. Besides the scattering concern, substreams also enable us to keep the output of each pass in cache so that it is ready for the next pass without hitting main memory excessively.

Our variant of radix sort first makes one pass through the input and accumulates four sets of counters, one for each radix digit. We are using radix-256, which means each digit is one byte. Next, we compute the prefix sums of the counters, giving us the final positions for each item. Finally, we make several passes through the input, one for each digit, and scatter the items into the correct order. The output of the scattering pass becomes the input to the next pass.

Radix sort was originally developed for integers since it relies on extracting parts of it using bit operations. Applying it directly to floating-point values works fine for positive numbers, but for negative numbers, the results are sorted in the wrong order. One common approach is to treat the most significant radix digit as a special case [Terdiman 2000]. However, that involves a test in the inner loop that we would like to avoid. A nice bit hack by [Herf 2001] solves this nasty problem for radix sort.

For efficient merging, we use an oldie but goodie called the *loser tree*. It is a lot more efficient than the common heap-based merger.

At the end we get a sorter that is two to six times faster than STL and has stable performance across a wide range of datasets and platforms.

28.1 Stream Splitting

This chapter discusses a class using Wright's [2006] sample as a base to obtain the cache and translation lookaside buffer description of various Intel and AMD CPUs. The basic code is very simple since it uses the cpuid instruction, as shown in Listing 28.1. Some of the detailed information is not available directly from the returned results. In the sample code on the website, a lot of the details based on Intel's application notes are manually entered. They are not needed for the purpose of the sample code, which only requires the size of the caches.

```
U32 CPU::CacheSize(U32 cachelevel)
{
    U32     ax, bx, cx, dx;

    cx = cachelevel;
```

```
    cpuid(kCacheParameters, &ax, &bx, &cx, &dx);

    if ((ax & 0x1F) == 0) return (0);

    U32 sets = cx + 1;
    U32 linesize = bx & 0x0FFF;              // [11:0]
    U32 partitions = (bx >> 12) & 0x03FF;    // [21:12]
    U32 ways = (bx >> 22) & 0x03FF;          // [31:22]
    return ((ways + 1) * (partitions + 1) * (linesize + 1) * sets);
}

static void GetCacheInfo(U32 *cacheLineSize)
{
    U32     ax, bx, cx, dx;

    cpuid(kProcessorInfo, &ax, &bx, &cx, &dx);
    *cacheLineSize = ((bx >> 8) & 0xFF) * 8;     // Intel only

    // For AMD Microprocessors, the data cache line size is in CL and
    // the instr cache line size is in DL after calling cpuid function
    // 0x80000005.

    U32 csize0 = CacheSize(0);      // L0
    U32 csize1 = CacheSize(1);      // L1
    U32 csize2 = CacheSize(2);      // L2
    U32 csize3 = CacheSize(3);      // L3
}
```

Listing 28.1. Returns the actual cache sizes on the CPU.

To determine our substream size, we have to consider the critical step for our radix sorter, the scatter phase. To perform scattering, the code has to access the count table using the next input radix digit to determine where to write the item to in the output buffer. The input stream access is sequential, which is good for the prefetcher and for cache hits. The count table and output buffer writes are both random accesses using indexing, which means they should both be in the cache. This is very important since we need to make several passes, one for each radix digit. If we are targeting the L1 cache, then we should reserve space for the counters (1–2 kB) and local temporaries (about four to eight cache lines). If we are targeting the L2 cache, then we might have to reserve a little more space in L2 since the set-associative mechanism is not perfect.

For the counting phase, the only critical data that should be in the cache is the counts table. Our use of radix-256 implies a table size of `256 *` `sizeof(int)`, which is 1 kB or 2 kB. For a GPU-based sort that has limited shared memory, one might consider a lower radix and a few extra passes.

28.2 Substream Sorting

At first glance, once we have partitioned the input, we can just use the "best" serial sort within each partition. Our interest is in first developing a good usable serial algorithm that is well suited for parallel usage. In a parallel context, we would like each thread to finish in roughly the same time given a substream of identical length. Keep in mind that one thread can stall the entire process since other threads must wait for it to finish before the merge phase can commence. Quicksort's worst-case behavior can be $O(n^2)$ and is data dependent. Radix sort is $O(n)$, and its run time is not dependent on the data.

Our radix sorter is a highly optimized descendent of Herf's code, which is built on top of Terdiman's. It is a radix-256 least-significant-bit sort. The main difficulty with using radix sort on game data is handling floating-point values. An IEEE floating-point value uses a sign-exponent-mantissa format. The sign bit makes all negative numbers appear to be larger than positive ones. For example, 2.0 is `0x40000000`, and −2.0 is `0xC0000000` in hexadecimal, while −4.0 is `0xC0800000`, which implies −4.0 > −2.0 > 2.0, just the opposite of what we want. The key idea is Herf's rule for massaging floats:

1. Always invert the sign bit.
2. If the sign bit was set, then invert the exponent and mantissa too.

Rule 1 turns 2.0 into `0xC0000000`, −2.0 into `0x40000000`, and −4.0 into `0x40800000`, which implies 2.0 > −4.0 > −2.0. This is better, but still wrong. Applying rule 2 turns −2.0 into `0x3FFFFFFF` and −4.0 into `0x3F7FFFFF`, giving 2.0 > −2.0 > −4.0, which is the result we want. This bit hack has great implications for radix sort since we can now treat all the digits the same. Interested readers can compare Terdiman's and Herf's code to see how much this simple hack helps to simplify the code.

The code in Listing 28.2 is taken from Herf's webpage. Like Herf's code, we build all four histograms in one pass. If you plan on sorting a large number of items and you are running on the CPU, you can consider Herf's radix-2048 opti-

```
static inline U32 FloatFlip(U32 f)
{
    U32 mask = -int32(f >> 31) | 0x80000000;
    return (f ^ mask);
}

static inline U32 IFloatFlip(U32 f)
{
    U32 mask = ((f >> 31) - 1) | 0x80000000;
    return (f ^ mask);
};
```

Listing 28.2. Herf's original bit hack for floats and its inverse (`IFloatFlip`).

mization, which reduces the number of scattering passes from four to three. The histogram table is bigger since $2^{11} = 2$ kB, and you need three of them. Herf reported a speedup of 40 percent. Keep in mind that the higher radix demands more L1 cache and increases the fixed overhead of the sorter. Our substream split strategy reduces the benefit of a higher radix since we strive to keep the output of each pass within the cache.

One small but important optimization for `FloatFlip` is to utilize the natural sign extension while shifting signed numbers. Instead of the following line:

```
U32 mask = -int32(f >> 31) | 0x80000000;
```

we can write

```
int32 mask = (int32(f) >> 31) | 0x80000000;
```

The right shift smears the sign bit into a 32-bit mask that is used to flip the input if the sign bit is set, hence implementing rule 2. Strictly speaking, this behavior is not guaranteed by the C++ standard. Practically all compilers we are likely to encounter during game development do the right thing. Please refer to the chapter "Bit Hacks for Games" in this book for more details. The same idea can be applied to `IFloatFlip`, as follows:

```
static inline void IFloatFlip(int32& f)
{
    f ^= (int32(f ^ 0x80000000) >> 31) | 0x80000000;
}
```

28.3 Stream Merging and Loser Tree

Our substreams are now sorted, and all we need is to merge them into one output stream. The most common approach is to use a priority queue (PQ). We insert the head of each substream into the PQ together with the `stream_id`. Then we remove the "min-member" and output it. We then take the next item from the substream in which the min-member was located, and insert it into the PQ. Finally, we re-establish the heap property and repeat. This approach was tried using one of the best available heap-based PQs, and the result was poor.

The best approach we found is to utilize an old idea reported by Knuth [1973] called the loser tree.[1] A loser tree is a kind of tournament tree where pairs of players engage in rounds of a single-elimination tournament. The loser is kept in the tree while the winner moves on, in a register. Our tree node consists of a floating-point key and a payload of `stream_id`, as shown in Listing 28.3. The loser tree is stored as a linearized complete binary tree, which enables us to avoid storing pointers. Navigating up and down the tree is done by shifts and adds.

The loser tree is initialized by inserting the head of each substream, as shown in Listing 28.4. At the end of this, the winner literally rises to the top. We remove the winner and output the key to the output buffer since this is the smallest among the heads of the sorted substreams. The winner node also carries the `stream_id` from which it came. We use that to pull the next item from the stream to take the place of the last winner. A key idea of a loser tree is that the new winner can only come from matches that the last winner had won—i.e., the path

```
struct Node
{
    float   key;      // our key
    int     value;    // always the substream index 'key' is from
};

// Returns our parent's index:
int Parent(int index) { return (index >> 1); }
int Left(int index) { return (index << 1); }
int Right(int index) { return ((index << 1) + 1); }
```

Listing 28.3. Data structures and abstractions for a loser tree.

---

[1] For an excellent graphical illustration of a loser tree in action, please refer to the course notes by Dr. Thomas W. Bennet, available at http://sandbox.mc.edu/~bennet/cs402/lec/losedex.html.

```
// Build the loser tree after populating all the leaf nodes:
int InitWinner(int root)
{
    if (root >= kStreams)
    {
        // leaf reached
        return (root);     // leaf index
    }
    else
    {
        int left = InitWinner(root * 2);
        int right = InitWinner(root * 2 + 1);

        Key lk = m_nodes[left].m_key;
        Key rk = m_nodes[right].m_key;

        if (lk <= rk)
        {
            // right subtree loses
            m_nodes[root] = m_nodes[right];    // store loser
            return (left);                     // return winner
        }
        else
        {
            m_nodes[root] = m_nodes[left];
            return (right);
        }
    }
}
```

Listing 28.4. Initialization for a loser tree.

from the root of the tree to the original position of the winner at the bottom. We repeat those matches with the new item, and a new winner emerges, as demonstrated in Listing 28.5.

The next little trick to speed up the merge is to mark each substream with an end-of-stream marker that is guaranteed to be larger than all keys. The marker stops the merger from pulling from that substream when it reaches the merger. We chose infinity() as our marker. This also allows us to handle the uneven

```
// Update loser tree after storing 'nextval' at 'slot':
int Update(int slot, Key newKey)
{
    m_nodes[slot].m_key = newKey;

    // should always be the same stream
    assert(m_nodes[slot].m_value == slot);

    int loserslot = Parent(slot);
    int winner = slot;
    while (loserslot != 0)
    {
        float loser = m_nodes[loserslot].m_key;    // previous loser
        if (newKey > loser)
        {
            // newKey is losing to old loser

            // new winner's key
            newKey = loser;

            // new winner's slot
            int newwinner = m_nodes[loserslot].m_value;

            // newKey is the new loser
            m_nodes[loserslot] = m_nodes[winner];
            winner = newwinner;
        }

        loserslot = Parent(loserslot);
    }

    return (winner);
}
```

Listing 28.5. Update method for a loser tree.

substream lengths when the input is not exactly divisible by the number of
substreams. In the sample code on the website, we copy the input and allocate the
extra space needed by the marker. With some careful coding, we only need the
extra space for the last substream. In a production environment, if one can be

sure the input buffer always has extra space at the end, then we can avoid the extra buffer and copying. For example, we can define the interface to the sorter as such or take an extra input argument that defines the actual size of the input buffer and copy only if no room is reserved.

Our merge phase is very cache friendly since all the memory operations are sequential. The sorted streams are accessed sequentially, and the output is streamed out from the merger. The merger is small since it only needs space to store one item from each stream. The size of the tree is `2 * kNumStreams` and, therefore, fits into the L1 cache easily. One can even consider keeping the merger entirely within registers for maximum speed. For small datasets, our sort is 2.1 to 3.5 times faster than STL sort. The relative disadvantage of quicksort is smaller when the dataset is smaller and fits nicely into the caches.

The loser tree might not be the best approach if you are writing GPU-based applications. An even–odd or bitonic merge network is probably a better way to exploit the wide SIMD parallelism. That being said, the merge phase is only a small fraction of the total processing time (~25%). The sorter is encapsulated in a class to hide details, like the stream markers and substream sizes, and to reuse temporary buffers for multiple sorts.

28.4 Multicore Implementation

The sorter is refactored to expose the initialization, substream sort, and merge as separate methods. In the sample code on the website, the initialization function is called from the main thread before firing off the worker threads, each of which sorts one of the substreams. Some care is taken to create the threads up front and to use auto-reset events to reduce the overhead with thread synchronization. The class is carefully designed to leave the threading logic outside of the sorter, without impacting performance. The merging is performed on the main thread.

The data set used for our performance tests consists of 0.89 million floating-point values that happen to be the x coordinates for the Stanford Dragon. The one-stream times in Table 28.1 show that our threading code is efficient, and only about 0.12 to 0.17 ms of overhead is introduced. One can see the per-core scalability is not great on the Q6600—using four threads only doubles the speed of the sort phase. This is somewhat understandable since radix sort is very bandwidth hungry. Keep in mind we are sorting substreams that have been spliced. If we directly sort the whole input array using a single radix-256 sort, then the runtime is 28 ms. The serial merge costs 5 to 6 ms, which gives us a 65 percent improvement. Most of that improvement can be had with as little as two cores. For a point of reference, STL sort runs in 76 ms, and while using four threads,

	Threaded (Q6600)	Serial (Q6600)	Threaded (I7)	Serial (I7)
1 stream	5.12	5	3.89	3.62
2 streams	6.90	10.04	4.20	7.1
3 streams	8.08	15.07	4.56	10.69
4 streams	10.97	20.55	4.86	14.2
4 + merge	16.4	26.01	9.61	19.0

Table 28.1. Threading performance, in milliseconds.

our sort runs in 16 to 17 ms. The picture is very different on an I7, which shows better scalability, where four threads is about three times faster than serially sorting the four substreams. The runtime for STL sort is 58 ms on an I7, which is a six-times speed-up for the hybrid sorter.

Parallel Loser Tree

The Update() operation of the loser tree is the key step during merging. Can we find a way to parallelize the Update() step? Superficially, in Listing 28.5, the loserslot appears to be dependent on the previous loop's result. However, if we remember our invariants for a loser tree, the matches between entries in the tree have already been played earlier. Hence, any new loser must be a new player—i.e., the one we inserted, triggering the call to Update(). We arrive at the critical insight that *all tournaments along the path from a new leaf to the root are independent of each other*. In addition, the path to the root for a given leaf is always the same. This means that all the addressing needed is known at compile time. This kind of parallelization can never be derived automatically by a compiler. Marin [1997] presented the following parallel Update() step for a binary tournament tree:

```
for n in [0..height-1] do in parallel
    parent = Parent(n)
    if (m_nodes[parent].m_key > loser)
        m_nodes[parent] = m_nodes[winner]
```

28.5 Conclusion

We now have a sorter that is fast in a single-threaded application and also functions well in a multicore setting. The sort is a stable sort and has a predictable run time, which is critical for real-time rendering environments. It can also be used as an external sorter, where inputs are streamed from the disk and each core performs the radix sort independently. The merging step would be executed on a single core, but the I/O is purely sequential. The merger is very efficient; it took 6 milliseconds to merge one million items on a single core of a 2.4-GHz Q6600. It can be further improved by simple loop unrolling since the number of tournaments is always equal to the height of the tree.

This chapter is heavily influenced by research in cache-aware and cache-oblivious sorting. Interested readers are encouraged to follow up with excellent works like Brodal et al. [2008], where funnel sort is introduced. While this work is independently conceived, a paper by Satish et al. [2009] shares a lot of the same ideas and shows great results by carefully using radix sort to get great parallelism on a GPU.

Appendix

Radix sort is a refinement of distribution sort. If we have a list of n input bytes, we need 256 bins of size n to distribute the inputs into, since in the worst case, all the inputs can be the same byte value. Of course, we could have dynamically grown each of the bins, but that would incur a large overhead in calls to the memory allocator and also fragments our heap. The solution is a two-pass approach, where we allocate 256 counters and only count how many items belong in each bin during the first pass. Next, we form the prefix sum within the array of counters and use that to distribute each input digit to an output buffer of size n. We have just discovered counting sort.

The most commonly used form of radix sort is called the *least-significant-digit sort*. That is, we start from the least-significant digit and work ourselves towards more significant digits. For a 32-bit integer, the most common practice is to use radix-256—i.e., we break the integer into four 8-bit digits and perform the sort in four passes. For each pass, we perform the above counting sort. Since counting sort is a stable sort, the ordering of a pass is preserved in the later passes.

Acknowledgements

We would like to thank Michael Herf for his permission to include his radix sort code and for inventing the bit hack for floating-point values. We would also like to thank Warren Hunt for supplying his excellent radix sort implementation for another project that uses negative indexing and the improved version of the bit hack.

References

[Brodal et al. 2008] Gerth Stølting Brodal, Rolf Fagerberg, and Kristoffer Vinther. "Engineering a Cache-Oblivious Sorting Algorithm." *Journal of Experimental Algorithms* 12 (June 2008).

[Herf 2001] Michael Herf. "Radix Tricks." 2001. Available at http://www.stereopsis.com/radix.html.

[Knuth 1973] Donald Knuth. *The Art of Computer Programming, Volume 3: Sorting and Searching.* Reading, MA: Addison-Wesley, 1973.

[Garanzha and Loop 2010] Kirill Garanzha and Charles Loop. "Fast Ray Sorting and Breadth-First Packet Traversal for GPU Ray Tracing." *Computer Graphics Forum* 29:2 (2010).

[Lin 2000] Ming C. Lin. "Fast Proximity Queries for Large Game Environments." Game Developers Conference Course Notes. Available at http://www.cs.unc.edu/~lin/gdc2000_files/frame.htm.

[Marin 1997] Mauricio Marin. "Priority Queues Operations on EREW-PRAM." *Proceedings of Euro-Par '97 Parallel Processing.* Springer, 1997.

[Nyberg et al. 1995] Chris Nyberg, Tom Barclay, Zarka Cvetanovic, Jim Gray, and Dave Lomet. "AlphaSort: A Cache Sensitive Parallel External Sort." *The VLDB Journal* 4:4 (October 1995), pp. 603–627.

[Patney et al. 2010] Anjul Patney, Stanley Tzeng, and John D. Owens. "Fragment-Parallel Composite and Filter." *Computer Graphics Forum* 29:4 (June 2010), pp. 1251–1258.

[Satish et al. 2009] Nadathur Satish, Mark Harris, and Michael Garland. "Designing Efficient Sorting Algorithms for Manycore GPUs." *Proceedings of the 2009 IEEE International Symposium on Parallel & Distributed Processing.*

[Terdiman 2000] Pierre Terdiman. "Radix Sort Revisited." April 1, 2000. Available at http://codercorner.com/RadixSortRevisited.htm.

[Wright 2006] Christopher Wright. "Using CPUID for SIMD Detection." 2006. Available at http://softpixel.com/~cwright/programming/simd/cpuid.php.

29

Thread Communication Techniques

Julien Hamaide

Fishing Cactus

Multithreaded systems are common nowadays. They usually involve synchronization and lock primitives implemented at the operating system level. Working with those primitives is far from trivial and can lead to problems such as deadlocks. This chapter introduces communication techniques that do not use such primitives and rely on simple concepts. They can be applied to a lot of common threading systems.

29.1 Latency and Threading

When writing multithreaded code, care should be taken to protect all data that could be accessed simultaneously by several threads. Most of the time, by using a mutex or a critical section, the problem is avoided. But this approach has a cost: threads may stall waiting for a resource to be released. If the resource is heavily accessed, it may put all threads to sleep, wasting cycles. On the other hand, this approach has a low latency. When working on protected data, the thread waits until the action can be executed and then executes it right away. But does the action really need it to be executed instantaneously? Latency of some actions is not critical (e.g., playing a sound). Fortunately, few problems in video games need a very low-latency treatment. While writing multithreaded code, don't try to solve the general problem. Solve your particular problem. For example, if we use a thread pool, we know the number of threads is limited and that a thread won't be destroyed.

Communication among threads is accomplished by simply sending data from one thread to another, whether by using an operating system primitive, such as a semaphore, or a piece of memory. A common practice is to send data through shared variables, variables that should be protected by primitives such as critical sections. But most of the time, those variables are falsely shared. If a first-in-first-out (FIFO) queue is used to communicate commands between threads, the item count is a shared variable. If only a single thread writes to the collection, and only a single thread is reading from the collection, should both threads share that variable? The item count can always be expressed with two variables—one that counts the inserted element and one that counts the removed element, the difference being the item currently in the queue. This strategy is used in the simple structures presented here.

29.2 Single Writer, Single Reader

This problem is the classic producer-consumer model. A thread produces items that are consumed by another thread. It is this example that is generally employed to teach how semaphores are used. The goal is to provide a single writer, single reader (SWSR) FIFO queue, without using synchronization objects. A fixed-size item table is preallocated. This table serves as memory storage for the data transfer. Two indices are used, one for the first object to be popped and the other for the storage of the next item to be pushed. The object is "templatized" with item type and maximum item count, as shown in Listing 29.1.

A variable must not be written to by two threads at the same time, but nothing prevents two threads from reading that variable concurrently. If the producer thread only writes to `WriteIndex`, and the consumer thread only writes to `ReadIndex`, then there should be no conflict [Acton 2009]. The code shown in Listing 29.2 introduces the `Push()` method, used by the producer to add an item at the end of the queue.

The method first tests if there is space available. If no, it simply returns false, letting the caller decide what to do (retry after some sleep, delay to next frame, etc.). If some space is available, the item is copied to the local item table using the `WriteIndex`, and then a write barrier is inserted. Finally, the `WriteIndex` is incremented. The call to `WriteBarrier()` is the most important step of this method. It prevents both the compiler and CPU from reordering the writes to

```
template <typename Item, int ItemCount> class ProducerConsumerFIFO
{
    Item                    ItemTable[ItemCount];
    volatile unsigned int   ReadIndex, WriteIndex;
};
```

Listing 29.1. FIFO class definition.

```
bool ProducerConsumerFIFO::Push(const Item& item)
{
    if (!IsFull())
    {
        unsigned int index = WriteIndex;
        ItemTable[index % ItemCount] = item;
        WriteBarrier();
        WriteIndex = index + 1;

        return (true);
    }

    return (false);
}
```

Listing 29.2. FIFO `Push()` method.

memory and ensures the item is completely copied to memory before the index is incremented. Let's imagine that the CPU had reordered the writes and the `WriteIndex` has already updated to its new value, but the item is not yet copied. If at the same time, the consumer queries the `WriteIndex` and detects that a new item is available, it might start to read the item, although it is not completely copied.

When the `WriteIndex` is incremented, the modulo operator is not applied to it. Instead, the modulo operator is applied only when accessing the item table. The `WriteIndex` is then equal to the number of items ever pushed onto the queue, and the `ReadIndex` is equal to the number of items ever popped off the queue. So the difference between these is equal to the number of items left in the queue. As shown in Listing 29.3, the queue is full if the difference between the indices is equal to the maximum item count. If the indices wrap around to zero, the difference stays correct as long as indices are unsigned. If `ItemCount` is not a power of two, the modulo operation is implemented with a division. Otherwise, the modulo can be implemented as `index & (ItemCount - 1)`.

```
bool ProducerConsumerFIFO::IsFull() const
{
    return ((WriteIndex - ReadIndex) == ItemCount);
}
```

Listing 29.3. FIFO `IsFull()` method.

```
bool ProducerConsumerFIFO::Pop(Item& item)
{
    if (!IsEmpty())
    {
        unsigned int index = ReadIndex;
        item = ItemTable[index % ItemCount];
        ReadBarrier();
        ReadIndex = index + 1;
        return (true);
    }

    return (false);
}

bool IsEmpty() const
{
    return (WriteIndex == ReadIndex);
}
```

Listing 29.4. FIFO `Pop()` method.

The `Push()` method only writes to `WriteIndex` and `ItemTable`. The write index is only written to by the producer thread, so no conflict can appear. But the item table is shared by both threads. That's why we check that the queue is not full before copying the new item.

At the other end of the queue, the `Pop()` method is used by the consumer thread (see Listing 29.4). A similar mechanism is used to pop an item. The `Is-Empty()` method ensures there is an item available. If so, the item is read. A read barrier is then inserted, ensuring `ReadIndex` is written to after reading the item. If incremented `ReadIndex` was stored before the end of the copy, the item contained in the item table might be changed by another `Push()`.

A question that may arise is what happens if the consumer has already consumed an item, but the read index is not updated yet? Nothing important—the producer still thinks the item is not consumed yet. This is conservative. At worst, the producer misses a chance to push its item. The latency of this system is then dependent on the traffic.

For this technique to work, a fundamental condition must be met: writing an index to the shared memory must be atomic. For x86 platforms, 32-bit and 64-bit writes are atomic. On PowerPC platforms, only 32-bit writes are atomic. But on the PlayStation 3 SPU, there are no atomic writes under 128 bytes! That means each index must use 128 bytes, even though only four bytes are really used. If some undefined behavior occurs, always verify the generated code.

We now have a simple SWSR FIFO queue using no locks. Let's see what can be built with this structure.

29.3 The Aggregator

By using the structure we have introduced, an aggregator can be created. This structure allows several threads to queue items for a single receiver thread. This kind of system can be used as a command queue for some subsystems (e.g., sound system, streaming system). While commands might not be treated in the order they are queued, commands inserted by a single thread are processed in order. For example, if you create a sound, then set its volume, and finally play it, the pushed commands are processed in order. As shown by Listing 29.5, the aggregator is built upon the FIFO queue described earlier.

Each pushing thread is assigned to a SWSR FIFO. This FIFO must be accessible to the thread when it pushes an item, and this can be accomplished through thread local storage or a simple table containing a mapping between thread ID and its assigned FIFO. Listing 29.6 shows the Push() method using a thread local storage. This code assumes that a FIFO has already been assigned to the thread. To assign it, a registration method is called from each producer thread. This registration method stores the pointer of the FIFO assigned to the caller thread in the thread local storage, as shown in Listing 29.7. The GetNextAvailableFifo() function returns the next available FIFO and asserts whether any are available.

Listing 29.8 presents the consumer code. It needs to select an item from all threads using the most equitable repartition. In this example, a simple round-robin algorithm is used, but this algorithm might be adapted to special situations. Once again, it depends on your problem and your data.

```
template <typename Item, int ItemCount, int ThreadCount>
class Aggregator
{
    ProducerConsumerFIFO<Item, ItemCount> ThreadFifoTable[ThreadCount];
};
```

Listing 29.5. Aggregator class definition.

```
bool Aggregator::Push(const Item& item)
{
    // From thread local storage.
    ProducerConsumerFIFO *fifo = ThreadFifo.GetValue();
    assert(fifo);
    return (fifo->Push(item));
}
```

Listing 29.6. Aggregator `Push()` method.

```
void Aggregator::InitializeWriterThread()
{
    int data_index = GetNextAvailableFifo();
    assert((index != -1) && (!"No buffer left"));

    ThreadFifo.SetValue(&ThreadFifoTable[data_index]);
}
```

Listing 29.7. Method used to register writer threads.

```
bool Aggregator::Pop(Item& item)
{
    assert(GetCurrentThreadId() == ReaderThreadIdentifier);
    int current_table_index = LastTableIndex;
    do
    {
        current_table_index = (current_table_index + 1) % ThreadCount;
        ProducerConsumerFIFO *fifo =
                &ThreadDataTable[current_table_index];
        if (fifo->Pop(item))
        {
            LastTableIndex = current_table_index;
            return (true);
        }
    } while (current_table_index != LastTableIndex);

    return (false);
}
```

Listing 29.8. Aggregator `Pop()` method.

If no element is available at all after checking each thread's FIFO, the function returns false. The function does not wait for an element, and lets the caller decide what to do if none are available. It can decide to sleep, to do some other work, or to retry if low latency is important. For example, in a sound system, the update might process all commands until there are no commands left and then update the system.

To prevent any other thread from reading from the aggregator, we also register the reader thread and store its identifier. The identifiers are then matched in Pop() in debug mode.

This implementation does not allow more than a fixed number of threads. But generally in a game application, the number of threads is fixed, and a thread pool is used, thus limiting the need for a variable-size architecture.

From a performance point of view, care should be taken with memory usage. Depending on the architecture, different FIFO structures should be placed on different cache lines. To ensure consistency, when a thread writes to a memory slot, the corresponding line cache is invalidated in all other processor caches. An updated version must be requested by the thread reading the same cache line. The traffic on the bus can increase substantially, and the waiting threads are stalled by cache misses. The presented implementation does not take this problem into account, but an ideal solution would be to align the FIFO items with cache lines, thus padding the item up to the cache line size so that neighbor items are not stored in the same cache line.

29.4 The Dispatcher

In the case of a single producer and multiple consumers, we can create a dispatcher. The dispatcher is a single writer, multiple readers queue. This system can be used to dispatch work to several threads, and it can be implemented using the SWSR FIFO. The implementation is similar to that of the aggregator.

29.5 The Gateway

The final structure presented here is the gateway. It is a multiple writers, multiple readers queue. This queue is not really a FIFO anymore since a thread must be used to pop items from the aggregator and push them in the dispatcher. Its role is quite similar to a scheduler, with the central task of distributing items. This distribution can process the data before dispatching it. Nothing prevents this central thread from reading every possible item, doing some work on them (e.g., sorting or filtering), and finally pushing them to the reader threads. As an example, in a job system, this thread can pop all jobs in local variables and sort them by priority before pushing them to each worker thread. It might also try to prevent two ALU-intensive jobs from being assigned to threads that shared the same ALU.

While implementing this structure, some important questions drive its details. Do all writer threads also read from the queue? Is the central thread one of the writer or reader threads? Or is it a special thread that only works on dispatching the items? No generic implementation of the gateway is practical. A custom implementation is recommended for each system.

29.6 Debugging

Multithreading and debugging frequently can't get along. Every queue we have presented is made from an SWSR queue. This means that a condition must be satisfied: for a single SWSR queue, only a single thread can read and only a single thread can write. The queue should be able to detect if an unexpected thread accesses its data. Comparing the ID of the accessing thread with the expected thread ID and asserting their equality is strongly advised. If some crash continues to occur despite the correct thread accessing the collection, it might be the barrier. Check compiler-generated code and validate that the correct instructions are issued to prevent reordering (e.g., `lwsync` on the PowerPC).

References

[Acton 2009] Mike Acton. "Problem #1: Increment Problem." CellPerformance. August 7, 2009. Available at http://cellperformance.beyond3d.com/articles/index.html.

30

A Cross-Platform Multithreading Framework

Martin Fleisz
Thinstuff s.r.o.

Over the last couple of years, a new trend in game engine design has started due to changing processor designs. With the increasing popularity of multicore CPUs, game engines have entered the world of parallel computing. While older engines tried to avoid multithreading at all costs, it is now a mandatory technique in order to be able to take full advantage of the available processing power.

To be able to focus on the important parts of a game engine, like the graphics engine or the networking code, it is useful to have a robust, flexible, and easy to use multithreading framework. This framework also serves as an abstraction layer between the platform-dependent thread and the platform-independent game engine code. While the C++0x standard provides support for threading, developers still have to refer to external libraries like Boost to gain platform-independent threading support. The design and interface of the Boost threading framework strongly resembles the POSIX Threads (or Pthreads) library, which is a rather low-level threading library. In contrast, our framework offers a higher-level interface for threading and synchronization. What we provide is a collection of easy to use synchronization objects, a flexible and simple way of handling threads, and additional features like deadlock detection. We also show a few examples of how to extend and customize our framework to your own needs.

30.1 Threading

ThreadManager

Let us first take a look at the core component of our threading framework, the `ThreadManager` class. As can be seen in the sample code on the website, this class is implemented using a simple singleton pattern [Meyers 1995]. We decided to use this pattern for two reasons. First, we want global access to the Thread-Manager in order to be able to start a thread from any location in our engine. The second reason is that we want to have just one instance of this class at a time because we use it to create snapshots of our application's current state.

The ThreadManager has several responsibilities in our framework. Whenever we start a new thread, the manager attaches a ThreadInfo object to the new thread instance. The ThreadManager also provides a unique ID for each synchronization object. Upon creation, a synchronization object registers with the ThreadManager, which keeps a complete list of all existing threads and synchronization objects. This information is used to create a snapshot of the application that shows us which threads exist and in what state the currently used synchronization objects are. Using this information, we can get an overview of our application run-time behavior, helping us to find deadlocks or performance bottlenecks.

ThreadInfo

The most important component in our framework, when working with threads, is the ThreadInfo class. In order to be able to uniquely identify a thread, our ThreadInfo class stores a thread name and a thread ID. This information can be used during logging in order to identify which thread wrote which log entry. ThreadInfo also offers methods to wait for a thread, to stop it, or to kill it, although killing a thread should usually be avoided.

Another important feature of ThreadInfo is the recording of wait object information. Whenever a synchronization object changes its wait state, it stores that information in the ThreadInfo's wait object information array. With this information, the ThreadManager is able to construct a complete snapshot of the threading framework's current state. Additionally, each thread is able to detect whether there are any pending wait operations left before it is stopped (e.g., a mutex might still be locked after an exception occurred). In this case, it can issue a warning as such a scenario could potentially lead to a deadlock.

Another feature that we have built into our ThreadInfo objects is active deadlock detection. Using a simple trigger system, we can easily create a watchdog service thread that keeps checking for locked threads. In order for this feature to work, a thread has to continuously call the TriggerThread() method of its ThreadInfo instance. If the gap between the last TriggerThread() call and the current timestamp becomes too large, a thread might be deadlocked. Do not use too small values (i.e., smaller than a second) in order to prevent accidental timeouts. If a thread is known to be blocking, you can temporarily disable the trigger system using the SetIgnoreTrigger() method. You can also use this method to permanently disable triggering in the current thread in order to avoid the overhead caused by `TriggerThread()`.

`ThreadInfo` objects are stored and managed by the `ThreadManager`. For faster access, a reference to each thread's own `ThreadInfo` is stored in thread local storage (TLS). The `ThreadManager` uses our `TLSEntry` class in order to allocate a storage slot. This storage slot is local to each thread, which means we only have to allocate one slot in order to store a reference for each thread's `ThreadInfo`. Another nice feature of the `ThreadInfo` class is that it provides a great way to easily extend the threading framework, as discussed later.

30.2 Synchronization Objects

In order to coordinate two or more threads, we need a set of tools that enable us to synchronize their execution, depending on certain criteria. There are various different synchronization mechanisms available, but they all have one thing in common—they all use atomic operations. An atomic operation guarantees that its effect appears to the rest of the system instantaneously. A few examples of atomic operations include test-and-set, compare-and-swap, and load-link/store-conditional. All higher-level synchronization methods, like mutexes or semaphores are implemented using these atomic primitives.

Our framework orients along the Windows API synchronization methods, including critical sections, mutexes, semaphores, and events. On other platforms, we use the Pthreads library to emulate the Windows API. The abstraction of the platform-dependent synchronization APIs can be found in the `BasicSync.cpp` source file. All the classes defined in this file are used later in higher-level synchronization classes in order to provide enhanced functionality like wait state tracking. In the following subsections, we take a closer look at how we implemented each synchronization object and what tweaks were necessary in order to provide as much functionality as possible on all the different platforms.

Critical Section (class `BasicSection`)

A critical section can be compared to a lightweight version of a mutex. Just like a mutex, a critical section is used to ensure mutually exclusive access to a shared resource. However, it lacks some advanced functionality like support for inter-process communication (IPC) or timed wait operations. If you don't need any of these features, then you should always prefer using a critical section over a mutex for performance reasons.

The Pthreads standard does not provide us with an exact critical section equivalent. In order to provide the same functionality, our `BasicSection` class can either use a mutex or a spin lock. Spin locks, depending on their implementation, can cause various problems or performance issues [Sandler 2009]. Therefore, we decided to go with the safer option and implemented the `BasicSection` class using a mutex. Of course, this doesn't result in any performance improvement compared to using the `BasicMutex` class.

Mutex (class `BasicMutex`)

As already described in the previous section, a mutex ensures mutually exclusive access to a shared resource. However, a mutex is able to provide this exclusive access on a system-wide basis, across process boundaries. Windows uses so-called named mutexes to enable this way of IPC. If two threads create a mutex with the same name, one is going to actually create the mutex object, whereas the other just opens a handle to it. The mutex implementation in Pthreads works very similarly to an unnamed mutex in the Windows API. Unfortunately, there is no support for IPC-capable mutexes in Pthreads.

In order to provide this functionality in our framework, we have two different code paths in our basic mutex implementation. One path implements a simple unnamed mutex using the Pthreads mutex API. The other path provides support for named mutexes using a binary semaphore. While a binary semaphore doesn't offer as much functionality as a mutex (i.e., the semaphore lacks recursivity), it still provides mutual exclusivity and can be used to synchronize across process boundaries.

Semaphore (class `BasicSemaphore`)

A semaphore can be seen as a protected variable that controls access by several processes to a single resource. A special form is the binary semaphore, which has either a value of one (free) or zero (locked) and exposes behavior similar to that of a mutex. The implementation of `BasicSemaphore` for Windows is quite simple and basically just wraps the semaphore API functions. On other platforms, we can choose between two, quite different APIs, POSIX semaphores and System V semaphores.

POSIX semaphores are an extension to the Pthreads standard and are similar to the semaphores in the Windows API. In order to create a semaphore, we can either use the `sem_init()` (unnamed) or the `sem_open()` (named) function. Both functions also atomically initialize the semaphore to a specified value. This is an important feature, as discussed later. To acquire a semaphore, we use the sem_wait() function, which decrements the semaphore value on success. If the count is zero, then the thread is suspended (unless sem_trywait() is used, which is non-blocking) until the semaphore value becomes greater than zero again. The sem_post() function is used to release the semaphore lock, and it simply increments the internal counter. Cleanup is performed using either the sem_destroy() (unnamed) or sem_unlink() (named) function, which both destroy the semaphore.

The System V semaphore interface is not as simple as its POSIX counterpart. For example, instead of just a single semaphore counter, System V works with sets of one to n semaphore counters, all identified by a single semaphore ID. To create such a set, we have to use the semget() function to specify the number of semaphore counters in the set and whether the set is unnamed or named. Instead of strings, keys are used to identify a semaphore set in other threads or processes. We use a simple hash function to generate a key value for a given semaphore name, and we use the constant IPC_PRIVATE for unnamed semaphores. Semaphore operations are performed using the semop() function, which is a very powerful tool, as discussed later.

Before a System V semaphore can be used, it has to be initialized to its initial value. This is a disadvantage of the System V API because it does not provide any way to create and initialize semaphores atomically. The two code snippets shown in Listings 30.1 and 30.2 illustrate how we create and initialize a System V semaphore set correctly.[1]

We start by creating a new semaphore or opening an existing semaphore using a precalculated hash of the semaphore name (hash) as the key value. As you can see in the code snippet, we create a semaphore set with three counters. Counter 0 contains the actual semaphore value, counter 1 tracks how many objects reference the semaphore set (required later for cleanup), and counter 2 serves as a binary lock semaphore used during creation and destruction. Next, we lock semaphore 2 in our set by checking whether the counter's value is zero. If it is, we increment it, locking out any other threads or processes. We need to execute semget() and semop() within a loop to avoid a race condition where another thread or process could delete the semaphore between those two function calls. This is also the reason why we ignore EINVAL and EIDRM errors returned by semop().

After successful creation, we retrieve the value of semaphore 1 in order to determine whether we created the set (semval equals zero) or whether it already

---

[1] See sem_pack.h, available on Joseph Kee Yin's home page at http://www.comp.hkbu. edu.hk/~jng/comp2320/sem_pack.h.

```
int     sem;

int hash = CalcHashFromString("SemaphoreName");
for (;;)
{
    // Create the semaphore.
    if ((sem = semget(p_Hash, 3, 0666 | IPC_CREAT)) == -1)
        THROW_LAST_UNIX_ERROR();

    // Lock the newly created semaphore set.
    if (semop(sem, OpBeginCreate,
        sizeof(OpBeginCreate) / sizeof(sembuf)) >= 0)
            break;
    if (errno != EINVAL && errno != EIDRM)
        THROW_LAST_UNIX_ERROR();
}
```

Listing 30.1. System V semaphore creation.

existed. In case a new semaphore set was created, we set the semaphore value to `p_InitialValue` and the semaphore reference count to a large integer value (we cannot count up from zero because we use this state to determine whether the set has been initialized). Finally, we decrement our reference counter and release the lock on semaphore 2 to complete the initialization.

In Listings 30.1 and 30.2, we can also see the power of the `semop()` function. Beside the semaphore set ID, this function also expects an array of semaphore operations. Each entry in this array is of type `sembuf` and has the data fields described in Table 30.1.[2]

```
int     semval;

if ((semval = semctl(sem, 1, GETVAL, 0)) < 0)
    THROW_LAST_UNIX_ERROR();

// If semaphore 1's value is 0 the set has not been initialized.
if (semval == 0)
```

---

[2] See also *The Single UNIX Specification, Version 2*, available at http://opengroup.org/onlinepubs/007908799/xsh/semop.html.

```
{
    // Initialize semaphore value.
    if (semctl(sem, 0, SETVAL, p_InitialValue) < 0)
        THROW_LAST_UNIX_ERROR();

    // Init semaphore object references counter.
    if (semctl(sem, 1, SETVAL, MaxSemObjectRefs) < 0)
        THROW_LAST_UNIX_ERROR();
}

// Decrement reference counter and unlock semaphore set.
if (semop(sem, OpEndCreate, sizeof(OpEndCreate) / sizeof(sembuf)) < 0)
    THROW_LAST_UNIX_ERROR();
```

Listing 30.2. System V semaphore initialization.

The operation member specifies the value that should be added to the semaphore counter (or subtracted in case it is negative). If we subtract a value that is greater than the semaphore's current value, then the function suspends execution of the calling thread. When we are using an operation value of zero, semop() checks whether the given semaphore's value is zero. If it is, the function returns immediately, otherwise, it suspends execution. Additionally, there are two flags that can be specified with each operation, IPC_NOWAIT and SEM_UNDO. If IPC_NOWAIT is specified, then semop() always returns immediately and never blocks. Each operation performed with SEM_UNDO is recorded internally in the semaphore set. In case the program terminates unexpectedly, the kernel uses this information to reverse all effects of the recorded operations. Using this mechanism, we avoid any situation in which a semaphore remains locked by a dead process.

Member	Description
sem_num	Identifies a semaphore within the current semaphore set.
sem_op	Semaphore operation.
sem_flg	Operation flags.

Table 30.1. System V semaphore operation fields.

Cleanup of our semaphore set works in a way similar to its creation. First, we acquire the lock of our binary semaphore with ID 2 and increment the reference count of semaphore 1. Then, we compare the current value of semaphore 1 to the constant integer we used to initialize it. If they have the same value, then we can go ahead and delete the semaphore from the system using semctl() with the IPC_RMID flag. If the values do not match, we know that there are still references to the semaphore set, and we proceed by freeing our lock on semaphore 2.

Our implementation of BasicSemaphore makes use of both POSIX and System V semaphores for various reasons. For named semaphores, we use the System V API because of its SEM_UNDO feature. While developing the framework, it so happened that an application with a locked POSIX semaphore crashed. After restarting the application and trying to obtain a lock to the same semaphore, we became deadlocked because the lock from our previous run was still persisting. This problem doesn't arise when we use System V semaphores because the kernel undoes all recorded operations and frees the lock when an application crashes. A problem that both solutions still suffer from is that semaphore objects remain alive in case of an application crash since there is no automatic cleanup performed by the operating system. POSIX semaphores are used for process-internal communication because they have a little smaller overhead than private System V semaphores.

Event (class BasicEvent)

The last type of synchronization mechanism we implemented in our framework is the event. An event can have two different states, signaled or nonsignaled. In a common scenario, a thread waits for an event until another thread sets its state to signaled. If a thread waits for an event and receives a signal, then the event may remain signaled; or in case of an auto-reset event, the event is returned to the nonsignaled state. This behavior is slightly different from the Pthreads counterpart, the condition variable. A condition variable has two functions to set its state to signaled. The first, pthread_cond_signal(), unblocks at least one waiting thread; the other, pthread_cond_broadcast(), unblocks all waiting threads. If a thread is not already waiting when the signal is sent, it simply misses that signal and blocks. Another difference with the Windows API is that each condition variable must be used together with a Pthreads mutex. Before a thread is able to wait on a condition variable, it first has to lock the associated mutex. The thread is granted the ownership of this mutex after it receives a signal and leaves the pthread_cond_wait() function.

Again, our implementation orients along the Windows API, and functions such as Wait(), Set(), and Reset() simply call WaitForSingleObject(),

SetEvent(), and ResetEvent(). To support other platforms, we once more have two different implementation paths. Because Pthreads provides no support for system-wide condition variables, we emulate this scenario using a System V semaphore. The other implementation uses a Pthread condition variable in conjunction with a boolean flag indicating the state of the event.

Let us now take a closer look at the Pthreads implementation, starting with the Set() method. After acquiring the lock on our mutex, we set the signaled flag to true and continue by signaling the condition variable. If our event has manual reset enabled, then we use pthread_cond_broadcast() to resume all threads waiting on our condition variable. In the other case, we use pthread_cond_signal() to wake up exactly one waiting thread.

Pseudocode for the implementation of the Wait() method, when specifying an infinite timeout, is illustrated in Listing 30.3. After locking the mutex, we check the state of the signaled flag. If it is true, we can leave the wait function immediately. If we do not use a manual reset event, then the signal flag is reset to false before leaving the method. In case the event state is nonsignaled, we enter a loop that keeps waiting on the condition variable until the event state changes to signaled or an error occurs.

```
Lock Mutex

if signal flag set to true
    if manual reset disabled
        Set signal flag to false
    Unlock Mutex and return true

Loop infinitely
    Wait on Condition Variable
    if return code is 0
        if signal flag set to true
            if manual reset disabled
                Set signal flag to false
            Unlock Mutex and return true
    else
        Handle error
End Loop

Unlock Mutex
```

Listing 30.3. Pseudocode for waiting for an event.

The pseudocode in Listing 30.3 is one of three separate cases that require distinct handling, depending on the timeout value passed to the `Wait()` method. It shows the execution path when passing an infinite timeout value. If we use a timeout value of zero, then we just test the current signal state and return immediately. In this case, we only execute the first block and skip the loop. If neither zero nor infinite is used as the timeout value, then we can use the same code as in the listing with two minor changes. First, we use `pthread_cond_timedwait()` to wait on the condition variable with a given timeout. The second adaption is that our `Wait()` method has to return false in case `pthread_cond_timedwait()` returns `ETIMEDOUT`.

The last method, which is rather easy to implement, is used to set an event's state to nonsignaled and is called `Reset()`. All we need to do is to obtain the lock on our mutex and set the event's signal state to false.

Thanks to the flexibility of System V semaphores, we can easily rebuild the functionality of an event using a binary semaphore. Construction and cleanup is performed exactly as described for the `BasicSemaphore` class. The value of our semaphore is directly mapped to the event state—a value of one indicates nonsignaled and a value of zero represents a signaled state. The `Set()` method is a simple call to `semop()` that decrements the semaphore value. We also specify the `IPC_NOWAIT` flag to avoid blocking in case the event is already in the signaled state. If the semaphore value is zero, then the `semop()` call simply fails with `EAGAIN`, which means that `Set()` can be called multiple times without any unexpected side effects. The `Reset()` method works almost as simply, but uses two operations. The first operation checks whether the event is currently in signaled state using the `IPC_NOWAIT` flag. In case the semaphore value is zero, the first operation succeeds and performs the second operation, which increments the value. In case the event is already reset, the semaphore value is already one, and the first operation fails with `EAGAIN`. In this case, the second operation is not executed, which ensures the coherence of our event status in the semaphore.

The implementation of the `Wait()` method for our semaphore-based event is also quite easy. Again, we have three different scenarios, depending on the specified timeout value. In order to check whether the event is in signaled state, we use `semop()` with either one or two operations, depending on whether the event is a manual reset event. The first operation simply checks whether the semaphore value equals zero, with or without (zero or infinite timeout specified, respectively) using the `IPC_NOWAIT` flag. If we are not using a manual reset event, we increment the semaphore value after testing for zero, thus setting the state back to nonsignaled. This is almost the same trick we used in the `Reset()` method, again using the flexibility and power of `semop()`. Some systems also support a timed version of `semop()`, called `semtimedop()`, which can be used for the remaining timeout scenario. Our implementation does not use `semtimedop()` in order to maintain a high level of portability. Instead, we poll the semaphore at a predefined interval using the `nanosleep()` function to put the thread to sleep between attempts. Of course, this is not the best solution and, if `semtimedop()` is available on your platform, you should favor using it.

Wait Objects

Now that we have the basic implementation of our synchronization mechanism, we have to extend our objects in order to add state recording functionality. To do so, we introduce a new class called `WaitObject` that implements state recording using the wait state array from our `ThreadInfo` class. This array is a simple integer array with a constant size, and it belongs to exactly one thread. Because only the owner thread is allowed to modify the content of the array, we do not have to use any synchronization during state recording.

The format of a single state record is shown in Table 30.2. The first field specifies the current state of the object, which can be either `STATE_WAIT` or `STATE_LOCK`. The next two fields specify the time at which the object entered the current state and, in case of a wait operation, the specified timeout. The object count field defines how many object IDs are to follow, which might be more than one in case of a `WaitForMultipleObjects()` operation. Finally, the last field specifies the total size of the record in integers, which in most cases is six. You might wonder why we put the state size field at the end and not at the beginning of a record. In most cases, we remove a record from the end of our wait state array, and it therefore makes sense that we start our search from the end of the integer array, which means the state size field is actually the first element in our record.

In order to set an object's state to wait, we use the `SetStateWait()` function from our `WaitObject` class, which simply fills the array with the required information. If an object's state becomes locked, then we need to call the `SetStateLock()` method. This method requires a parameter that specifies whether the object immediately entered the locked state or whether it made a

Field	State	Timestamp	Timeout	Object Count	Object IDs	State Size
Integers	1	1	1	1	1 – object count	1

Table 30.2. Wait state record.

state transition from a previous wait. In the first case, again we can simply fill the array with the lock state information. However, in the second case, we have to find the wait state record of our object in the wait state array, change the state to locked, and set the timestamp value. Thanks to the state size field, scanning the array for the right entry can be done very quickly. RemoveStateWait() and RemoveStateLock() work in a similar fashion. First, we search for the wait or lock entry of the current object in the array. Then we delete it by setting all fields to zero. Of course, it might happen that the entry is not the last in the array, in which case we have to relocate all of the following entries.

WaitObject() also defines four important abstract methods that need to be implemented by derived classes: Lock(), Unlock(), LockedExtern(), and UnlockedExtern(). We provide classes derived from WaitObject for each of our basic synchronization objects. Each class implements Lock() and Unlock() by forwarding the call to the underlying object and setting the according wait and lock states. LockedExtern() and UnlockedExtern() are called if a WaitObject's internal synchronization object is locked or unlocked outside the class. These methods are helper functions used by our WaitForMultipleObjects() implementation in order to keep each WaitObject's state correctly up-to-date. In order to prevent someone else from messing up our wait states with these methods, we declared them private and made the WaitList class a friend of WaitObject.

With the recorded wait state information, our ThreadManager is now able to construct a complete snapshot of the current program state as in the example shown in Listing 30.4. The output shows us that we have three different threads. We start with the first one with the ID 27632, which is our main program thread. Because our ThreadManager did not create this thread but only attached itself to it, no thread name is given. We can see that our main thread owns a lock (since

```
Thread WaitObject Information:
    27632 : Attached Thread
       - Locked (500ms) wait object 1013 : WaitMutex
    28144 : WaitThread
       - Waiting (500ms/INFINITE) for multiple wait objects (2)
       - 1012 : WaitEvent
       - 1013 : WaitMutex
    28145 : DumpThread
       - No Locks
```

Listing 30.4. Example dump of wait state information.

the timestamp is 500 ms) on a wait object with ID 1013 named `WaitMutex`. `WaitThread` is the name of the next thread in the list, and this thread is blocked in a `WaitForMultipleObjects()` call. It waits for an event named `WaitEvent` with ID 1012, and it waits for the same mutex that our main thread has locked. We can also see that we have already been waiting for 500 ms and that the specified timeout value is infinite. Finally, the last thread in the list is called `Dump-Thread`. This is the thread that was used to output the status information, and it has neither waits nor locks.

Lock Helper Classes

An important practice when programming in multithreaded environments is to release object ownership once a thread is done with its work. Of course, ownership must also be released in case of an error, which is actually not that easy. Common pitfalls include functions that do not have a single point of exit and unhandled exceptions. To better illustrate this problem, take a look at the code in Listing 30.5, which contains one obvious error where it fails to release the mutex lock. In the case that an error occurs, we simply return false but forget to release our lock. While this issue can be easily fixed by adding the missing `Unlock()` call before the `return` statement, we still have another error hiding in the sample code. Imagine that somewhere between the `Lock()` and `Unlock()` calls, an exception occurs. If this exception does not occur within a try-catch block, we exit the function, leaving the mutex in a locked state.

A simple and elegant solution to this problem is to provide RAII-style[3] locking using a little helper class. Our framework offers two different versions of this class called `LockT` and `WaitLock`. `LockT` is a simple template class that works with all synchronization objects that implement a `Lock()` and `Unlock()` method. On construction, this class receives a reference to the guarded synchronization object and, if requested, immediately locks it. `LockT()` also offers explicit `Lock()` and `Unlock()` functions, but the most important feature is that it unlocks a locked object in its destructor. Listing 30.6 shows how to use `LockT()` to fix all issues in Listing 30.5.

The `WaitLock` class implementation is almost identical to `LockT`'s—the only difference is that it only works with synchronization objects derived from `WaitObject`. There are two reasons why we do not use `LockT` for `WaitObjects`. The first is that these objects are reference counted and need special handling when passing a raw pointer to our helper class. The other reason is that `WaitObjects`

---

[3] RAII stands for resource acquisition is initialization. See Stroustrup, *The Design and Evolution of C++*, Addison-Wesley, 1994.

```
BasicMutex     mtx;

mtx.Lock();
... // do some work here
if (errorOccured) return (false);
mtx.Unlock();
return (true);
```

Listing 30.5. A bad example for using the mutex class.

```
BasicMutex     mtx;

LockT<BasicMutex> mtxLock(mtx, true);  // true means lock the Mutex
... // do some work here

if (errorOccured) return (false);
return (true);
```

Listing 30.6. Example for using a mutex with the `LockT` helper class

introduce a certain overhead due to the additional wait state recording. Therefore, you should carefully choose what synchronization object type to use when performance matters.

Waiting for Multiple Objects

A nice feature on Windows platforms is the ability to wait for several different objects to enter the signaled state. These objects can be synchronization objects (mutex, semaphore, event), as well as other object types like sockets or threads. Unfortunately, the POSIX standard doesn't specify a similar API. In order to provide the same functionality on UNIX platforms, we introduce the concept of wait lists. A `WaitList` is a simple container that holds an array of `WaitObjects` that we want to wait for. It provides two wait methods, `WaitForSingleObject()` and `WaitForMultipleObjects()`, where the first method waits for a specific wait object to enter the signaled state, and the second method waits for any object in the list to become signaled. On Windows platforms, we can simply call the corresponding Windows API functions, `WaitForSingleObject()` and `WaitForMultipleObjects()`. However, our UNIX implementation is a bit more complex.

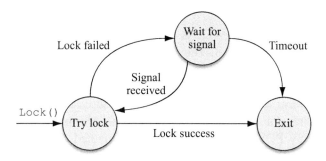

Figure 30.1. WaitForMultipleObjects() states.

Of course, the easiest solution is to simply poll our wait object list and try to lock each object with a zero timeout (nonblocking). If one of our attempts succeeds, then we can return true immediately. Otherwise, if all attempts fail, then we put our thread to sleep for a short period. This solution is not the most elegant one since we keep waking up the thread and polling the wait objects even though none of their states have changed. Therefore, our implementation uses a different approach, slightly based on [Nagarajayya and Gupta 2000].

Figure 30.1 shows the different states and the respective transitions in our WaitForMultipleObjects() implementation. We start off by trying to lock one of the wait objects in our list. If this step is not successful, then we put our thread to sleep, waiting on a condition variable. We are using a single global condition variable that notifies us whenever a synchronization object has changed its state to signaled. Every synchronization object in our framework sets this variable when its Unlock() or Reset() method is called. Only if such a state transition has occurred do we go through our list of wait objects again and try to lock each one. The advantage of this solution is that it only resumes the thread and checks whether a lock can be obtained after an object has actually changed to a signaled state. The overhead of our solution is also relatively small because we only use one additional condition variable in the whole framework. Of course, it can still happen that the thread performs unnecessary checks in the case that an object is unlocked but is not part of the WaitList. Therefore, this function should be used with caution on UNIX systems.

30.3 Limitations

Now that we have provided an overview of our threading framework and its features, it is time to discuss some limitations and potential pitfalls. The first thing we want to discuss is the possibility of a Mac OS X port. The source code provided on the website is basically ready to be built on a Mac, but there are a few things to consider. The first one is that the Mac OS X kernel only supports a total of 10 System V semaphores, using the SEM_UNDO flag, in the whole system [Huyler 2003]. This means that if no other application is using a System V semaphore, we can have a maximum of three named semaphores, mutexes, or events with our framework on Mac OS X. A simple solution to this problem is to use only POSIX semaphores for both named and unnamed synchronization objects on Mac OS X platforms. However, a problem that is introduced by this fix is that the named auto-reset event implementation in our framework requires the flexibility of System V's semop() function. A POSIX implementation could be made similar to the one for an unnamed event, but it has to keep the signal flag in a shared memory segment. Another problem we experienced on Mac OS X was that the sem_init() function always failed with an ENOSYS error. This is because the POSIX semaphores implementation on Mac OS X only supports named semaphores and does not implement sem_init() [Jew 2004]. Apart from these limitations, the framework should already work perfectly on Mac OS X, without requiring any code changes.

In this chapter, we also showed how to implement a wait function similar to the Windows API WaitForMultipleObjects() function. However, our implementation currently only works with unnamed events. This limitation is caused by the condition variable used to signal that an object has been unlocked, which is only visible to a single process. This means that if a named mutex is unlocked in process A, our wait function in process B won't be notified of the state change and will remain blocked. A possible solution to this problem is to use a System V semaphore to signal a state change across process boundaries. However, this should be used with care since our processes might end up receiving many notifications, resulting in a lot of polling in the WaitList. On Windows, you can also specify whether you want to wait for any or all objects in the WaitList. Our current implementation only supports the first case, but adding support for the second case should be straightforward. Finally, if you are really concerned about performance, you might want to add a BasicWaitList class that works with the basic synchronization classes, instead of WaitObject-derived ones.

The last potential pitfall that we want to highlight occurs when using named mutexes. On Windows, mutexes can be locked by the same thread multiple times without blocking. With Pthreads, you can specify whether the mutex should be recursive using the PTHREAD_MUTEX_RECURSIVE type. Unfortunately, our implementation does not offer this functionality using our System V semaphore without introducing additional overhead to keep track of the current thread and its lock count. Therefore, if you really require recursive mutexes, then you should try to avoid using named mutex instances. One more thing left to do when integrating the framework into your own projects is to implement logging. If you search through the code, you will find some comments containing TRACE statements. These lines should be replaced with your engine's own logging facilities.

30.4 Future Extensions

We believe that our framework can be extended in many useful ways. In the code on the website, we show two examples in the WaitObjectEx.* files. The first is a small class that derives from WaitObject and encapsulates the thread stop event found in ThreadInfo. Using this class, we can construct a thread wait object from a ThreadInfo instance, which can be used in a WaitForMultiple-Objects() call to wait for a thread to exit. Additionally, it also enables wait state tracking for thread events. The second example is a simple last-in-first-out queue that is in the signaled state when elements are in the queue and in the nonsignaled state when the queue is empty. The class is derived from WaitEvent, which means that the queue can also be used with a WaitList.

Another possibility to enhance the threading framework is offered by the ThreadInfo class. In our system, we are using an exception-based error handling framework that is tightly integrated into the threading framework. If an exception occurs in a thread, then we have a generic exception handler in our thread start procedure (in the ThreadInfo class) that collects the exception information for any uncaught exception and stores it in the thread's ThreadInfo. The parent thread is then able to retrieve the exact error information from that same Thread-Info and use it for proper error reporting.

References

[Meyers 1995] Scott Meyers. *More Effective C++: 35 New Ways to Improve Your Programs and Designs*. Reading, MA: Addison-Wesley, 1995.

[Sandler 2009] Alexander Sandler. "pthread mutex vs pthread spinlock." Alex on Linux, May 17, 2009. Available at http://www.alexonlinux.com/pthread-mutex-vs-pthread-spinlock.

[Nagarajayya and Gupta 2000] Nagendra Nagarajayya and Alka Gupta. "Porting of Win32 API WaitFor to Solaris." September 2000. Available at http://developers.sun.com/solaris/articles/waitfor_api.pdf.

[Huyler 2003] Christopher Huyler. "SEM_UNDO and SEMUME kernel value issues." osdir.com, June 2003. Available at http://osdir.com/ml/macosx.devel/2003-06/msg00215.html.

[Jew 2004] Matthew Jew. "semaphore not initialized - Question on how to implement." FreeRADIUS Mailing List, October 28, 2004. Available at http://lists.cistron.nl/pipermail/freeradius-devel/2004-October/007620.html.

31

Producer-Consumer Queues

Matthew Johnson
Advanced Micro Devices, Inc.

31.1 Introduction

The producer-consumer queue is a common multithreaded algorithm for handling a thread-safe queue with first-in-first-out (FIFO) semantics. The queue may be bounded, which means the size of the queue is fixed, or it may be unbounded, which means the size can grow dynamically based on available memory. Finally, the individual items in the queue may be fixed or variable in size. Typically, the implementation is derived from a circular array or a linked list data structure. For simplicity, this chapter describes bounded queues with elements of the same size.

Multithreaded queues are a common occurrence in existing operating system (OS) APIs. One example of a thread-safe queue is the Win32 message model, which is the main communication model for applications to process OS and user events. Figure 31.1 shows a diagram of a single-producer and single-consumer model. In this model, the OS produces events such as WM_CHAR, WM_MOUSEMOVE, WM_PAINT, etc., and the Win32 application consumes them.

Figure 31.1. Single producer and single thread.

The applications for producer and consumer queues extend beyond input messaging. Imagine a real-time strategy game where the user produces a list of tasks such as "build factory," "move unit here," "attack with hero," etc. Each task is consumed by a separate game logic thread that uses pathfinding knowledge of the environment to execute each task in parallel. Suppose the particular consumer thread uses an A* algorithm to move a unit and hit a particularly worst-case performance path. The producer thread can still queue new tasks for the consumer without having to stall and wait for the algorithm to finish.

The producer-consumer queue naturally extends to data parallelism. Imagine an animation engine that uses the CPU to update N animated bone-skin skeletons using data from static geometry to handle collision detection and response. As pictured in Figure 31.2, the animation engine can divide the work into several threads from a thread pool by producing multiple "update bone-skin" tasks, and the threads can consume the tasks in parallel. When the queue is empty, the animation thread can continue, perhaps to signal the rendering thread to draw all of the characters.

In Figure 31.3, only one item at a time is consumed from the queue (since the queue is read atomically from the tail); however, the actual time the consumer thread starts or finishes the task may be out of order because the OS thread scheduler can preempt the consumer thread before useful work begins.

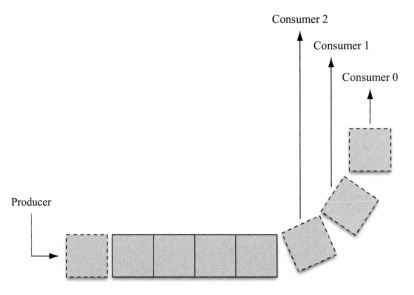

Figure 31.2. Single producer and multiple consumers.

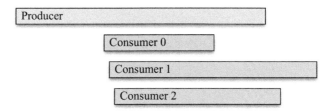

Figure 31.3. Sample timeline for producer-consumer model from Figure 31.2.

31.2 Multithreading Overview

Since the producer pushes items onto the head of the queue and the consumer pops items off of the tail, the algorithm must serialize access between these memory locations atomically. An atomic operation is an operation that appears to all processors in the system as occurring instantly and is uninterruptible by other processes or threads. Although C++ does not natively support atomic data types (until C++0x), many libraries have other mechanisms to achieve the same results.

Serialization is typically supported by the OS and its API in various forms. At the process level, a critical section can be used to synchronize access among multiple threads, whereas a mutex can synchronize access among multiple processes. When a thread enters a critical section, no other thread can execute the code until that thread leaves the critical section. Finally, synchronization primitives can be avoided in the design of a lock-free algorithm by exclusively using atomic instructions to serialize resources. This requires detailed working knowledge of the compiler, memory model, and processor.

31.3 A First Approach: Using Win32 Semaphores and Critical Sections

Since this chapter covers bounded queues, where each item is fixed in size, we need a thread-safe way to internally manage the item count in the queue. Semaphores are an appropriate way to handle this.

Our first implementation of a FIFO, shown in Listing 31.1, uses a standard circular queue, which is not thread safe. Note that the head precedes the tail but, if wrapped around zero for an unsigned integer, continues to work because the modulus is used for each array access. A common optimization is to use a power of two for the queue size, allowing the modulus operation to be replaced with a bitwise AND with one less than the queue size. To keep track of the number of

```cpp
template <typename T> class Fifo
{
    public:

        Fifo(UINT maxItems) : m_maxSize(maxItems), m_head(0), m_tail(0)
        {
            m_items = new T[m_maxSize];
        }

        ~Fifo()
        {
            delete m_items;
        }

        bool IsEmpty()
        {
            return (m_tail == m_head);
        }

        bool IsFull()
        {
            return (m_tail == m_head + m_maxSize);
        }

        void Insert(const T& item)
        {
            assert(!IsFull());
            m_items[m_tail++ % m_maxSize] = item;
        }

        T Remove()
        {
            assert(!IsEmpty());
            return (m_items[m_head++ % m_maxSize]);
        }

    private:

        T               *m_items;
```

```
        UINT            m_head;
        UINT            m_tail;

        const UINT m_maxSize;
};
```

Listing 31.1. A simple FIFO queue.

elements currently in the queue, an m_size member variable can be added. It is incremented for each insert at the tail and decremented for each removal at the head.

Next, a producer-consumer class can be created for thread-safe access to the FIFO queue. The code in Listing 31.2 demonstrates a thread-safe queue using a semaphore to maintain a count of the maximum number of items allowed in the queue. The semaphore is initialized to zero by default. Calling ReleaseSemaphore() causes the OS to atomically increment the semaphore count by one. Calling WaitForSingleObject() waits if the semaphore count is zero; otherwise, the OS atomically decrements the count by one and returns from the function.

```
template <typename T> class ProducerConsumerQueue
{
    public:

        enum
        {
            MaxItemsInQueue = 256
        };

        ProducerConsumerQueue() : m_queue(MaxItemsInQueue)
        {
            InitializeCriticalSection(&m_cs);

            m_sm = CreateSemaphore(NULL,          // security attributes
                    0,                            // initial semaphore count
                    MaxItemsInQueue,              // maximum semaphore count
                    NULL);                        // name (useful for IPC)
        }
```

```cpp
    void Insert(const T& item)
    {
        for (;;)
        {
            EnterCriticalSection(&m_cs);

            if (m_queue.IsFull())
            {
                LeaveCriticalSection(&m_cs);
                SwitchToThread();
                continue;           // Queue full
            }
            else if (SUCCEEDED(ReleaseSemaphore(m_sm, 1, NULL)))
            {
                m_queue.Insert(item);
            }

            LeaveCriticalSection(&m_cs);
            break;
        }
    }

    T Remove()
    {
        T   item;

        for (;;)
        {
            if (WAIT_OBJECT_0 != WaitForSingleObject(m_sm, INFINITE))
                break;

            EnterCriticalSection(&m_cs);

            if (!m_queue.IsEmpty())
            {
                item = m_queue.Remove();

                LeaveCriticalSection(&m_cs);
                break;
            }
            else
```

```
            {
                LeaveCriticalSection(&m_cs);     // Queue empty
            }
        }

        return (item);
    }

    DWORD LastError()
    {
        return (::GetLastError());
    }

private:

    Fifo<T>              m_queue;

    CRITICAL_SECTION     m_cs;
    HANDLE               m_sm;
};
```

Listing 31.2. A simple producer-consumer queue.

When the semaphore count is zero, the consumer(s) enter into a low-overhead wait state until a producer inserts an item at the head of the queue. Note that in this implementation, the `Insert()` function can still return without inserting an item, for example, if the `ReleaseSemaphore()` function fails. Finally, the `Remove()` function can still return without removing an item, for example, if the `WaitForSingleObject()` function fails. These failure paths are rare, and the application can call the `LastError()` function to get the Win32 error result.

Some designs may want to defer to the application to retry the operation on the producer side if the queue is full. For example, one may want to return false if an item can't be inserted at that moment in time because the queue is full. In the implementation of a thread-safe queue, one needs to be careful how to proceed if the queue is empty or full. For example, when removing an object, the code in Listing 31.2 first requests exclusive access to check if the queue has at least one item, but backs out and tries again if it has no items to remove. If the check was done before entering the critical section, a thread context switch could occur and another consumer could remove the last item right before entering the critical section, which would result in trying to remove an item from an empty queue!

This particular class has a few limitations. For example, there is no way to signal to all the consumers to stop waiting for the producer(s) to insert more data. One way to handle this is by adding a manual reset event that is initially non-signaled and calling the `WaitForMultipleObjects()` function on both the stop event and semaphore, with the `bWaitAll` set to `FALSE`. After a `SetEvent()` call, all consumers would then wake up, and the return value of the `WaitForMultipleObjects()` function indicates whether it was the event that signaled it.

Finally, the semaphore object maintains an exact physical count of the objects, but this is only internal to the OS. To modify the consumer to get the number of items currently in the queue, we can add a member variable that is incremented or decremented whenever the queue is locked. However, every time the count is read, it would only be a logical snapshot at that point in time since another thread could insert or remove an item from the queue. This might be useful for gauging a running count to see if the producer is producing data too quickly (or the consumer is consuming it too quickly), but the only way to ensure the count would remain accurate would be to lock the whole queue before checking the count.

31.4 A Second Approach: Lock-Free Algorithms

The goal of every multithreaded algorithm is to maximize parallelism. However, once a resource is requested for exclusive access, all threads must stall while waiting for that resource to become available. This causes contention. Also, many OS locks (such as mutexes) require access to the kernel mode and can take hundreds of clock cycles or more to execute the lock. To improve performance, some multithreaded algorithms can be designed not to use any OS locks. These are referred to as *lock-free* algorithms.

Lock-free algorithms have been around for decades, but these techniques are seeing more and more exposure in user-mode code as multicore architectures become commonplace and developers seek to improve parallelism. For example, Valve's Source Engine internally uses lock-free algorithms for many of the multithreaded components in their game engine [Leonard 2007].

The design of new lock-free algorithms is notoriously difficult and should be based on published, peer-reviewed techniques. Most lock-free algorithms are engineered around modern x86 and x64 architectures, which requires compiler and processor support around a specific multiprocessor memory model. Despite the complexity, many algorithms that have high thread contention or low latency requirements can achieve measurable performance increases when switching to a lock-free algorithm. Lock-free algorithms, when implemented correctly, generally do not suffer from deadlocks, starvation, or priority inversion. There is, however, no guarantee that a lock-free algorithm will outperform one with locks, and the additional complexity makes it not worth the headache in many cases. Finally, another disadvantage is that the range of algorithms that can be converted to a lock-free equivalent is limited. Luckily, queues are not among them.

One of the challenges when accessing variables shared among threads is the reordering of read and write operations, and this can break many multithreaded algorithms. Both the compiler and processor can reorder reads and writes. Using the `volatile` keyword does not necessarily fix this problem, since there are no guarantees in the C++ standard that a memory barrier is created between instructions. A memory barrier is either a compiler intrinsic or a CPU instruction that prevents reads or writes from occurring out of order across the execution point. A full barrier is a barrier that operates on the compiler level and the instruction (CPU) level. Unfortunately, the `volatile` keyword was designed for memory-mapped I/O where access is serialized at the hardware level, not for modern multicore architectures with shared memory caches.

In Microsoft Visual C++, one can prevent the compiler from reordering instructions by using the `_ReadBarrier()`, `_WriteBarrier()`, or `_ReadWriteBarrier()` compiler intrinsics. In Microsoft Visual C++ 2003 and beyond, volatile variables act as a compiler memory barrier as well. Finally, the `MemoryBarrier()` macro can be used for inserting a CPU memory barrier. Interlocked instructions such as `InterlockedIncrement()` or `InterlockedCompareAndExchange()` also act as full barriers.

31.5 Processor Architecture Overview and Memory Models

Synchronization requirements for individual memory locations depend on the architecture and memory model. On the x86 architecture, reads and writes to aligned 32-bit values are atomic. On the x64 architecture, reads and writes to aligned 32-bit and 64-bit values are atomic. As long as the compiler is not optimizing away the access to memory (use the `volatile` keyword to ensure this), the value is preserved and the operation is completed without being interrupted. However, when it comes to the ordering of reads and writes as it appears to a particular processor in a multiprocessor machine, the rules become more complicated.

Processors such as the AMD Athlon 64 and Intel Core 2 Duo operate on a "write ordered with store-buffer forwarding" memory model. For modern x86 or x64 multiprocessor architectures, reads are not reordered relative to reads, and writes are not reordered relative to writes. However, reads may be reordered with a previous write if the read accesses a different memory address.

During write buffering, it is possible that the order of writes across all processors appears to be committed to memory out of order to a local processor. For example, if processor A writes A.0, A.1, and A.2 to three different memory locations and processor B writes B.3, B.4, and B.5 to three other memory locations in parallel, the actual order of the writes to memory could appear to be written ordered A.0, A.1, B.3, B.4, A.2, B.5. Note that neither the A.0, A.1, and A.2 writes nor the B.3, B.4, and B.5 writes are reordered locally, but they may be interleaved with the other writes when committed to memory.

Since a read can be reordered with a write in certain circumstances, and writes can be done in parallel on multiprocessor machines, the default memory model breaks for certain multithreaded algorithms. Due to these memory model intricacies, certain lock-free algorithms could fail without a proper CPU memory barrier such as Dekker's mutual exclusion algorithm, shown in Listing 31.3.

```
static declspec(align(4)) volatile bool    a = false, b = false,
                                           turn = false;
void threadA()
{
    a = true;

    // <- write a, read b reorder possible, needs MemoryBarrier()

    while (b)
    {
        if (turn)
        {
            a = false;
            while (turn) { /* spin */ };
            a = true;
        }
    }

    // critical section

    turn = true;
    a = false;
}
```

```
void threadB()
{
   b = true;

   // <- write a, read b reorder possible, needs MemoryBarrier()

   while (a)
   {
      if (!turn)
      {
         b = false;
         while (!turn) { /* spin */ };
         b = true;
      }
   }

   // critical section

   turn = false;
   b = false;
}
```

Listing 31.3. Dekker's mutual exclusion algorithm.

Assuming that each thread is running in a different process, Figure 31.4 shows a parallel operation that fails with the write-ordered, store-buffer forwarding memory model. Since the read (cmp) can be reordered relative to the previous write on each processor, it is possible for threadA to read b as false on processor A and threadB to read a as false on processor B, before any instructions to set them to true are finished executing. This causes both threads to enter into the critical section at the same time, which breaks the algorithm. The way to solve this is by inserting a CPU memory barrier before the write and read.

Processor A	Processor B
mov dword ptr [a], 1	mov dword ptr [b], 1
cmp dword ptr [b], 0	cmp dword ptr [a], 0

Figure 31.4. On modern x86 or x64 processors, a read instruction can be reordered with a previous write instruction if they are addressing different memory locations.

Microsoft Visual C++ 2008 has a compiler-specific feature to treat access to volatile variables with acquire and release semantics [Dawson 2008]. In order to enforce acquire and release semantics completely on modern x86/64 processors, the compiler would need to enforce compiler memory barriers up the call stack in addition to using atomic or lock instructions. Unfortunately, when disassembling Dekker's algorithm on a Core2 Duo platform, locking instructions were not used, thus leaving the possibility of a race condition. Therefore, to be safe, aligned integers that are shared across threads should be accessed using interlocked instructions, which provide a full barrier when compiled under Microsoft Visual C++. A cross-platform implementation may require the use of preprocessor defines or different code paths depending on the target compiler and target architecture.

31.6 Lock-Free Algorithm Design

On x86 and x64 architectures, lock-free algorithms center around the compare-and-swap (CAS) operation, which is implemented atomically when supported as a CPU instruction. Pseudocode for a 32-bit CAS operation is shown in Listing 31.4. If implemented in C++, this operation would not be atomic. However, on x86 and x64 processors, the 32-bit CMPXCHG instruction is atomic and introduces a processor memory barrier (the instruction is prefixed with a LOCK). There are also 64-bit (CAS2) and 128-bit (CAS4) versions on supported platforms (CMPXCHG8B and CMPXCHG16B). In Win32, these are available as the InterlockedCompareExchange() and InterlockedCompareExchange64() functions. At the time of this writing, there is no InterlockedExchange128() function, but there is a _InterlockedCompareExchange128() intrinsic in Visual C++ 2010.

Support for the 64-bit CMPXCHG4B instruction has been available since the Pentium, so availability is widespread. The 128-bit CMPXCHG8B instruction is available on newer AMD64 architectures and Intel 64 architectures. These features can be checked using the __cpuid() intrinsic function, as shown in Listing 31.5.

```
UINT32 CAS(UINT32 *dest, UINT32 old, UINT32 new)
{
    UINT32 cur = *dest;
    if (cur == old) *dest = new;
    return (old);
}
```

Listing 31.4. CAS pseudocode.

```
struct ProcessorSupport
{
    ProcessorSupport()
    {
        int cpuInfo[4] = {0};
        __cpuid(cpuInfo, 1);

        supportsCmpXchg64 = ((cpuInfo[2] & 0x200) != 0);
        supportsCmpXchg128 = ((cpuInfo[3] & 0x100) != 0);
    }

    bool    supportsCmpXchg64;
    bool    supportsCmpXchg128;
};


ProcessorSupport        processorSupport;

if (processorSupport.supportsCmpXchg64) { /*...*/ }
```

Listing 31.5. Verifying CAS processor support.

This chapter uses the 32-bit CAS and 64-bit CAS2 in its design. The advantage of using a CAS operation is that one can ensure, atomically, that the memory value being updated is exactly as expected before updating it. This promotes data integrity. Unfortunately, using the CAS operation alone in the design of a lock-free stack and queue isn't enough. Many lock-free algorithms suffer from an "ABA" problem, where an operation on another thread can modify the state of the stack or queue but not appear visible to the original thread, since the original item appears to be the same.

This problem can be solved using a version tag (also known as a reference counter) that is automatically incremented and stored atomically with the item when a CAS operation is performed. This makes the ABA problem extremely unlikely (although technically possible, since the version tag could eventually wrap around). The implementation requires the use of a CAS2 primitive to reserve room for an additional version tag.

31.7 Lock-Free Implementation of a Free List

The AllocRef data structure in Listing 31.6 contains the union shared by both the stack and the queue algorithm. It is a maximum of 64 bits wide to accommo-

```
declspec(align(8)) union AllocRef
{
    enum
    {
        NullIndex = 0x000FFFFF
    };

    AllocRef(UINT64 idx, UINT64 ver)
    {
        arrayIndex = idx;
        version = ver;
    }

    AllocRef(UINT64 idx, UINT64 si, UINT64 v)
    {
        arrayIndex = idx;
        stackIndex = si;
        version = v;
    }

    AllocRef() : arrayIndex(NullIndex), version(0)
    {
    }

    AllocRef(volatile AllocRef& a)
    {
        val = a.val;
    }

    struct
    {
        UINT64      arrayIndex : 20;
        UINT64      stackIndex : 20;
        UINT64      version : 24;
    };

    UINT64      val;
};
```

Listing 31.6. A 64-bit allocation block reference.

date the CAS2 primitive. In addition, it includes a version tag to prevent the ABA problem.

The `Allocator` class in Listing 31.7 implements a free list, which is an algorithm that manages a preallocated pool of memory. The code uses an array-based lock-free implementation of a stack algorithm by Shafiei [2009]. Since this algorithm also requires a stack index, we pack the stack index, stack value (array index), and version number in one 64-bit value. This limits the maximum number of items in the stack and queue to 1,048,575 entries ($2^{20} - 1$) and the version number to 24 bits. The 20-bit index `0x000FFFFF` is reserved as a null index.

```cpp
__declspec(align(8)) class Allocator
{
  public:

      Allocator(UINT maxAllocs) : m_maxSize(maxAllocs + 1)
      {
          m_pFreeList = new AllocRef[m_maxSize];
          assert(m_pFreeList != NULL);

          m_pFreeList[0].arrayIndex = AllocRef::NullIndex;
          m_pFreeList[0].stackIndex = 0;
          m_pFreeList[0].version = 0;

          for (UINT i = 1; i < m_maxSize; ++i)
          {
              m_pFreeList[i].arrayIndex = i - 1;
              m_pFreeList[i].stackIndex = i;
              m_pFreeList[i].version = i;
          }

          m_top.val = m_pFreeList[m_maxSize - 1].val;
      }

      Allocator::~Allocator()
      {
          delete[] m_pFreeList;
      }

      inline bool IsFull() const
      {
```

```
            return (m_top.stackIndex == m_maxSize - 1);
        }

        inline bool IsEmpty() const
        {
            return (m_top.stackIndex == 0);
        }

        void Free(UINT32 arrayIndex);
        UINT32 Alloc();

    private:

        volatile AllocRef    *m_pFreeList;
        volatile AllocRef    m_top;

        const UINT           m_maxSize;
};

UINT32 Allocator::Alloc()
{
    for (;;)
    {
        AllocRef top = m_top;
        AllocRef stackTop = m_pFreeList[top.stackIndex];

        CAS2(&m_pFreeList[top.stackIndex].val,
            AllocRef(stackTop.arrayIndex, stackTop.stackIndex,
                top.version - 1).val,
            AllocRef(top.arrayIndex, stackTop.stackIndex,
                top.version).val);

        if (top.stackIndex == 0) continue; // Stack Empty?

        AllocRef belowTop = m_pFreeList[top.stackIndex - 1];

        if (CAS2(&m_top.val, AllocRef(top.arrayIndex, top.stackIndex,
                top.version).val,
            AllocRef(belowTop.arrayIndex, top.stackIndex - 1,
                belowTop.version + 1).val))
        {
```

```
                return (top.arrayIndex);
            }
        }
    }

void Allocator::Free(UINT32 arrayIndex)
{
    for (;;)
    {
        AllocRef top = m_top;

        AllocRef stackTop = m_pFreeList[top.stackIndex];

        CAS2(&m_pFreeList[top.stackIndex].val,
            AllocRef(stackTop.arrayIndex, stackTop.stackIndex,
                top.version - 1).val,
            AllocRef(top.arrayIndex, stackTop.stackIndex,
                top.version).val);

        if (top.stackIndex == m_maxSize - 1) continue;    // Stack full?

        UINT16 aboveTopCounter = m_pFreeList[top.stackIndex + 1].version;

        if (CAS2(&m_top.val, AllocRef(top.arrayIndex, top.stackIndex,
                top.version).val,
            AllocRef(arrayIndex, top.stackIndex + 1,
                aboveTopCounter + 1).val))
        {
            return;
        }
    }
}
```

Listing 31.7. A free list memory block allocator.

31.8 Lock-Free Implementation of a Queue

The implementation shown in Listing 31.8 is based on the lock-free, array-based queue algorithm by Shann and Haung [2000] with bug fixes by Colvin and Groves [2005]. In the original algorithm, the complete item in the queue is atomically swapped. Unfortunately, due to the absence of a hardware CAS*n* instruction, this limits an item size to 32 bits (further limited to 20 bits since we share the same index value in the stack algorithm). Therefore, the algorithm is extended to use a circular queue of index elements that are offsets to the specific item in an array. When a producer wants to insert an item, the index is atomically retrieved from the free list, and it is not inserted into the queue until the data is finished being copied into the array. Finally, the index can't be reused until a consumer retrieves the index from the queue, copies the data locally, and returns the index back to the free list.

```
/*   Lock-free Array-based Producer-Consumer (FIFO) Queue

Producer: Inserting an item to the tail of the queue.

- The allocator atomically retrieves a free index from the free pool.
- This index points to an unused entry in the item array.
- The item is copied into the array at the index.
- The index is inserted atomically at the queue's tail and is now visible.
- The index will remain unavailable until consumer atomically removes it.
- When that happens, the index will be placed back on the free pool.

Consumer: Removing an item from the head of the queue.

- The index is atomically removed from the head of the queue.
- If successfully removed, the head is atomically incremented.
- The item is copied from the array to a local copy.
- The index is placed back on the free pool and is now available for reuse.
*/

template <typename T> class Queue
{
    public:

        Queue(UINT maxItemsInQueue) : m_maxQueueSize(maxItemsInQueue),
                m_allocator(maxItemsInQueue),
                m_head(0), m_tail(0)
        {
            // Each value references a unique array element.
            m_pQueue = new AllocRef[maxItemsInQueue];
            assert(m_pQueue != NULL);
```

```cpp
            // By default, the queue is empty.
            for (UINT i = 0; i < m_maxQueueSize; ++i)
            {
                m_pQueue[i].arrayIndex = AllocRef::NullIndex;
                m_pQueue[i].version = 0;
            }

            // Array of Items
            m_pItems = new T[maxItemsInQueue];
            assert(m_pItems != NULL);
        }

        ~Queue()
        {
            delete[] m_pItems;
            delete[] m_pQueue;
        }

        void Insert(const T& item);
        T Remove();

    private:

        Allocator           m_allocator;      // Free list allocator.
        T                   *m_pItems;        // Array of items.
        volatile AllocRef   *m_pQueue;        // FIFO queue.
        volatile UINT       m_head;           // Head of queue.
        volatile UINT       m_tail;           // Tail of queue.
        const UINT          m_maxQueueSize;   // Max items in queue.
};

template <typename T> void Queue<T>::Insert(const T& item)
{
    UINT32      index;

    do
    {
        // Obtain free index from free list.
        index = m_allocator.Alloc();
    } while (index == AllocRef::NullIndex);     // Spin until free index.
```

```
    m_pItems[index] = item;

    for (;;)   // Spin until successfully inserted.
    {
        UINT tail = m_tail;
        AllocRef alloc = m_pQueue[tail % m_maxQueueSize];

        UINT head = m_head;
        if (tail != m_tail) continue;

        if (tail == m_head + m_maxQueueSize)
        {
            if (m_pQueue[head % m_maxQueueSize].arrayIndex !=
                AllocRef::NullIndex)
                    if (head == m_head) continue;   // Queue is full.

            CAS(&m_head, head, head + 1);
            continue;
        }

        if (alloc.arrayIndex == AllocRef::NullIndex)
        {
            if (CAS2(&m_pQueue[tail % m_maxQueueSize].val, alloc.val,
                AllocRef(index, alloc.version + 1).val))
            {
                CAS(&m_tail, tail, tail+1);
                return;
            }
        }
        else if (m_pQueue[tail % m_maxQueueSize].arrayIndex !=
            AllocRef::NullIndex)
        {
            CAS(&m_tail, tail, tail+1);
        }
    }
}

template <typename T> T Queue<T>::Remove()
{
    for (;;)   // Spin until successfully removed.
    {
```

```
        UINT32 head = m_head;
        AllocRef alloc = m_pQueue[head % m_maxQueueSize];

        UINT32 tail = m_tail;
        if (head != m_head) continue;

        if (head == m_tail)
        {
            if (m_pQueue[tail % m_maxQueueSize].arrayIndex ==
                AllocRef::NullIndex)
                    if (tail == m_tail) continue;    // Queue is empty.

            CAS(&m_tail, tail, tail + 1);
        }

        if (alloc.arrayIndex != AllocRef::NullIndex)
        {
            if (CAS2(&m_pQueue[head % m_maxQueueSize].val, alloc.val,
                AllocRef(AllocRef::NullIndex, alloc.version + 1).val))
            {
                CAS(&m_head, head, head+1);
                T item = m_pItems[alloc.arrayIndex];

                // Release index back to free list.
                m_allocator.Free(alloc.arrayIndex);
                return (item);
            }
        }
        else if (m_pQueue[head % m_maxQueueSize].arrayIndex ==
            AllocRef::NullIndex)
        {
            CAS(&m_head, head, head+1);
        }
    }
}
```

Listing 31.8. A lock-free, array-based producer-consumer queue.

When inserting an item, both the stack and queue continue spinning until they are not full. When removing an item, both the stack and queue continue spinning until they are not empty. To modify these methods to return an error code instead, replace `continue` with `return` at the appropriate locations. This is a better strategy when data is being inserted and removed from the queue sporadically, since the application can handle the best way to wait. For example, the application can call `SwitchToThread()` to allow the OS thread scheduler to yield execution to another thread instead of pegging the processor at 100% with unnecessary spinning.

31.9 Interprocess Communication

Both the semaphore-based model and the lock-free model can be extended for use for interprocess communication. Win32 memory-mapped files can act as shared storage for all the volatile and constant memory used by the algorithms. To extend to the semaphore-based queue, replace the critical sections with named mutexes and the semaphores with named semaphores. The queue will be managed using kernel locks, and the head and tail offsets can be maintained as reserved space in the memory-mapped file.

The lock-free implementation is handled similarly, except no locks are needed, even across processes. However, just like the semaphore-based queue, the lock-free queue requires all producers and consumers to use the same lock-free code to ensure thread safety.

From the perspective of both algorithms, there are no differences between an array in one process and the same array remapped in different processes, since each algorithm is designed to operate using array indices instead of pointers.

References

[AMD 2010] AMD. *AMD64 Architecture Programmer's Manual Volume 2: System Programming*. Advanced Micro Devices, 2010. Available at http://support.amd.com/us/Processor_TechDocs/24593.pdf.

[Colvin and Groves 2005] Robert Colvin and Lindsay Groves. "Formal Verification of an Array-Based Nonblocking Queue." *Proceedings of Engineering of Complex Computer Systems*, June 2005.

[Dawson 2008] Bruce Dawson. "Lockless Programming Considerations for Xbox 360 and Microsoft Windows." MSDN, June 2008. Available at http://msdn.microsoft.com/en-us/library/ee418650%28VS.85%29.aspx.

[Intel 2010] Intel. *Intel 64 and IA-32 Architectures. Software Developer's Manual. Volume 3A: System Programming Guide*. Intel, 2010. Available at http://www.intel.com/Assets/PDF/manual/253669.pdf.

[Leonard 2007] Tom Leonard. "Dragged Kicking and Screaming: Source Multicore." *Game Developers Conference*, 2007. Available at http://www.valvesoftware.com/publications/2007/GDC2007_SourceMulticore.pdf.

[Newcomer 2001] Joseph M. Newcomer. "Using Semaphores: Multithreaded Producer/Consumer." The Code Project, June 14, 2001. Available at http://www.codeproject.com/KB/threads/semaphores.aspx.

[Sedgewick 1998] Robert Sedgewick. *Algorithms in C++, 3rd Edition*. Reading, MA: Addison-Wesley, 1998.

[Shafiei 2009] Niloufar Shafiei. "Non-blocking Array-Based Algorithms for Stacks and Queues." *Proceedings of Distributed Computing and Networking*, 2009.

[Shann et al. 2000] Chien-Hua Shann, Ting-Lu Huang, and Cheng Chen. "A Practical Nonblocking Queue Algorithm using Compare-and-Swap." *Proceedings of Parallel and Distributed Systems*, 2000.

Contributor Biographies

Rémi Arnaud
remi@acm.org

Rémi Arnaud is Chief Software Architect at Screampoint International, working on interoperable 5D digital models. Rémi's involvement with real-time graphics started at Thales Simulation where he designed the Space Magic real-time visual system and finalized his PhD. In 1996, Rémi relocated to California to join Silicon Graphics's IRIS Performer team before cofounding Intrinsic Graphics, an advanced technology cross-platform middleware company for PlayStation 2, Xbox, GameCube, and PC. In 2003, he joined Sony Computer Entertainment as Graphics Architect, working on the PlayStation 3 SDK, and joined the Khronos Group, creating the COLLADA standard. More recently, Rémi was building a game technology team for the Larrabee project at Intel.

Wessam Bahnassi
wbahnassi@inframez.com

Wessam Bahnassi is a software engineer with a background in building architecture. This combination is believed to be the reason behind Wessam's passion for game engine design. He has written and dealt with a variety of engines throughout a decade of game development. Currently, he is in charge of animation and rendering engineering at Electronic Arts Inc., and he is a supervisor of the Arabic Game Developer Network.

Tolga Çapın
tcapin@cs.bilkent.edu.tr

Tolga Çapın is an assistant professor in the Department of Computer Engineering at Bilkent University. Before joining Bilkent, he worked at the Nokia Research Center as a Principal Scientist. He has contributed to various mobile graphics standards, including Mobile SVG, JCP, and 3GPP. Tolga received his PhD at EPFL (Ecole Polytechnique Federale de Lausanne), Switzerland in 1998. He has published more than 20 journal papers and book chapters, 50 conference papers, and a book. He has 4 patents and 10 pending patent applications. His current research interests include perceptually aware graphics, mobile graphics platforms, human-computer interaction, and computer animation.

Patrick Cozzi
pjcozzi@siggraph.org

Patrick Cozzi is a senior software developer on the 3D team at Analytical Graphics, Inc. (AGI) and coauthor of the book *3D Engine Design for Virtual Globes*. His interests include real-time rendering, GPU programming and architecture, OpenGL, and software architecture. Patrick has an M.S. in Computer and Information Science from the University of Pennsylvania and a BS in Computer Science from Penn State. He is also a contributor to Siggraph. Before joining AGI in 2004, he participated in IBM's Extreme Blue internship program at the Almaden Research Lab, interned with IBM's z/VM operating system team, and interned with the chipset validation group at Intel.

Michał Drobot
hello@drobot.org

Michał Drobot has been working in game development for five years. He started as Technical Artist, switching later to Effect Programmer and Visual Technical Director at Reality Pump. Currently, he is creating some cutting edge rendering technology as Senior Tech Programmer at Guerrilla Games. In his spare time, Michał frequently speaks on the topic of game development and graphics programming, having presented at GDC, GCDC, SFI, and having given series of lectures at universities. Moreover, he is the author of several articles on real-time rendering published in books and magazines. He is also the main consultant and active teacher on EGA, the European Games Academy in Krakow, Poland. He loves eating pixels for breakfast.

Richard Egli
richard.egli@usherbrooke.ca

Richard Egli has been a professor in the Department of Computer Sciences at the University of Sherbrooke since 2000. He received his BSc in Computer Science and his MSc in Computer Science at the University of Sherbrooke. He received his PhD in Computer Science from the University of Montréal in 2000. He is the director of the centre MOIVRE (MOdélisation en Imagerie, Vision et RÉseaux de neurones). His research interests include computer graphics, physical simulations, and digital image processing.

Martin Fleisz
martin.fleisz@kabsi.at

Martin Fleisz received an MS degree in Computer Science from the University of Edinburgh, Scotland and a BS degree in Computer Science from the University of Derby, England. He has worked as a professional C++ developer for more than six years. Martin spends most of his spare time enlarging his knowledge about game and graphics related programming techniques. In his remaining time, he plays drums in a band, tries to learn to play the guitar, and sometimes enjoys playing a video game.

Simon Franco
simon_franco@hotmail.com

Simon Franco began his love of computer programming shortly after receiving a Commodore Amiga, and has been coding ever since. He joined the games industry in 2000 after completing a degree in Computer Science. He started at The Creative Assembly in 2004, where he has been to this day. During his spare time, Simon can be found playing the latest PC strategy game or writing assembly code for the ZX spectrum. His baby girl Astrid was recently born.

Marco Fratarcangeli
marco@fratarcangeli.net

Marco Fratarcangeli is a Senior Software Engineer at Taitus Software Italia, developing cross-platform visual tools for the representation of space mission data like planet rendering and information visualization. In 2009, he obtained a PhD in Computer Engineering from University of Rome Sapienza, Italy, jointly with the Institute of Technology of Linköping, in Sweden. During his academic activities, Marco researched mainly novel methods for automatic rigging of facial animation through physically-based animation and motion retargeting. His earliest memories are of programming Basic on the ZX Spectrum 48K.

Holger Grün
holger.gruen@amd.com

Holger Grün ventured into 3D real-time graphics right after university, writing fast software rasterizers in 1993. Since then, he has held research and development positions in the middleware, games, and simulation industries. He began working in developer relations in 2005 and now works for AMD's product group. Holger, his wife, and his four kids live in Germany, close to Munich and near the Alps.

Julien Hamaide
julien.hamaide@fishingcactus.com

Julien Hamaide graduated as a multimedia electrical engineer at the Faculté Polytechnique de Mons, Belgium. After two years working on speech and image processing at TCTS/Multitel and three years leading a team on next-generation consoles at Elsewhere Entertainment (www.elsewhereentertainment.com), Julien started his own studio called Fishing Cactus (www.fishingcactus.com) with three associates. He has published several articles in the *Game Programming Gems* series and in *AI Programming Wisdom 4*. He spoke about *10Tacle*'s movement and interaction system at the Game Developers Conference in 2008 and about multithreading applied to AI in 2009. His experience revolves around multithreaded and scripted systems, mostly applied to AI systems. He is now leading development at Fishing Cactus.

Daniel Higgins
dan@lunchtimestudios.com

Over ten years ago, Dan Higgins began his career in games at Stainless Steel Studios, where he was one of the original creators of the Titan game engine. As one of the chief AI programmers on *Empire Earth*, *Empires: Dawn of the Modern World* and *Rise & Fall: Civilizations at War*, he spent years designing and innovating practical AI solutions for difficult problems. Later, he enjoyed working for Tilted Mill Entertainment on *Caesar IV* and *SimCity Societies*.

Dan's coding domain extends well beyond AI, as he enjoys all aspects of game engine development and is often called on for his optimization skills both inside and outside of the games industry. Today, along with his wife, he is owner and manager of Lunchtime Studios, LLC.

Jason Hughes
jhughes@steelpennygames.com

Jason Hughes is an industry veteran game programmer of 16 years and has been actively coding for a decade longer. His background runs the gamut from modem drivers in 6502 assembly and fluid dynamics on the Wii to a proprietary multiplatform 3D engine and tools suite. When not working as a hired gun for game developers, Jason tinkers with exotic data structures, advanced compression algorithms, and various tools and technology relating to the games industry. Prior to founding Steel Penny Games, he spent several years at Naughty Dog, developing pipeline tools and technology for the ICE and Edge libraries on the PlayStation 3.

Matthew Johnson
matt.johnson@amd.com

Matthew Johnson is a Software Engineer at Advanced Micro Devices, Inc. with over 12 years of experience in the computer industry. He wrote his first game as a hobby in Z80 assembly language for the TI-86 graphic calculator. Today, he is a member of the DirectX 11 driver team and actively involved in developing software for future GPUs. Matthew currently lives in Orlando, Florida with his lovely wife.

Linus Källberg
linus.kallberg@mdh.se

Linus Källberg is a teacher in Computer Science at Mälardalen University in Sweden. He has previously worked with program analysis as a research engineer on the ALL-TIMES project. His MSc in Computer Science was completed in March 2010.

Frank Kane
fkane@sundog-soft.com

Frank Kane is the owner of Sundog Software, LLC, makers of the SilverLining SDK for real-time rendering of skies, clouds, and precipitation effects (see http://www.sundog-soft.com/ for more information). Frank's game development experience began at Sierra On-Line, where he worked on the system-level software of a dozen classic adventure game titles including *Phantasmagoria*, *Gabriel Knight II*, *Police Quest: SWAT*, and *Quest for Glory V*. He's also an alumnus of Looking Glass Studios, where he helped develop *Flight Unlimited III*. Frank developed the C2Engine scene-rendering engine for SDS International's Advanced Technology Division, which is used for virtual reality training simulators by every branch of the US military. He currently lives with his family outside Seattle.

Manny Ko
mannyk90@yahoo.com

Manny Ko is currently working in the Rendering Group for DreamWorks Animation. Prior to that, he worked for Naughty Dog as a member of the ICE team, where he worked on next-generation lighting and GPU technologies.

Balor Knight
Balor.Knight@Disney.com

Balor Knight is a Senior Principal Programmer at Black Rock Studio in Brighton, UK. He has been writing games for over 20 years, having started on the ZX Spectrum in the late 1980s. Since then, he has worked on many titles, including *Re-Volt*, the *Moto GP* series, *Pure*, and most recently, *Split/Second*. His main area of expertise lies in console game rendering engines.

Thomas Larsson
thomas.larsson@mdh.se

Thomas Larsson is an Assistant Professor at Mälardalen University in Sweden where he has been teaching Computer Science since 1996. His PhD thesis about efficient intersection queries was completed in January 2009. Currently, he gives courses in C/C++, multimedia, and computer graphics.

Eric Lengyel
lengyel@terathon.com

Eric Lengyel is a veteran of the computer games industry with over 16 years of experience writing game engines. He has a PhD in Computer Science from the University of California, Davis, and he has an MS in Mathematics from Virginia Tech. Eric is the founder of Terathon Software, where he currently leads ongoing development of the C4 Engine.

Eric was the Lead Programmer for *Quest for Glory V* at Sierra Online, he worked on the OpenGL team for Apple, and he was a member of the Advanced Technology Group at Naughty Dog, where he designed graphics driver software used on the PlayStation 3. Eric is the author of the bestselling book *Mathematics for 3D Game Programming and Computer Graphics* and several chapters in other books including the *Game Programming Gems* series. His articles have also been published in the *Journal of Game Development*, in the *journal of graphics tools*, and on *Gamasutra.com*.

Noel Llopis
noel@snappytouch.com

Noel Llopis is following his lifelong dream of being an indie developer. He founded Snappy Touch to focus exclusively on iPhone development and released *Flower Garden* and *Lorax Garden*. He writes about game development regularly, from a monthly column in *Game Developer Magazine* to the *Game Programming Gems* series or his book *C++ for Game Programmers*. Some of his past games include *The Bourne Conspiracy*, *Darkwatch*, and the *MechAssault* series. He earned an MS in Computer Science from the University of North Carolina at Chapel Hill and a BS in Computer Systems Engineering from the University of Massachusetts Amherst.

Curtiss Murphy
cmmurphy@alionscience.com

Curtiss Murphy is a Senior Project Engineer at Alion Science and Technology. He manages the game-based training and 3D visualization development efforts for the Norfolk-based AMSTO Operation of Alion. He is responsible for the serious game efforts for a variety of Marine, Navy, Air Force, and Joint DoD customers. He is an author and frequent speaker at conferences and leads the game development team that created the award winning Damage Control Trainer. He has been developing and managing software projects for 18 years and currently works in Norfolk, VA. Curtiss holds a BS in Computer Science from Virginia Tech.

George Parrish
George.Parrish@Disney.com

George Parrish is a Senior Engine Programmer at Black Rock Studio. George started programming on his Commodore 64 when he was 13. He then went on to a career in IT, developing network servers and web applications for clients such as the National Health Service and the Ministry of Defence. He finally entered the games industry in 2002 and has worked on a number of games, including *Pure* and *Split/Second*.

Michael Ramsey
mike@ramseyresearch.com

Mike Ramsey is the principle programmer on the GLR AI Engine. Mike has developed core technologies for the Xbox 360, PC, and Wii at various companies. He has also shipped a variety of games, including *World of Zoo* (PC and Wii), *Men of Valor* (Xbox and PC), *Master of the Empire*, several *Zoo Tycoon 2* products, and other titles. Mike has contributed multiple articles to both the *Game Programming Gems* and *AI Game Programming Wisdom* series, and he has presented at the AIIDE conference at Stanford on uniform spatial representations for dynamic environments. Mike has a BS in Computer Science from Metropolitan State College of Denver, and his publications can be found at http://www.masterempire.com/. He also has a forthcoming book entitled *A Practical Cognitive Engine for AI*. When Mike isn't working, he enjoys playing speedminton, drinking mochas, and having thought-provoking discussions with his fantastic wife and daughter, Denise and Gwynn!

Matthew Ritchie
Matthew.Ritchie@Disney.com

Matthew Ritchie is a Graphics Engine Programmer at Black Rock Studio. He joined the games industry when he was 18. Since then, he has worked in the areas of networking, sound, and tools, but he currently specializes in graphics. His most recent titles are the *MotoGP07* series and *Split/Second*.

Sébastien Schertenleib
sscherten@bluewin.ch

Sébastien Schertenleib has been involved in academic research projects creating 3D mixed-reality systems using stereoscopic visualization while completing his PhD in Computer Graphics at the Swiss Institute of Technology in Lausanne. In 2006, he joined Sony Computer Entertainment Europe R&D, and since then, he has been trying to support game developers on all PlayStation platforms by providing technical training, presenting at various games conferences, and working directly with game developers via on-site technical visits and code sharing.

Olivier Vaillancourt
olivier.vaillancourt@usherbrooke.ca

Olivier Vaillancourt is an MSc student in the Department of Computer Science at University of Sherbrooke. His research interests include computer graphics and scientific visualization. He is a lecturer for undergraduate students in interaction design and computer graphics for the University of Sherbrooke. He has contributed to multiple handheld game productions as a software engineer at Artificial Mind and Movement, where he worked on high-profile licenses such as *The Sims*, *Lord of The Rings*, and *Ironman*. When he's not working on his LCD tan, he's spending time on various home improvement projects and enjoying the Canadian great outdoors.

Jon Watte
jwatte@gmail.com

Jon Watte started programming at the age of 10 on a Z80-based computer named "ABC-80." After evolving through a variety of micro- and minicomputer platforms through the 1980s and attending the Master of Computer Science program at KTH, Stockholm, Sweden, Jon moved from the arctic winter to the sweltering summer of Austin, Texas, where he worked on CodeWarrior products for game consoles and the alternative BeOS operating system.

One thing led to another, and Jon combined his love of music and video with his love of systems programming by taking the reins of the media group at Be, building a foundation for low-latency, high-throughput multiprocessing computing on desktop computers.

After Be was successfully sold to Palm, Jon moved to help build avatar, networking, and simulation technology for the There.com virtual world, a job that then led to a position as CTO of enterprise virtual world provider Forterra Systems, where he contributed to the future of virtual worlds for entertainment and business.

Currently, Jon is enjoying adding some exciting new technology to power the spectacular growth of the connected entertainment company IMVU, Inc., which is powered by an amazing collection of user-generated 2D and 3D content.

Jon also moderates online forums on game development and networking for Microsoft XNA Creators and the independent game developer site GameDev.net, and he has published both code and articles that help power several successful games, both independent and published.

M. Adil Yalçın
yalcin@umd.edu

M. Adil Yalçın is a PhD student in the Department of Computer Science at University of Maryland, College Park. He received his MS in Computer Engineering at Bilkent University, Ankara, Turkey in June 2010 with a thesis focused on deformations of height field structures. If something is about computer graphics and programming, more specifically real-time shading techniques, graphic engines, GPU programming or physically based animations, it will probably catch his attention. He tries to experiment with new stuff and organize most of his studies under an open-source modern rendering engine development effort.

Index

Printed and bound by CPI Group (UK) Ltd, Croydon, CR0 4YY

23/10/2024

01777689-0001